Algorithms for RPN Calculators

Algorithms for RPN Calculators

John A. Ball

Center for Astrophysics
Cambridge, Massachusetts

A Wiley-Interscience Publication
JOHN WILEY & SONS
New York Chichester Brisbane Toronto

Library of Congress Cataloging in Publication Data

Ball, John A. 1935-
Algorithms for RPN calculators.

"A Wiley-Interscience publication."
Bibliography: p.
Includes index.
1. Calculating-machines. 2. Algorithms. I. Title.
II. Title: RPN calculators.

QA75.B33 510'.28 77-14977
ISBN 0-471-03070-8

Printed in the United States of America

10 9 8 7 6 5 4 3 2

**For Professor A. E. Lilley
who helped keep the wolves away
while I wrote it.**

PREFACE

The enjoyment of the tools one works with is, of course, an essential ingredient of successful work.*

—Donald E. Knuth (1938–)

Reverse Polish notation (RPN) calculators are used by Nobel laureates and grade school children, but few users are aware of the power beneath their fingers. One can, for example, calculate the distance and heading between points on the earth using an algorithm from Section A.7.2 (without knowing anything about spherical trigonometry). Section A.4.2 shows how to integrate a second-order differential equation using fourth-order Runge-Kutta. Section A.12 illustrates the calculation of the interest rate of an annuity using a very rapidly converging iterative algorithm. But this book is intended primarily for those who want to write their own concise and precise algorithms.

Many of the slide-rule tricks developed by engineers and many of the techniques of classical numerical analysis and modern computer programming are herein adapted to RPN. But RPN is different and needs special attention if the resulting algorithms are to be optimum. RPN is an elegant calculator system, and those who are willing to study it carefully can write powerful RPN algorithms. But the difficulties are not as formidable as they might seem, for RPN is a logical system—easy to remember and use.

I have written for readers with a diversity of backgrounds. Much of this book should be accessible to anyone who has an RPN calculator and has read the accompanying instruction booklet. High school mathematics through logarithms and trigonometry, but rarely calculus or advanced mathematics, is needed as background for the first parts of this book. Later

*From Knuth (1969), p. 204.

parts deal with more advanced concepts such as differential equations. Sections 1.3.4 and 2.4, and Chapters 3, 4, and 5 use, at least to some extent, concepts from calculus.

Although not really intended as a textbook—I know of no courses in which it would so serve—I include exercises and other pedagogical devices because I believe that they may be useful, even fun. Answers for all the exercises are given in Appendix B.

Most of this book applies to all RPN calculators, programmable or not. Contrary to popular opinion, I find the differences between programmable and nonprogrammable RPN calculators less significant than, for example, the differences discussed in Section 1.4 between RPN and algebraic entry system (AES) calculators. Examples in the text and in Appendix A are algorithms for specific calculators. I try to choose the simplest calculator that will do the problem easily. Using the minimum necessary power for the job is aesthetically pleasing, and converting upward is clearly easier. That is, converting an HP-35 algorithm to run on an HP-67 is easier than the converse.

In 1973 I began using an RPN calculator extensively both in my job (radio astronomy) and for other calculations. No books of RPN algorithms were available then, and I began writing some algorithms for my own use. A few friends got photocopies, and in August 1974 several hundred copies of "Algorithms for the HP-45 and HP-35" were printed as a technical report. A later edition of this report was dated March 9, 1975. These technical reports, which are similar to Appendix A in this book, are collections of algorithms with enough explanatory material to enable the reader to use them. As other collections of algorithms became available in 1974 (*HP-35 Math Pac*, and *HP-45 Applications Book*; see references in Appendix C), I noted that my algorithms for the same problems were frequently more compact and more elegant. This prompted me to analyze carefully my techniques for writing algorithms and to try to describe these techniques in a coherent manner. This book is the result.

Norman Brenner of MIT and IBM, and George Rybicki and Hays Penfield of the Center for Astrophysics (CfA) wrote some of the algorithms in Appendix A as noted. Norman Brenner also carefully read and commented on the typescript. I also thank my other colleagues at the CfA for help and encouragement. I thank the following persons and organizations for useful ideas or information: C. V. Briggs, III, of MIT; Win Chan, Roy Martin, Jeff Nagle, J. Peter Nelson, and Sharon Northrup of Hewlett-Packard Company; D. Jividen of Compucorp; Ron Ames of Monroe; Bruce Balick of the University of Washington; Georgene H. Berglund of Novus Consumer Products Division of National Semiconductor Corporation; R. C. Vanderburgh, Dayton, Ohio; and T. A. Bates, Montague,

Massachusetts. Although I adopted only some of their recommendations, I sincerely appreciate the comments and criticisms of Jon M. Smith and Richard W. Hamming, who read a draft of this book. I thank my wife Audrey for typing, and typing, and typing, and my daughter Fifi for checking many of the algorithms.

This book and this author have given John Wiley & Sons rather more trouble than usual. But the appearance of the result is, in my view, outstanding. I thank the editorial and production people at Wiley, and especially the compositor, for taking the time and effort to make it so.

I would sincerely appreciate having your comments and criticisms, especially about any errors you may find. Thank you.

JOHN A. BALL

January 1978
Harvard, Massachusetts
Harvard College Observatory
Harvard, Massachusetts 01451

CONTENTS

Algorithms for RPN Calculators

1
Introduction to RPN

I do not say: science is useful because it allows us to construct machines; I do say: machines are useful, for by working for us they permit us more time to study science.

—*Jules Henri Poincaré (1854–1912)*

It is unworthy of excellent men to lose hours like slaves in the labor of calculation.

—*Gottfried Wilhelm, Baron von Leibniz (1646–1716)*

As the available computational power increases, some among us, perhaps less-excellent men, take on progressively more complex problems, so that the hours per day, lost in the labor of calculation, are nearly constant.

—*Anomalous*

...when you are declaiming, declaim; and when you are calculating, calculate.

—*Samuel Johnson (1709–1784)*

1.1 INTRODUCTION TO RPN: HISTORY

RPN stands for "reverse Polish notation" and the "Polish" refers to Jan Łukasiewicz (1878–1956), a great Polish logician and mathematician. "Polish" is easier to pronounce than "Łukasiewicz" (wū-ka-shā'-vich) and "reverse Polish" is much easier to pronounce than "Zciweisakuł."*

In their advertisements and also in a letter to me, Hewlett-Packard Company (HP), the best-known manufacturer of RPN calculators,† says that RPN is based on a suggestion by Łukasiewicz, and that RPN was invented and is patented by HP. Aside from the apparent contradiction in these two statements, I do not think that either of them is quite true. My first experience with RPN involved a nice old Friden EC-130 desktop electronic calculator, circa 1964. The EC-130 has RPN with a push-down stack (defined in Section 1.2) of four registers, all visible simultaneously on a cathode ray tube display. Furthermore, they are shown upside down, that is, the last-in–first-out register is at the bottom. The same orientation is used in instruction booklets for HP calculators, perhaps by coincidence.

Around 1966, the Monroe Epic calculator offered RPN with a stack of four, a printer, and either 14 or 42 step programmability. The instruction booklets with these two calculators make no mention of RPN or Jan Łukasiewicz.

In his book *Aristotle's Syllogistic from the Standpoint of Modern Formal Logic* (1951), p. 78, and in some of his other publications, Łukasiewicz recommends a parenthesis-free algebraic notation in which the operation symbols (e.g., +, −, ×, ÷), which Łukasiewicz calls "functors," *precede* their parameters. Thus in his notation + 5 3 would add to 8. Compactness is the motivation for this notation; parentheses are eliminated, and the number of symbols that must be written is minimized. Since with a calculator, each symbol represents a keystroke, Łukasiewicz's system has obvious interest. Placing the functors *after* the parameters gives *reverse* Polish notation, RPN.

But remember the old mechanical adding machine. One clears it to zero, keys a number, and presses [+] to "enter" the number into the machine. What has happened is that the number was added to zero, already in the machine. Then one can key another number, press [+] again, and see the sum of the two numbers, and so on. Note that the operation [+] comes after the number. Using exactly this sequence, one can add numbers on an

*I am indebted to Peter Collins and Jerome Cherniack for pointing this out to me.

†The terminology is sometimes confusing. I try to be consistent and use "calculator" to mean a device. A calculator operator is a human who presses the keys of a calculator and performs other useful functions such as buying this book.

RPN calculator. RPN represents a reasonable extrapolation of this mechanical calculator scheme, and would, and probably did, occur to many people who might never have heard of Jan Łukasiewicz.

Hewlett-Packard Company is to be commended for the beautiful design of the original HP-35 calculator (c. 1972). They avoided many pitfalls that lesser minds would have become mired in. HP provided many features in a truly concinnate way. But ignoring the foundations on which this accomplishment was built serves no purpose. In my view, the RPN calculator owes about as much to the venerable mechanical calculator, and to a number of anonymous designers, some of whom worked for Friden, as it does to Jan Łukasiewicz.

More than a dozen models of scientific RPN pocket calculators are now available from five manufacturers, plus other models intended for business and financial calculations or for desktop rather than portable use. RPN with its push-down stack of numbers and its post operators is also being used extensively in computers. The FORTH software system by Charles H. Moore, for example, is designed around RPN. And minicomputers now exist with hardware that incorporates RPN. The motivation for using RPN is the same in calculators and computers—terseness.

1.2 PUSHING AND POPPING; EXCHANGING AND ROLLING

A push-down stack of numbers is named by analogy with the device that holds plates in some cafeterias. A stack of plates rests on a mechanism with a spring under it such that the weight of each added plate presses all the other plates downward, usually into a hole in a table surface. Only the top plate is accessible, and it is called the last-in–first-out plate. Adding a plate to the stack is called "pushing," and removing a plate from the stack is called "popping" (the plate is not supposed to break).

In RPN calculators, the push-down stack usually has room for four numbers, each of which can be up to ten digits long with a sign and exponent. The last-in–first-out number is in the X register, which is, always and only, displayed. The other three registers are called Y, Z, and T, in order. If one pushes five numbers, the first-in number is lost. On most RPN calculators, however, T does not change on pop; thus a number pushed into T can be reused indefinitely. This feature is remarkably useful.

Some calculators have room for only three or even only two numbers in their push-down stack. Two registers are the minimum possible to have a calculator at all, and perhaps such a calculator should not be called RPN. Less than a four-register stack, regardless of what it is called, is very undesirable.

The registers in the push-down stack are usually shown upside down in calculator instruction booklets (and in this book), giving one, in effect, a push-up stack. This may seem illogical at first, but with the upside-down arrangement, the registers are oriented so that [÷] and [−] are in the correct order: [÷] causes the number in Y to be *divided by* the number in X, and [−] causes the number in X to be *subtracted from* the number in Y, rather than the other way round. This convention in turn allows numbers to be entered in the logical order; thus A [↑] B [−] gives $A-B$ rather than $B-A$, and A [↑] B [÷] gives A/B rather than B/A. The answer is always displayed in the X register.

Operations such as [+], [−], [×], and [÷] are called dyadic functors—two numbers popped from the stack are combined to give one answer, which is pushed back onto the stack.

Figure 1.2.1 illustrates the effect of stack manipulators and functors. Each part of this figure begins with a number A in X, B in Y, and so on. The calculator is usually in a state called "auto-enter enabled," "push pending," or "auto stack lift"; a more exact name might be "auto-push." A new number keyed in this state pushes itself into X, the previous contents of X go into Y, and so forth. The functor [↑], called "enter," duplicates the

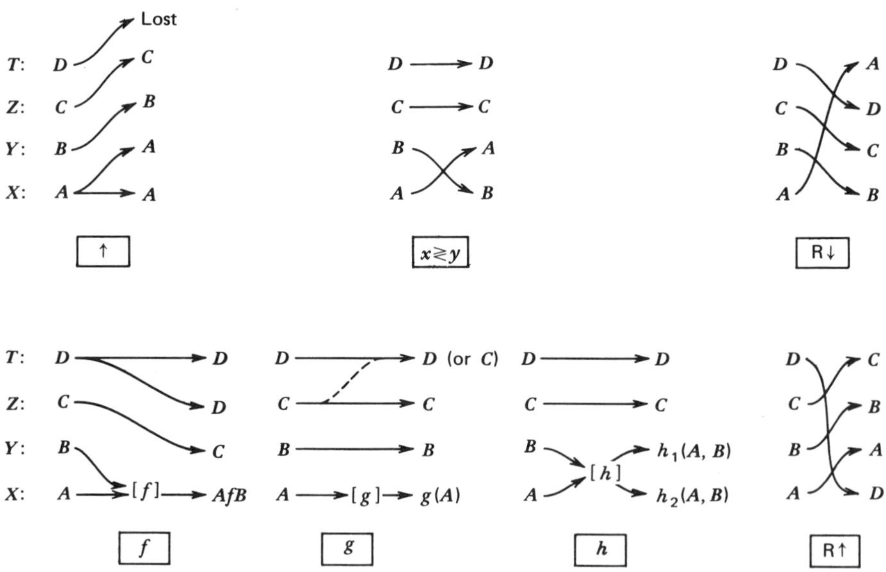

Figure 1.2.1. The effect of various stack manipulators and functors: [f] represents any dyadic functor, [g] any monadic functor, and [h] any bifid functor. On an HP-35, trigonometric functors lose the contents of T; C is duplicated into T as shown by the dashed line. The bottom half of this figure does not apply to Novus calculators.

T:														
Z:					225	225								
Y:		15		225	54	54	225			3	15		12	
X:	15	15	225	54	54	4	216	9	3	15	3	12	2	6
Key:	15	[↑]	[×]	54	[↑]	4	[×]	[−]	[$\sqrt{x}$]	15	[$x \gtrless y$]	[−]	2	[÷]

Figure 1.2.2 Example: $(15-\sqrt{15^2-54\times 4})/2=6$.

contents of X into Y as shown in Figure 1.2.1, but also *disables* auto-enter; therefore a number keyed following [↑] overwrites X—no push.

In Figure 1.2.1, [f] represents a dyadic functor such as [+], [−], [×], or [÷], and [g] a monadic functor such as [$\sqrt{x}$] or [log]. This figure can be taken to define dyadic, monadic, and bifid functors; an extended discussion of these terms is in Section 1.3.2. The symbol [R↓] is called "roll down," [R↑] is "roll up," and [$x \gtrless y$] is "exchange."

Figure 1.2.2 shows the contents of the stack after each step of a particular problem written as

15 [↑] [×] 54 [↑] 4 [×] [−] [$\sqrt{x}$] 15 [$x \gtrless y$] [−] 2 [÷] (see 6).

This gives $(15-\sqrt{15^2-54\times 4})/2$ which is one of the roots of $x^2-15x+54=0$. The [↑] [×] after 15 gives $15^2=225$. Some calculators have an [x^2] key, which would save a keystroke. The [$x \gtrless y$] before [−] gives $15-\sqrt{\cdots}$ rather than $\sqrt{\cdots}-15$. Try it. Another algorithm for this problem is

15 [↑] [x^2] 54 [↑] 4 [×] [−] [$\sqrt{x}$] [−] 2 [÷] (see 6).

This section and these examples are intended to give just a hint of what RPN calculating is all about. All these topics are more thoroughly discussed in later sections: Section 1.3.1 describes most of the keys on RPN calculators, Section 1.3.2 discusses the functors in Figure 1.3.1 and considers auto-enter in more detail, and Section 1.3.3 compares two methods for translating algebraic expressions into RPN keystrokes.

1.3 INTRODUCTION TO RPN CALCULATING

"Curiouser and curiouser," cried Alice.

—*Lewis Carroll (1832–1898)*

1.3.1 Key to Keys

Figures 1.3.1 through 1.3.9 present a selection of available or recently available RPN pocket calculators. This selection is obviously incomplete; many other RPN calculators are available, and a plethora of non-RPN

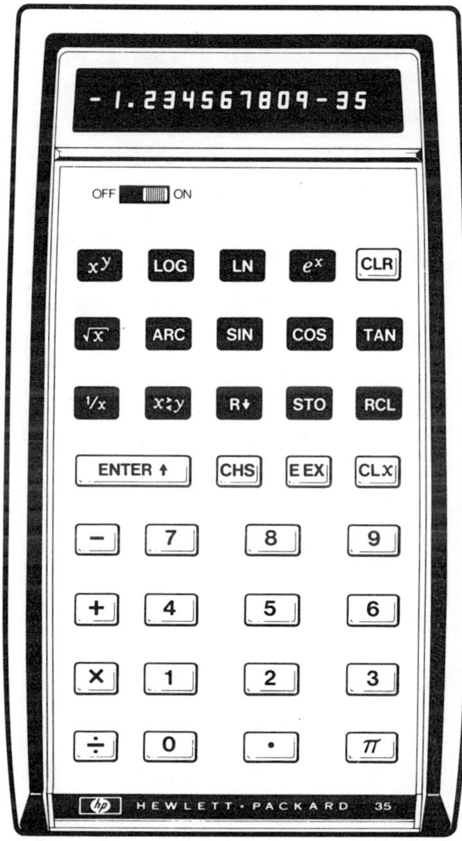

Figure 1.3.1. The original RPN pocket calculator, the HP-35, was introduced in 1972. (A later version is illustrated. On the top half of the keyboard of the 1972 model, the keys were more nearly square and were labeled on the land area above the keys rather than on the keys themselves.) Figure courtesy of Hewlett-Packard Company.

calculators (discussed in Section 1.4) are now on the market. The following tabulation shows most of the possible keystrokes from the devices illustrated, with an explanation of the function. For this description, x is the number initially in X, y in Y, and so on. Symbols inside dashed boxes (e.g., y^x) require a prefix or shift keystroke. In the auto-enter column, E means enable, D means disable, and U means unaffected. Auto-enter is discussed in more detail in Section 1.3.2.

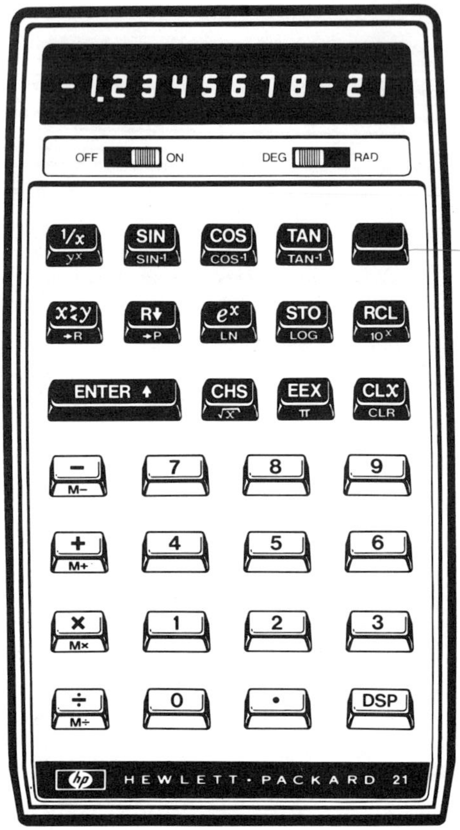

Figure 1.3.2. The HP-21 calculator, which replaced the HP-35, has all the features of the HP-35 and more. However, the HP-21 is smaller, the display is somewhat harder to read, and prefix or shift keystrokes are required more often than with the HP-35. The unlabeled shift key in the upper right-hand corner is blue and is indicated in the text by B. Figure courtesy of Hewlett-Packard Company.

Figure 1.3.3. The Novus Mathematician is a simple and very inexpensive RPN calculator. However, it has only three registers in its push-down stack and does not have scientific notation (for very large and very small numbers). Figure courtesy of National Semiconductor Corporation.

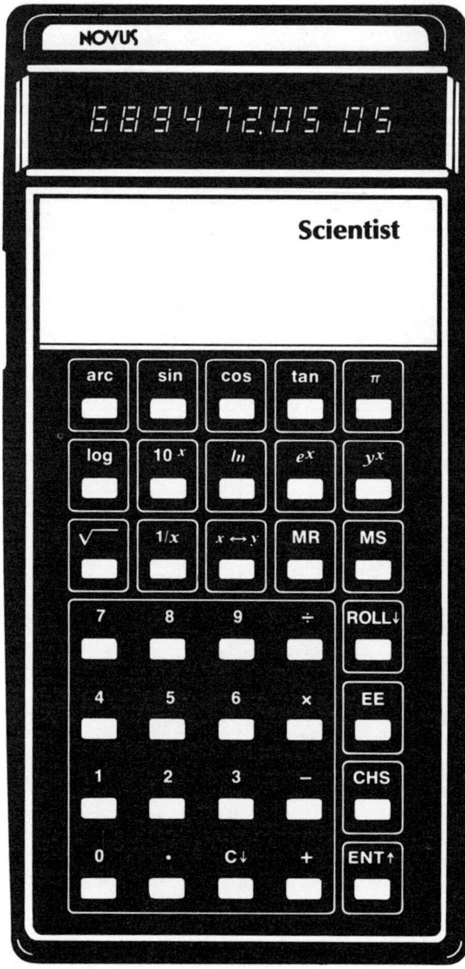

Figure 1.3.4. The Novus Scientist is also inexpensive and avoids some of the limitations of the Mathematician. Figure courtesy of National Semiconductor Corporation.

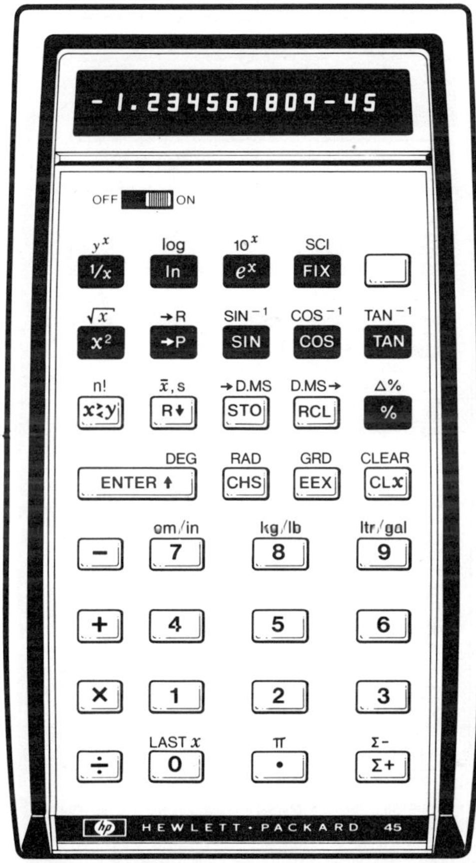

Figure 1.3.5. The HP-45 is the big brother of the HP-35 and has many additional features. The HP-45 was introduced in 1973 and, like the HP-35, has been discontinued. The unlabeled shift key in the upper right-hand corner is gold (or yellow) and is indicated in the text by [G]. Figure courtesy of Hewlett-Packard Company.

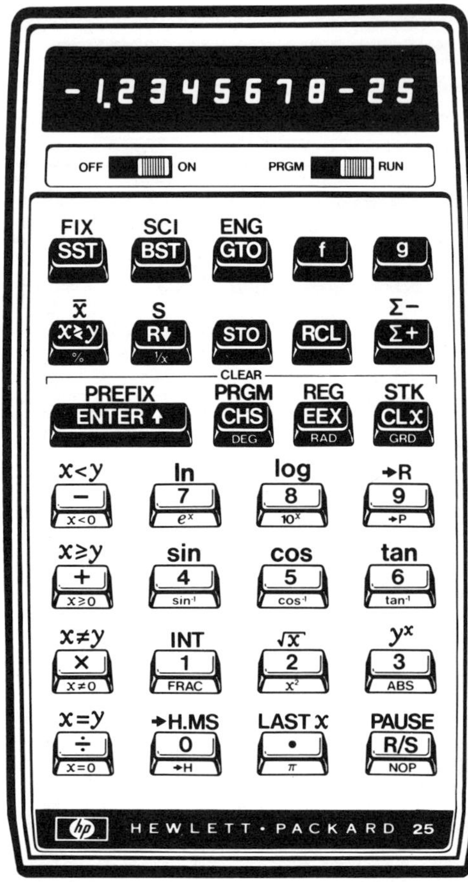

Figure 1.3.6. The HP-25, which replaced the HP-45, has almost all the features of the HP-45 and is also programmable with room to store 49 program steps. Figure courtesy of Hewlett-Packard Company.

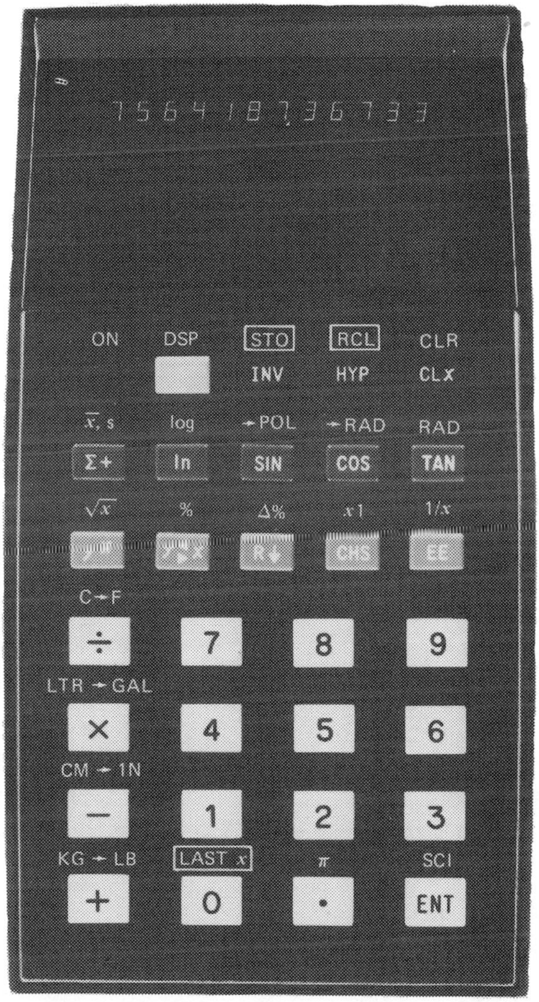

Figure 1.3.7. The Corvus 500 (which is internally identical to the APF Mark 55 and Omron 12-SR) is closely modeled after the HP-45 and has most of the same features, as well as some additional features. The unlabeled shift key in the top center is gold (or yellow) and is indicated in the text by $\boxed{G}$.

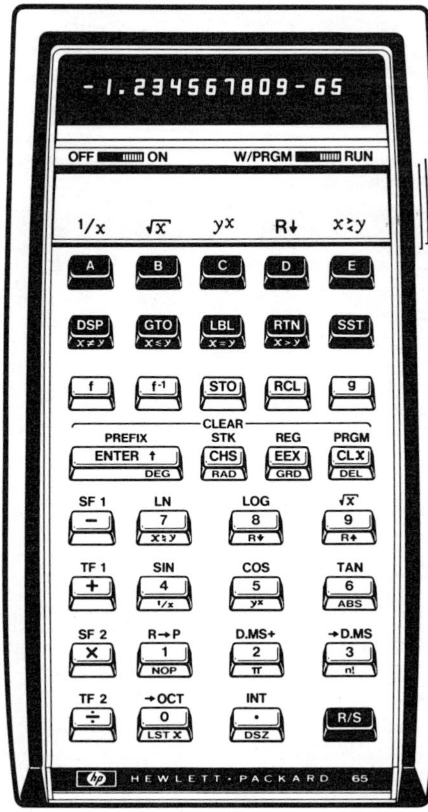

Figure 1.3.8. The HP-65 is an expensive programmable RPN calculator with room for 100 program steps that can be written onto and read from a tiny (not tinny) magnetic card. An impressive library of these programs is available (see Appendix C). Figure courtesy of Hewlett-Packard Company.

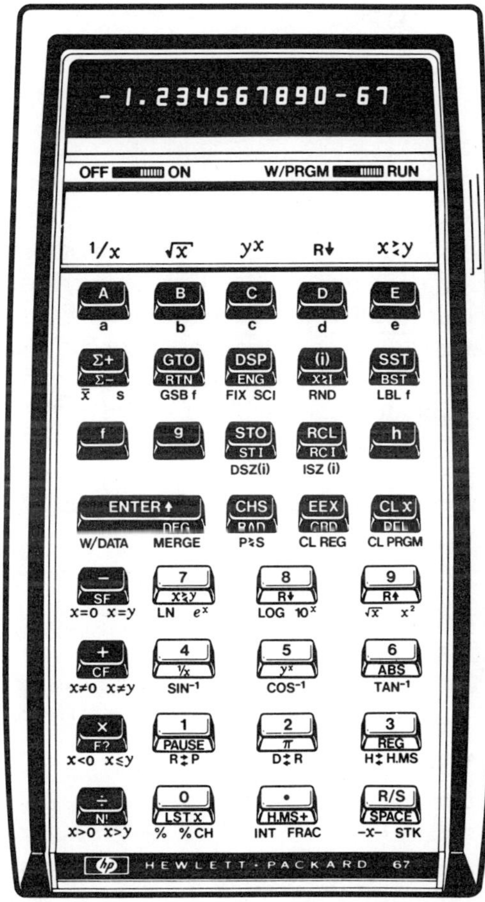

Figure 1.3.9. The HP-67 calculator, which replaced the HP-65, has all the features of the HP-65 and more. The HP-67 has room for 224 program steps. Figure courtesy of Hewlett-Packard Company.

Keystroke	Type	Effect on Auto-enter
[ENTER↑] [ENT] [ENT↑] [EN] [SAVE↑]	Stack manipulator	D

Pushes $x \to Y$, $y \to Z$, and $z \to T$ and leaves x in X (see Figure 1.3.1). [ENTER↑] is abbreviated herein as [↑]. This key performs two functions and can lead to some confusion. The old Friden EC-130 has two keys labeled [ENTER] and [REPEAT]. [↑] is most frequently used as a number separator, that is, to signify the end of one number and the start of the next. But the repeat or duplicate function allows [↑] to be used also to duplicate numbers on the stack.

Keystroke	Type	Effect on Auto-enter
[$x \gtrless y$] [$x \gtrless y$] (dashed) [$x \leftrightarrow y$] [$y \gtrless x$]	Stack manipulator	E

Exchanges x and y. See Figure 1.2.1.

Keystroke	Type	Effect on Auto-enter
[R↓] [R↓] (dashed) [ROLL↓] [ROLL]	Stack manipulator	E

Rolls the stack downward; that is, $x \to T$, $y \to X$, $z \to Y$, and $t \to Z$. See Figure 1.2.1.

Keystroke	Type	Effect on Auto-enter
[R↑] [R↑] (dashed)	Stack manipulator	E

Rolls the stack upward; that is, $x \to Y$, $y \to Z$, $z \to T$, and $t \to X$. See Figure 1.2.1.

Keystroke	Type	Effect on Auto-enter
[0] through [9] and [·]	Number	E

Places the corresponding digit into X at the next available position. The first number key pressed pushes the stack if auto-enter was enabled, or overwrites x if not. A number is terminated by any keystroke other than [0] through [9], [·], [CHS] or [EEX] (certain extraneous prefix keystrokes

Keystroke	Type	Effect on Auto-enter

are ignored). If [·] is pressed twice within the same number, the second [·] is ignored on some calculators but is taken to be the decimal point on other calculators. [·] after [EEX] clears the exponent to zero on some calculators and undoes the effect of [EEX] on other calculators. If more than two digit keys are pressed after [EEX], the last two are taken to be the exponent.

[EEX] [EE]	Number	E

"Enter exponent" tells the calculator that the following number keys are the exponent (power of 10). If no number keys were pressed before [EEX], "1" is assumed on most calculators.

[π] [π]	Number	E

Puts $\pi = 3.141592654$ into X, terminates and pushes a previous number if any, pushes the stack if auto-enter was enabled, or overwrites x if not. [π] is an example of a special number key for frequently used constants to avoid keying each digit of the number. Some calculators have other special number keys as well as [π].

[CHS] [+/−]	Monadic functor *or* minus sign	U

Negates x; that is, replaces x by $-x$. This key ("change sign") like [↑], performs two functions. [CHS] can be used to key a negative number (push [CHS] *after* some or all of the digits are keyed on all calculators except an HP-35), a negative exponent (push [CHS] any time after [EEX]), or to negate a number already in X from a previous calculation. But on most RPN calculators [CHS] is *not* a number terminator, does not affect the status of auto-enter, and thus is very different from the ordinary monadic functors. Section 1.3.2 discusses [CHS] more fully.

[STO] [MS]	Storage operation	?

Stores x into a storage (memory) register, overwriting anything that was in there before. On most RPN calculators [STO] does *not* pop, and it has no effect on the stack. Although [STO] is always a number terminator, it may or may not enable or disable auto-enter; [STO] disables auto-enter on the

Keystroke	Type	Effect on Auto-enter

HP-35, does not affect auto-enter on the HP-45, HP-55, HP-65, Corvus 500, and APF Mark 55, and enables auto-enter on the HP-21 and HP-25. Check the fine print in the instruction booklets for other calculators.

[RCL] [MR]	Storage operation	E

Recalls the number from a storage register and puts it into *X*; [RCL] terminates and pushes a previous number if any, pushes the stack if auto-enter was enabled, or overwrites *x* if not.

Some RPN calculators have more than one storage register, and a number key pressed after [STO] or [RCL] designates the register to be used. Some RPN calculators also have storage register arithmetic; that is, a sequence such as [STO] [÷] 5 divides the contents of storage register number 5 by the contents of *X* and leaves the answer in the storage register. The stack registers are unchanged. The [÷] in this example could also be [+], [−], or [×]. On some calculators storage register arithmetic is performed by keys labeled [M+], [M−], and so on. Similarly, the HP-45 and HP-46 allow sequences such as [RCL] [−] 6, which subtracts the contents of storage register number 6 from the contents of *X* and leaves the answer in *X*. The storage register, and the stack registers other than *X*, are unchanged. If auto-enter was enabled, the sequence [RCL] [−] 6 (pop then push) differs from [RCL] 6 [−] (push then pop) in not losing the contents of *T*. If auto-enter was disabled, [RCL] [−] 6 uses, but [RCL] 6 [−] loses, the contents of *X*. Figure 1.3.10 gives examples of such sequences. These capabilities are valuable for certain problems.*

Some calculators allow the sequence [STO] [$x \gtrless y$] *n* or [MS] [$x \gtrless y$] *n*, where *n* is a storage register number. This sequence exchanges the contents of *X* with the contents of the designated storage register.

[CLX]	Special	D

Clears *X* to zero, overwriting *x* regardless of auto-enter. This key is often used to correct a miskeyed number, and so must disable auto-enter. However [R↓] will do as well for this purpose.

*Actually the situation is even more complicated. The very first HP-45s had [RCL] arithmetic but not [STO] arithmetic. Then beginning officially in January 1974 (unofficially in August 1973), HP-45s featured both [STO] and [RCL] arithmetic. HP-45s with an S serial number higher than 1301S2000 or an A serial number higher than 1336A00000 are the new model. Whenever "HP-45" is used herein, I assume the new model.

Before	Keystrokes	After
T: *D*		*C*
Z: *C*	RCL 4 ÷	*C*
Y: *B*		*B*
X: *A*		A/K
Auto-enter: E		E
R_4: *K*		*K*
T: *D*		*D*
Z: *C*	RCL 4 ÷	*D*
Y: *B*		*C*
X: *A*		B/K
Auto-enter: D		E
R_4: *K*		*K*
T: *D*		*D*
Z: *C*	RCL ÷ 4	*C*
Y: *B*		*B*
X: *A*		A/K
Auto-enter: either		E
R_4: *K*		*K*

Figure 1.3.10 Examples of the effect of recall arithmetic on the HP-45.

Keystroke	Type	Effect on Auto-enter
CLR CLR CLEAR CA	Stack operation	E

Clears the stack registers, and, on some calculators, some or all of the storage registers. Except in connection with Σ+, the CLR key is not very useful; CLR between problems is almost never necessary.

Keystroke	Type	Effect on Auto-enter
C↓ C	Stack operation	E

Performs the equivalent of CLX +. This would be a valuable function except that on the Novus calculators on which C↓ appears, *T* (or *Z*) clears on pop.

Keystroke	Type	Effect on Auto-enter
[DSP] [FIX] [FIX] [SCI] [SCI] [DS] [ENG] [ENG]	Display manipulator	U

These keys affect the format of the displayed number. See the instruction booklet with the individual calculator.

Keystroke	Type	Effect on Auto-enter
[G] [B] [Shift] [f] [f^{-1}] [g] [h] [F] [ARC] [INV] [HYP]	Prefix or shift	U

Affects a following keystroke by changing its effect to the shifted mode. For example, [ARC] [SIN] gets $\sin^{-1}$ and [INV] [HYP] [SIN] gets $\sinh^{-1}$ rather than sine. These shifted functions are very valuable, of course, and the shift keys save the cost and size of additional keys on the calculator, but many errors occur through accidental misuse of shift keys. Do *not* hold the shift key down while pressing the following keystroke.

[G] represents an unlabeled gold-colored key and [B] an unlabeled blue key. [g] is pronounced "glue" because it is a blue key with a "g" on it, [f] is "fold," [f^{-1}] is "dolf," and [h] is "hack." With just a little effort, calculator manufactures could think of a more logical color scheme.

Keystroke	Type	Effect on Auto-enter	
[+]	Dyadic functor	E	Pops x and y, adds them $(y + x)$, and pushes the answer.

Keystroke	Type	Effect on Auto-enter	
[−]	Dyadic functor	E	Pops x and y, subtracts them $(y-x)$, and pushes the answer.
[×]	Dyadic functor	E	Pops x and y, multiplies them $(y \times x)$, and pushes the answer.
[÷]	Dyadic functor	E	Pops x and y, divides them $(y \div x)$, and pushes the answer.
[$1/x$] [$1/x$]	Monadic functor	E	Replaces x by $1/x$ for $x \neq 0$.
[$\sqrt{x}$] [$\sqrt{x}$] [$\sqrt{\ }$]	Monadic functor	E	Replaces x by $\sqrt{x}$ for $x \geqslant 0$ or $x > 0$.
[x^2] [x^2]	Monadic functor	E	Replaces x by x^2.
[SIN] [SIN]	Monadic functor	E	Replaces x by $\sin(x)$.
[COS] [COS]	Monadic functor	E	Replaces x by $\cos(x)$.
[TAN] [TAN]	Monadic functor	E	Replaces x by $\tan(x)$.
[SIN^{-1}]	Monadic functor	E	Replaces x by the principal value ($-90°$ to $+90°$) of $\sin^{-1}(x)$ for $-1 \leqslant x \leqslant 1$.

Keystroke	Type	Effect on Auto-enter	
COS^{-1}	Monadic functor	E	Replaces x by the principal value (0 to 180°) of $\cos^{-1}(x)$ for $-1 \leqslant x \leqslant 1$.
TAN^{-1}	Monadic functor	E	Replaces x by the principal value ($-90°$ to $+90°$) of $\tan^{-1}(x)$.

For the six trigonometric functors above, the units of x and the restrictions on the range of x depend on the individual calculator. Many calculators allow a choice of degrees or radians as noted below.

DEG RAD GRD	Mode change	U

Changes the units of the angle to degrees, radians, or grads for all following trigonometric functors (a "grad" is a four hundredth of a circle or a hundredth of a right angle). These keys affect direct and inverse trigonometric functors, rectangular-to-polar-to-rectangular coordinate conversions, and also, on some calculators, the degrees-minutes-seconds conversions below.

→DMS →D.MS →H.MS	Monadic functor	E

Converts the angle or time in x into the format DD.MMSS (degrees, minutes, seconds) or HH.MMSS (hours, minutes, seconds).

D.MS→ →H →D	Monadic functor	E

Converts the angle or time in x from DD.MMSS or HH.MMSS format into decimal degrees or hours (or radians or grads).

Keystroke	Type	Effect on Auto-enter	
[→R] [→P] [→P] [→POL] [R→P]	Bifid functor	E	

Some RPN calculators have rectangular-to-polar ([→P]) and polar-to-rectangular ([→R]) coordinate converters. Starting with x in X and y in Y, [→P] puts $R=\sqrt{x^2+y^2}$ into X and $\theta=\tan^{-1}(y/x)$ into Y. The angle θ is in the correct quadrant ($-180°$ to $+180°$) even if x or y is negative. Or starting with R in X and θ in Y, [→R] puts $x=R\cos\theta$ into X and $y=R\sin\theta$ into Y. On some calculators, θ and R are exchanged in the stack from this description. With either [→P] or [→R], Z and T are unchanged. These functors are extremely useful for many problems in which trigonometric functions play a role. Appendix A contains examples in plane and spherical trigonometry and complex numbers.

[log] [LOG]	Monadic functor	E	Replaces x by $\log_{10}(x)$ for $x>0$
[ln] [LN]	Monadic functor	E	Replaces x by $\ln(x)$; that is, $\log_e(x)$, where $e = 2.718281828$, for $x>0$.
[e^x]	Monadic functor	E	Replaces x by e^x; that is, $\exp(x)$.
[10^x]	Monadic functor	E	Replaces x by 10^x.
[x^y] [y^x] [y^x]	Dyadic functor	E	

Pops x and y, computes x^y for $x>0$ or y^x for $y>0$ and pushes the answer. Neither x nor y needs to be an integer. Some RPN calculators can also do [y^x] for $y\leqslant 0$ and x an integer. Most RPN calculators other than the HP-35 have a [y^x] rather than an [x^y] key. Each has some advantages. For

Keystroke	Type	Effect on Auto-enter	

example, A [↑] B [y^x] gives A^B, which seems more logical than A [↑] B [x^y], which gives B^A. But the stack as usually drawn has Y above X so [x^y] might seem more reasonable. Sometimes [x^y] saves keystrokes over [y^x]. The cube root of A, for example, would be 3 [1/x] A [x^y] or A [↑] 3 [1/x] [y^x]. But other examples favor [y^x]. On some calculators [y^x] requires a prefix or shift key, thus adding another keystroke.

Keystroke	Type	Effect on Auto-enter	
[n!] [x!]	Monadic functor	E	Replaces x by $x!$ for x a positive integer.
[Σ+] [Σ−]	Special functor	D	

Sums Σx_i, Σx_i^2, $\Sigma 1$ (increment), and, on some calculators, Σy_i, $\Sigma x_i y_i$, and Σy_i^2 also. In these sums x_i is the quantity in X and y_i in Y when [Σ+] is pressed. These sums are accumulated in designated storage registers to be used later to calculate means, standard deviations from the means, and perhaps linear regressions. Since [Σ−] accumulates negatively, it can be used to delete an incorrect (x_i, y_i) pair from the sums. On most RPN calculators [Σ+] overwrites x with the latest $\Sigma 1$ but disables auto-enter so a subsequent number overwrites $\Sigma 1$. [CLEAR] is used to zero all the appropriate registers before the first [Σ+]. Check the instruction booklet for details. [Σ+] is discussed further in Section A.1.2; see especially Table A.1.1.

Keystroke	Type	Effect on Auto-enter
[$\bar{x}$,s] [$\bar{x}$] [s] [SD]	Special functor	E

Computes means and standard deviations from the sums previously accumulated by [Σ+].

Keystroke	Type	Effect on Auto-enter
[%] [%]	Bifid functor	E

Replaces x by $xy/100$; Y is unchanged.

Keystroke	Type	Effect on Auto-enter
Δ% %CHS	Bifid functor	E

Replaces x by $100(x-y)/y$ for $y \neq 0$; Y is unchanged.

LASTX LSTX	Storage operation	E

Recalls the contents of the LASTX register; which was previously set from X by any functor that "uses up" x. LASTX terminates and pushes a previous number if any, pushes the stack if auto-enter was enabled, or overwrites x if not.

INT	Monadic functor	E

Replaces x by its integer part. That is, x is truncated (*not* rounded) to an integer. The sign is retained.

FRAC	Monadic functor	E

Replaces x by its fractional part. That is, the integer part of x is subtracted off. The sign is retained.

ABS	Monadic functor	E

Replaces x by the absolute value of x. That is, a minus sign on x, if present, is deleted.

RND	Monadic functor	E

Replaces x by x rounded off as selected by the display format.

The following keystrokes are on programmable RPN calculators. Using a program on such a calculator is a two-step process: first one keys or reads in the program, and then runs it, repeatedly if necessary, to give numerical answers. A switch on the calculator determines whether keystrokes are stored and saved (W/PRGR or PRGM) or executed (RUN).

Some quite significant differences exist among programmable RPN calculators, even from the same manufacturer, and the descriptions below are only approximations in some cases. Consult the instruction booklet for details.

Keystroke

[R/S] (run/stop)
[START]
[HALT]

In RUN mode, these keystrokes start or stop the operation of a program. An internal program pointer determines which instruction will be executed next. On encountering [R/S] or [HALT], a running program stops to allow operator intervention (e.g., keying a number).

[GTO]

In RUN mode, this keystroke sets the program pointer to the instruction corresponding to the keystroke or keystrokes following, so that a subsequent [R/S] starts there. On encountering [GTO] in a running program, the program pointer jumps (forward or backward) to the instruction corresponding to the keystroke or keystrokes following, and the program runs on from there.

There are two systems for labeling instructions within programs. The HP-65, HP-67, and HP-97 use formal labels (A through E and 1 through 9) which can be inserted anywhere in the program. This is a very versatile system, but these labels occupy instruction locations. Other calculators use a two-digit number for each instruction location.

[LBL]

This keystroke in a stored program defines the keystroke following to be the label of this location, so that a [GTO] can be used to start execution at this point. The keystroke following [LBL] must be [A] through [E], [1] through [9], or [a] through [e].

[A]
[B]
[C]
[D]
[E]
[a]
[b]
[c]
[d]
[e]

Keystroke

In RUN mode, these keystrokes start program execution at the corresponding label and set the subroutine return address to the keyboard. In a stored program, following [LBL], these keystrokes label the start of subroutines. In a stored program not following [LBL], these keystrokes cause program execution to transfer to the appropriate subroutine (a subroutine "call") and the calling address is saved so that

[RTN]

at the end of the subroutine transfers control back to the instruction following the subroutine call. If a subroutine is called from the keyboard, [RTN] is equivalent to [R/S] and control returns to the keyboard. Subroutine calls and [RTN]s are very valuable for certain problems.

[SF1]
[SF2]
[SF]
[CF]
[STF]
[CLF]

The HP-65 has two and the HP-67 and HP-97 have four internal flags that can be set or reset by these keystrokes.

[$x<y$]
[$x<0$]
[$x\geqslant y$]
[$x\geqslant 0$]
[$x\neq y$]
[$x\neq 0$]
[$x=y$]
[$x=0$]
[$x\leqslant y$]
[$x>y$]
[$x>0$]
[TF1]
[TF2]
[F?]

These are conditional skip instructions. When encountered in a running program, they cause the instruction following (two instructions on the HP-65) to be skipped if the indicated condition is *false* (i.e., no skip if *true*).

Keystroke

[DSZ]
[DSZ]
[ISZ]

"Decrement and skip on zero" or "increment and skip on zero" adds or subtracts 1 from the contents of a storage register and skips the next instruction (or two) if the result is then zero.

[SST]
[SKIP]

"Single step" moves the program pointer ahead by one step in PRGM mode to allow editing of a program. In RUN mode, [SST] causes one instruction of a program to be executed.

[BST]
[BST]
[DEL]

"Back step" or "delete" allows editing in PRGM mode by backspacing so an incorrect instruction can be overwritten. In RUN mode [BST] resets the program pointer to zero on the HP-55; [f] [PRGM] performs this function on the HP-25, [RTN] on the HP-65.

[PAUSE]

This instruction in a running program pauses about a half-second—almost long enough to be able to read the number in the display.

[NOP]

"No operation" does just that; it is used as a filler after skips or to wipe out an instruction without affecting the locations of the other instructions.

Omitted from the foregoing list are some very specialized functors that are not mentioned in this book. In any case the reader should peruse the instruction booklet that accompanies the calculator. The instruction booklets with HP calculators, especially recent models, are very well prepared. The instruction booklets with National Semiconductor and Novus calculators are not quite as good, and the instruction booklet with the Corvus-500 calculator is just awful. (Hence Corvus-500 owners will better appreciate this book.)

Everything You've Always Wanted To Know About RPN But Were Afraid To Pursue—Comprehensive Manual for Scientific Calculators (see Appendix C) is, in effect, an instruction booklet for the Corvus 500. I recommend *Everything...* for Corvus-500 owners, even though it has at least 86 errors (mostly typographical), including one on the cover.

1.3.2 Functors and Flags

Functors take at least one parameter or argument off the stack and put at least one answer onto the stack. A monadic functor (e.g., [SIN]) pops a number, calculates the function, and pushes the answer. A dyadic functor (e.g., [+]) pops two numbers, calculates the function, and pushes the answer. A bifid functor (e.g., [→P]) pops two numbers, calculates the functions, and pushes the two answers. Other possibilities also occur. For example, the functor [%] found on some RPN calculators *uses* two numbers off the stack but *pops* only one of them; Y is unchanged. This is logically equivalent to a bifid functor that pushes back as its first answer the same number as its second parameter, that is, $h_1(A,B)=B$ in Figure 1.2.1.

Three internal flags (one-bit numbers) in a typical RPN calculator indicate the status. Auto-enter can be enabled or disabled, a number can be in the process of being entered, or an exponent can be in the process of being entered. Some keystroke operations check and others change the status of these three flags. Only the auto-enter flag seems to cause any confusion. When a new number goes into X, there are two possibilities: if auto-enter is enabled, the new number pushes the stack as described previously; but if auto-enter is disabled, the new number overwrites the old number in X and leaves the contents of Y, Z, and T unchanged. The new number can come from number keys, from [RCL], or from a special key such as [π]. To save the previous contents of X, the first possibility above (push, or auto-enter) is needed.

The [CLX] key writes zero into X (no push) but disables auto-enter, so that a number following [CLX] overwrites the zero rather than pushing it. And [↑] performs a "duplicate" function (push the contents of X into Y also, etc.) but disables auto-enter so that if [↑] is followed by a new number, X is overwritten and only one copy of the previous number remains on the stack. Thus [↑] can be either a number separator or a duplicator.

The minus symbol ($-$) in algebra serves *three* separate functions that are performed by *two* keys—[−] and [CHS]—on a typical RPN calculator. [−] is unambiguous; it is the dyadic functor "subtract." However, [CHS] can correspond either to a minus sign on a number or to the monadic

T:								
Z:						A		
Y:		A	A	A		$-A$		A
X:	A	A	$-A$	B	*not*:	B	*nor*:	$-B$
Key:	A	↑	CHS	B				

Figure 1.3.11 A common error on calculators other than the HP-35: CHS following ↑ gets lost.

functor "negate." In its minus-sign role CHS is different from other number keys in that it cannot begin a number (except on the HP-35) as it logically would; and in its functor role, CHS differs from other monadic functors on most RPN calculators in not terminating a number and not enabling auto-enter. Nevertheless, this combination of functions in a single key is probably desirable because it reduces the number of keys on the keyboard and may reduce confusion in some problems.

But confusion can arise on some calculators if ↑ precedes CHS because auto-enter will still be disabled after the CHS. The negated number will then be overwritten (the CHS will be lost) if a number or RCL follows ↑ CHS. This error is illustrated in the sequence in Figure 1.3.11. On some RPN calculators, other keystrokes such as STO can occur between CHS and B in Figure 1.3.11, without affecting auto-enter.

T:				
Z:				
Y:		A	A	A
X:	A	A	$-A$	$-B$
Key:	A	↑	CHS	B

T:						
Z:						
Y:		A	A			$A+B$
X:	A	A	B	$A+B$	$-(A+B)$	$-C$
Key:	A	↑	B	+	CHS	C

STO:				$-A$	$-A$
T:					
Z:					
Y:		A	A	A	A
X:	A	A	$-A$	$-A$	B
Key:	A	↑	CHS	STO	B

Figure 1.3.12 On an HP-35, CHS never gets lost, but sometimes it attaches itself to an unlikely number.

On an HP-45, for example, the sequence A [↑] [CHS] [STO] 4 B puts $-A$ into storage register 4 but not onto the stack! But this is not necessarily an error—it might be what the user wanted. If a functor follows [↑] [CHS], the negated number is correctly used, and the functor will enable auto-enter.

On an HP-35, [CHS] is even curiouser. If [CHS] precedes a number entry, the [CHS] first negates the number in X (as it must, for the calculator cannot know which key will be pressed next). Then, when a number key is pressed, the number in X is negated again (becomes what it was originally), and the new number is negated and either pushed onto the stack if auto-enter was enabled, or just written into X if not. Three such sequences are illustrated in Figure 1.3.12.

1.3.3 Translating into RPN

In ordinary algebra the expression A/BC is ambiguous;* it might mean $(A/B)C$, but more likely means $A/(BC)$. In algebraic computer languages such as FORTRAN and BASIC, the expression $A/B*C$ is unambiguous; it means $(A/B)*C$ (* means multiply). In translating A/BC and similar expressions into RPN (which is *not* ambiguous), one may have to guess whether the author of the expression intended A [↑] B [÷] C [÷] or A [↑] B [÷] C [×]. In really ambiguous cases, try working out the units (meters, kilograms, seconds, etc.). This ambiguity sometimes occurs when authors translate their equations into serial form (all on one line) at the urging of the printer of a book or journal. The horizontal bar in a fraction in nonserial form is a vinculum and implies parentheses around both numerator and denominator.

Figure 1.3.13 is a flow chart of a method for evaluating arithmetic expressions as recommended by Hewlett-Packard in instruction booklets for some of their early RPN calculators. To use this method, first write the expression to be evaluated on one line (serial form), adding parentheses as necessary. This method has the advantage that one proceeds from left to right through the expression, keying each number (but *not* each operation) in turn. This method can be done mechanically with little thought, but it would work for any expression only if the push-down stack were arbitrarily long. The principal disadvantage of this approach is that it rarely yields the shortest possible keystroke sequence.

I recommend, instead, a method that requires more thought, at least at

*"Please Excuse My Dear Aunt Sallie" is a mnemonic for the algebraic hierarchy: parentheses, exponentiation (involution or evolution), multiplication, division, addition, and subtraction. Interpreted literally, this hierarchy would make A/BC into $A/(BC)$, but it would also make $A-B+C$ into $A-(B+C)$!

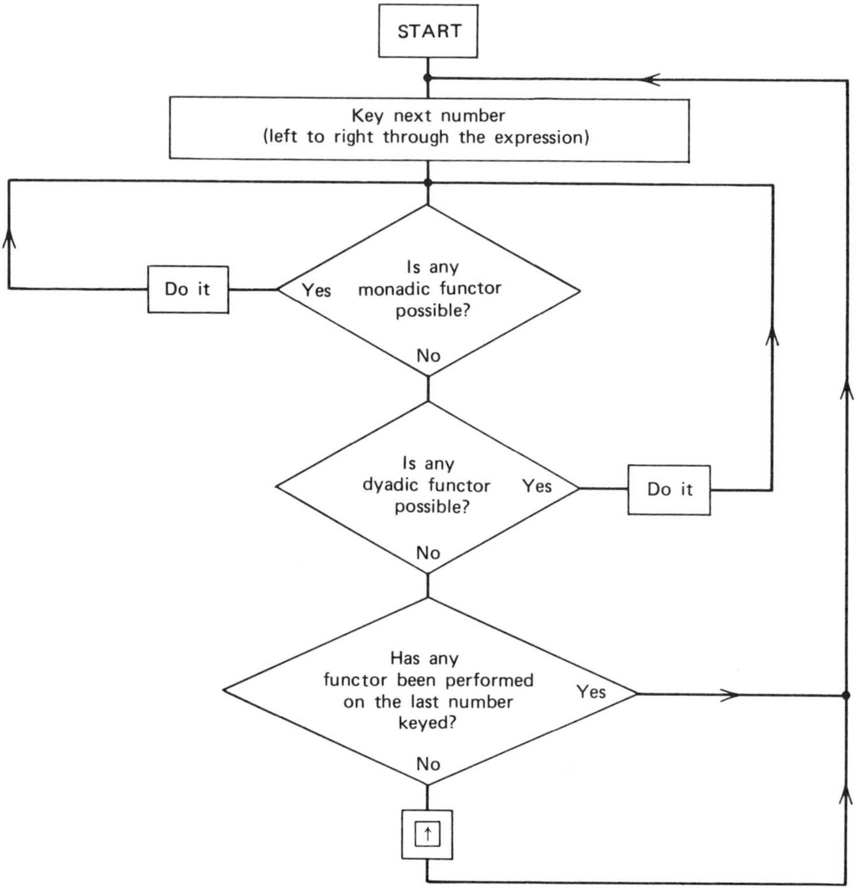

Figure 1.3.13. A method for translating expressions in serial form into RPN, as recommended by Hewlett-Packard.

first, but usually yields more nearly optimum keystroke sequences and works for much more complicated expressions than the method above. Essentially one works from inside out, starting inside the innermost parentheses and proceeding outward. Products and fractions have implied parentheses around them, and the arguments of functions usually must be evaluated first of all. This method is just about what would be done with pencil and paper or with slide rule and abacus, but intermediate answers can be left on the stack for later use (unless it fills up). This second method, or something very similar, is now also recommended by HP (Martin, R., private communication). Perhaps the best way to show these methods is through a series of examples. In the following, algorithm 1 is from HP's method in Figure 1.3.13; algorithm 2 is from my method.

Expression	Algorithms
$A-BC$	1. $A \boxed{\uparrow} B \boxed{\uparrow} C \boxed{\times} \boxed{-}$.
	2. $B \boxed{\uparrow} C \boxed{\times} A \boxed{x \gtrless y} \boxed{-}$.

Note that BC has implied parentheses around it and that the two methods yield algorithms of the same length. In method 2, the $\boxed{x \gtrless y}$ could be omitted if a $\boxed{\text{CHS}}$ were appended at the end, but with no change in the number of keystrokes.

Expression	Algorithms
$A/(B+C)$	1. $A \boxed{\uparrow} B \boxed{\uparrow} C \boxed{+} \boxed{\div}$.
	2. $B \boxed{\uparrow} C \boxed{+} A \boxed{x \gtrless y} \boxed{\div}$.

In method 2, the $\boxed{x \gtrless y}$ could be omitted if a $\boxed{1/x}$ were appended at the end, but with no change in the number of keystrokes. These examples show that method 2 sometimes needs an $\boxed{x \gtrless y}$ before $\boxed{-}$ or $\boxed{\div}$ (or a $\boxed{\text{CHS}}$ or $\boxed{1/x}$ afterward) because the parameters can be in the wrong order on the stack. This never occurs with method 1.

Expression	Algorithms
$A+BC$	1. $A \boxed{\uparrow} B \boxed{\uparrow} C \boxed{\times} \boxed{+}$.
	2. $B \boxed{\uparrow} C \boxed{\times} A \boxed{+}$.
$A(B-CD)$	1. $A \boxed{\uparrow} B \boxed{\uparrow} C \boxed{\uparrow} D \boxed{\times} \boxed{-} \boxed{\times}$.
	2. $C \boxed{\uparrow} D \boxed{\times} B \boxed{x \gtrless y} \boxed{-} A \boxed{\times}$.
$(B-CD)A$	1. $B \boxed{\uparrow} C \boxed{\uparrow} D \boxed{\times} \boxed{-} A \boxed{\times}$.
	2. $C \boxed{\uparrow} D \boxed{\times} B \boxed{x \gtrless y} \boxed{-} A \boxed{\times}$.

Rewriting the expression algebraically sometimes shortens the resulting algorithm. This occurs frequently with method 1, and occasionally with method 2.

Expression	Algorithms
$(A-B)(C-D)$	1. $A \boxed{\uparrow} B \boxed{-} C \boxed{\uparrow} D \boxed{-} \boxed{\times}$.
	2. $A \boxed{\uparrow} B \boxed{-} C \boxed{\uparrow} D \boxed{-} \boxed{\times}$.
$AB-CD$	1. $A \boxed{\uparrow} B \boxed{\times} C \boxed{\uparrow} D \boxed{\times} \boxed{-}$.
	2. $A \boxed{\uparrow} B \boxed{\times} C \boxed{\uparrow} D \boxed{\times} \boxed{-}$.

Note the similarity of these two forms—product of sums and sum of products—both algebraically and in RPN.

Expression	Algorithms
$A(B+C(D-E))$	1. No go: a five-level stack is needed.
	2. $D \boxed{\uparrow} E \boxed{-} C \boxed{\times} B \boxed{+} A \boxed{\times}$.
$((D-E)C+B)A$	1. $D \boxed{\uparrow} E \boxed{-} C \boxed{\times} B \boxed{+} A \boxed{\times}$.
	2. $D \boxed{\uparrow} E \boxed{-} C \boxed{\times} B \boxed{+} A \boxed{\times}$.

To use method 1 efficiently, try rewriting the expression algebraically such that all the opening parentheses are at the beginning.

Expression	Algorithms
$1+\sin^2(2A)$	1. 1 [↑] 2 [↑] A [×] [SIN] [x^2] [+].
	2. A [↑] [+] [SIN] [x^2] 1 [+].

On a calculator without [x^2], use [↑] [×] instead.* To calculate $2A$, A [↑] [+] uses fewer keystrokes than 2 [↑] A [×] or A [↑] 2 [×]. Again method 1 would yield a shorter algorithm if the expression were rewritten as $\sin^2(2A)+1$.

1.3.4 Errors and Error Propagation†

When an approximate calculation is done, we usually need to know how precise the answer is. Relative or absolute errors can be expressed in several ways. If A is an approximate value and T is the true value, then

$$\xi = A - T \tag{1.3.1}$$

is the *absolute error*, and

$$\varepsilon = \frac{A-T}{T} = \frac{\xi}{T} \tag{1.3.2}$$

is the *relative error* in A. The percentage error is 100 times the relative error, that is, ε expressed as a percentage. Some authors define $-\xi$ as the error, $|\xi|$ as the absolute error, and $|\varepsilon|$ as the relative error. Usually ξ and ε are unknown (because T is unknown), but one may be able to estimate a value or an upper limit for $|\xi|$ and $|\varepsilon|$.

The number of *decimal digits* of precision in A is the number of correct digits to the right of the decimal point in the standard decimal expression for A. The number of *significant figures* in A is the number of correct digits regardless of the decimal point. But until one specifies how precise a digit must be to be "correct," these definitions are unusable. Will ± 1 or ± 0.5 do? And does it matter whether the digit is 1, 5, or 9?

Rather than these imprecise definitions, define the number of decimal digits of precision as

$$DD = -\log_{10}(|\xi|), \tag{1.3.3}$$

*But [↑] [×] loses the contents of T. This is not important in this example.

†This section is somewhat more difficult and specialized than other material in Chapter 1 and may be omitted on first reading.

and the number of significant figures as

$$SF = -\log_{10}(|\varepsilon|) = DD + \log_{10}(|T|). \tag{1.3.4}$$

These definitions accord approximately with the imprecise definitions just given. Specifically, 1 ± 0.1, 5 ± 0.5, and 9 ± 0.9, each corresponds to one significant figure, $SF = 1$.

The absolute error of the *sum* or *difference* of several numbers is at most equal to the sum of the absolute values of the absolute errors of the individual numbers, and, if the errors in the numbers are *uncorrelated*, is more probably approximately the square root of the sum of the squares of the absolute errors of the individual numbers. Similarly, the relative error of the *product* or *quotient* of several numbers is at most approximately equal to the sum of the absolute values of the relative errors of the individual numbers, and, if the errors in the numbers are uncorrelated, is more probably approximately the square root of the sum of the squares of the relative errors of the individual numbers. Symbolically, if

$$B = A_1 \pm A_2 \pm A_3 \pm \cdots \pm A_n, \tag{1.3.5}$$

then

$$|\xi_1| + |\xi_2| + |\xi_3| + \cdots + |\xi_n| \geqslant |\xi_B| \cong \left[\xi_1^2 + \xi_2^2 + \xi_3^2 + \cdots + \xi_n^2\right]^{1/2}, \tag{1.3.6}$$

or if

$$C = A_1 * A_2 * A_3 * \cdots * A_n, \tag{1.3.7}$$

where $*$ represents either $\times$ or $\div$, then

$$|\varepsilon_1| + |\varepsilon_2| + |\varepsilon_3| + \cdots + |\varepsilon_n| \gtrsim |\varepsilon_C|$$

$$\cong \left[\varepsilon_1^2 + \varepsilon_2^2 + \varepsilon_3^2 + \cdots + \varepsilon_n^2\right]^{1/2}. \tag{1.3.8}$$

In these formulas, ξ_1 represents the absolute error, and ε_1 the relative error in A_1, and so on.

For operations more complex than sums or products, one can use the approximate theory of linear error propagation with first derivatives. If D is any function of A,

$$D = f(A), \tag{1.3.9}$$

then

$$\xi_D \cong f'(A)\xi_A. \tag{1.3.10}$$

This is only an approximate value for ξ_D since higher-order derivatives are neglected. In general, if

$$E = f(A_1, A_2, A_3, \cdots, A_n), \tag{1.3.11}$$

and if we write

$$\rho_1 = \frac{\partial f}{\partial A_1}\xi_1, \ \rho_2 = \frac{\partial f}{\partial A_2}\xi_2, \text{ etc.,} \tag{1.3.12}$$

then

$$|\rho_1| + |\rho_2| + |\rho_3| + \cdots + |\rho_n| \gtrsim |\xi_E| \cong \left[\rho_1^2 + \rho_2^2 + \rho_3^2 + \cdots + \rho_n^2\right]^{1/2}. \tag{1.3.13}$$

As an example, suppose we average n numbers A_i,

$$F = \frac{1}{n}\sum_{i=1}^{n} A_i. \tag{1.3.14}$$

If the ξ_i are approximately equal and uncorrelated, using either equation 1.3.6 or 1.3.13, we get

$$|\xi| \geqslant |\xi_F| \cong \frac{|\xi|}{\sqrt{n}}. \tag{1.3.15}$$

In the denominator $\sqrt{n-1}$, rather than $\sqrt{n}$, is a better estimate in most cases.

As another example, suppose we raise a number A to a power $\alpha > 1$, or a root $0 < \alpha < 1$,

$$G = A^{\alpha}. \tag{1.3.16}$$

Then

$$\varepsilon_G \cong \frac{\xi_G}{G} \cong \frac{\alpha A^{\alpha-1}\xi_A}{A^{\alpha}} = \frac{\alpha\xi_A}{A} \cong \alpha\varepsilon_A. \tag{1.3.17}$$

Thus the relative error in G is approximately α times the relative error in

A. Or if

$$H = \ln A;\ A = e^H, \tag{1.3.18}$$

then

$$\xi_H \cong \frac{\xi_A}{A} \cong \varepsilon_A. \tag{1.3.19}$$

Therefore the relative error in A is approximately equal to the absolute error in H.

The overused example of what not to do involves subtracting two nearly equal numbers or adding two numbers approximately equal in magnitude but opposite in sign. From equation 1.3.6, the absolute error is not too bad, but the relative error, with its small denominator, can be huge. "Avoid subtracting nearly equal numbers" is an admonition sometimes unnecessary and sometimes impossible to follow. The relative error in an approximate value for zero is necessarily large.

If more than one algorithm is available for a problem, as is usually the case, investigate the error in the answer with each algorithm, using the formulas in this section, and bearing in mind that intermediate answers can be expressed only to the calculator accuracy. This rule of thumb is more difficult to follow but more reliable than the naïve admonition above. An example appears in Section A.9.5.

Some of the ideas in this section are unconventional. For a detailed discussion of the conventional viewpoint, see Chapter 1 in Demidovich and Maron (1973) or Chapter 1 in Scarborough (1962).

1.4 WHY RPN?

Although the first scientific pocket calculators employed RPN, many more models are available today with some form of an algebraic entry system (AES). RPN calculators have an [ENTER↑], [ENT], or [SAVE↑] key, AES calculators an [=] key. Advertisements for the two systems make what seem to be contradictory claims: each system is said to be easier to learn, remember, and use. This section attempts to separate sense from nonsense in this area.

Simple AES calculators have two internal registers for numbers and do each operation when the following operation key is pressed. Some expressions can be keyed directly in this system. For example,

$$\frac{AB}{C} - E \tag{1.4.1}$$

can be done on a simple AES calculator by

$$A \boxed{\times} B \boxed{\div} C \boxed{-} E \boxed{=}. \tag{1.4.2}$$

Note that $\boxed{=}$ is always needed at the end to complete the last operation. But some expressions need to be rewritten to work in this system. For example,

$$A - BC \tag{1.4.3}$$

can be done on a simple AES calculator by

$$B \boxed{\times} C \boxed{+/-} \boxed{+} A \boxed{=}. \tag{1.4.4}$$

The $\boxed{+/-}$ key corresponds to $\boxed{\text{CHS}}$ in RPN and is a post operator. If, instead, one keys

$$A \boxed{-} B \boxed{\times} C \boxed{=}, \tag{1.4.5}$$

the result is $(A-B)C$.

Another AES scheme, referred to as AESH, has three internal registers for numbers and an operational *h*ierarchy according to which $\times$ and $\div$ are done before $+$ and $-$. This system is similar to the ordinary algebraic hierarchy without parentheses. Any serial-form arithmetic expression without parentheses and with the foregoing convention can be done in AESH just by keying from left to right. For example, in AESH

$$A \boxed{-} B \boxed{\times} C \boxed{=} \tag{1.4.6}$$

gets $A-BC$ rather than $(A-B)C$. For this problem the AESH calculator saves A and "$-$" internally and does this operation after the multiply; both operations in this example actually take place when $\boxed{=}$ is pressed at the end.

Still another AES scheme, referred to as AESP, has four internal registers for numbers and keys for two-level *p*arentheses: $\boxed{(}$ and $\boxed{)}$. Any serial-form arithmetic expression with no more than two-level parentheses can be done in AESP just by keying from left to right. For example,

$$A \boxed{-} \boxed{(} B \boxed{\times} C \boxed{)} \boxed{=} \tag{1.4.7}$$

gets $A-BC$. On some (but not all) AESP calculators, the closing parenthesis before $\boxed{=}$ can be omitted.

The most elaborate AES scheme is the algebraic operating system (AOS), which combines AESH and AESP with several levels of parentheses and automatically closes all parentheses on $\boxed{=}$.

As one possible method to compare these systems, I wrote algorithms for all possible problems involving up to four parameters and the four basic dyadic operations (+, −, ×, ÷). Table 1.4.1 contains a reasonably systematic listing of this class of problems and what I believe to be the shortest algorithms for the five systems. To simplify the notation, I use ± for either + or −, ∗ for either × or ÷, and · to stand for any of these four operations. Categories of problems are used to shorten the table whenever this can be done without conferring an advantage on any of the systems. In some cases a particular category can be done without knowing which operation · represents. These cases are indicated in Table 1.4.1, but the results are not included in Table 1.4.2. For the AESP scheme, I assume that all parentheses need to be closed, although this is not necessary before $\boxed{=}$ on some AESP calculators. For the AES scheme, I assume that $\boxed{=}$ has no effect on the $\boxed{\text{STO}}$ register. A bar over an operation in Table 1.4.1 represents the complement of the original operation. For example, $\boxed{\overline{-}} = \boxed{+}$, and $\boxed{\overline{\times}} = \boxed{\div}$.

As a criterion to apply in choosing among these five systems, the total number of keystrokes needed for each system appears in Table 1.4.2. These tables contain some arbitrariness, particularly in the scoring system. Each line in Table 1.4.1 represents a category of problems and has a weight associated with it to approximate the frequency of occurrence of such problems. Other scoring methods are possible, but the final outcome would probably be the same. Clearly RPN wins *if* this type of problem is being done and *if* minimizing the number of keystrokes is the criterion.

The problems in Table 1.4.1 are only a small subset of the arithmetic problems encountered in life. But RPN wins also for more complex problems. Since the AES and AESH calculators have already used up their $\boxed{\text{STO}}$ register for some types of problem in Table 1.4.1, the $\boxed{\text{STO}}$ register is not available for more complex problems. The additional internal registers enable RPN, AESP, and AOS calculators to do more complex problems that would require writing down and rekeying intermediate answers on AES or AESH calculators with only one $\boxed{\text{STO}}$ register. Consider, for example,

$$\frac{A+B}{C+D}+\frac{E+F}{G+H}. \tag{1.4.8}$$

An RPN algorithm for this expression is

$$A\boxed{\uparrow}B\boxed{+}C\boxed{\uparrow}D\boxed{+}\boxed{\div}\ E\boxed{\uparrow}F\boxed{+}G\boxed{\uparrow}H\boxed{+}\ \boxed{\div}\ \boxed{+}, \tag{1.4.9}$$

an AESP algorithm is

$$A \boxed{+} B \boxed{\div} \boxed{(} C \boxed{+} D \boxed{)} \boxed{+} \boxed{(} \boxed{(} E$$
$$\boxed{+} F \boxed{)} \boxed{\div} \boxed{(} G \boxed{+} H \boxed{)} \boxed{)} \boxed{=}, \tag{1.4.10}$$

and an AOS algorithm is

$$A \boxed{+} B \boxed{=} \boxed{\div} \boxed{(} C \boxed{+} D \boxed{)} \boxed{+} \boxed{(} E \boxed{+} F \boxed{)} \boxed{\div} \boxed{(} G \boxed{+} H \boxed{=}. \tag{1.4.11}$$

This problem cannot be done in AES or AESH without either writing down and rekeying an intermediate answer, or rewriting the expression and rekeying some of the parameters.

Although RPN and AESP calculators usually have the same number of

Table 1.4.2 Keystroke Summary

Sum of Wins × Weights from Table 1.4.1			
	RPN	AESP	Ties
RPN vs. AESP	28	0	328
	RPN	AES	Ties
RPN vs. AES	54	0	302
	RPN	AOS	Ties
RPN vs. AOS	122	38	196
	RPN	AESH	Ties
RPN vs. AESH	156	30	170
	AESP	AES	Ties
AESP vs. AES	16	0	340
	AESP	AOS	Ties
AESP vs. AOS	110	40	206
	AESP	AESH	Ties
AESP vs. AESH	144	30	182
	AES	AOS	Ties
AES vs. AOS	110	48	198
	AES	AESH	Ties
AES vs. AESH	122	31	203
	AOS	AESH	Ties
AOS vs. AESH	44	0	312

Sum of Keystrokes × Weights from Table 1.4.1	
RPN	1484
AESP	1512
AES	1528
AOS	1568
AESH	1626

working registers (four), the additional control through the stack manipulator keys ([x≷y], [R↓], and [↑]) gives RPN an advantage over AESP in many more complex problems. AOS calculators with more than two-level parentheses have more than four working registers, which sometimes is an advantage.

Monadic functions such as sines and logarithms are not considered in this comparison because they are handled identically in the five schemes; AES, AESP, AESH, and AOS calculators in fact use RPN!

This study has not really answered the question of which system is easier because "easy" is not easy to define. I can only comment that I found filling out the RPN column in Table 1.4.1 easier, I made fewer mistakes in the RPN column, and I selected an RPN calculator for my personal use and to write a book about.

Some of the material in this section is based on Ball (1975a).

1.5 EXERCISES

1.5.1 (*a*) List all the calculator keys that undo themselves; that is, if pushed twice, they are equivalent to no operation (ignore auto-enter). (*b*) List all the calculator keys that can be pushed twice or more and have the same effect as if they had been pushed only once.

For each part of the following two problems, assume that the calculator starts with A in all four stack registers (e.g., from A [↑] [↑] [↑]):

1.5.2 Calculate each of the following in two keystrokes or fewer: (*a*) $2A$, (*b*) A^2, (*c*) $2A^2$, (*d*) A^3, (*e*) $A(1-A)$.

1.5.3 What quantity is in X after each of the following sequences: (*a*) [+] [×] [×]; (*b*) [×] [×] [+]; (*c*) [×] [+] [×]; (*d*) [×] [−] [+]; (*e*) [+] [×] [−]; (*f*) [×] [−] [÷]; (*g*) [+] [+] [÷]?

1.5.4 Suppose the diameter of the earth were known precisely (it is approximately 12,740 km). Estimate the inaccuracy in the circumference calculated using only a 10-digit calculator.

1.5.5 An approximation for π is 355/113. How many significant figures and how many correct decimal digits are in this approximation?

1.5.6 Suppose the diameter of the earth were known precisely. Estimate the inaccuracy in the circumference calculated using 355/113 instead of π.

1.5.7 A railroad track a mile long is rigidly fixed at both ends. As a result of a temperature change, the track increases in length by an inch and buckles up in the middle. Assume that the track forms an approximate triangle with the ground and estimate how high the buckled track is off the ground at its center.

1.5.8 The golden ratio ϕ is the first of a series of (positive) numbers each characterized by being equal to the reciprocal of its own fractional part. Calculate the first four numbers of this series.

1.5.9 The average distance between the earth and the sun is about 93,000,000 miles, from the earth to the moon about 239,000 miles, and from the sun to the nearest star about 4.3 light years or 2.5×10^{13} miles. The diameter of the sun is about 865,000 miles, of the earth about 7900 miles, and of the moon about 2160 miles. A scale model is to be built with the sun represented by a ball one foot in diameter. On this scale (*a*) what is the diameter of the earth? (*b*) the diameter of the moon? (*c*) the distance from the sun to the earth? (*d*) the distance from the earth to the moon? (*e*) the distance from the sun to the nearest star? What is (*f*) the angular diameter of the sun as seen from the earth? (*g*) of the moon as seen from the earth?

1.5.10 As his reward for inventing the game of chess, Sissa asked his Rajah for one grain of wheat for the first square, two grains for the second, four for the third, eight for the fourth, and so on, up to the 64 squares of the chessboard. (*a*) If this request had been granted, how many grains of wheat would Sissa have received? (*b*) If Sissa had put his wheat into a granary 40×80 feet, and if 200 grains of wheat occupy a cubic inch, how deep a layer would have resulted?

1.5.11 Two hundred years ago you could buy a pound of butter, two pounds of coffee, a pound of cheese, and a five-pound sack of flour for less than a dollar. The same items in 1976 cost about \$6.95. If "less than a dollar" means \$0.99, what is the average inflation rate over the two centuries?

1.5.12 On April 26, 1976, the length of the shadow of a tower near Boston was 93 feet when the shadow was shortest (local noon). How tall is the tower? *Hints:* Use the "Sun Ephemeris" algorithm in Section A.7.14 to find the declination δ of the sun. Then at local noon, the elevation of the sun is just $90° - \text{latitude} + \delta$. The latitude of Boston is about 42°20′.

1.5.13 Calculate or look up (*a*) the reciprocal speed of sound in seconds per mile, (*b*) the reciprocal speed of light (or radio waves) in microseconds per mile, (*c*) the number of cubic inches in a gallon, and (*d*) the number of fluid ounces in a cubic foot.

2 Terser and Tighter Algorithms

How do I love thee? Let me count the [keystrokes]. I love thee to [calculate] the depth and breadth and height...

—Corrupted from Elizabeth Barrett Browning (1806–1861)

Omit needless words. Vigorous writing is concise. A sentence should contain no unnecessary words, a paragraph no unnecessary sentences, [and an algorithm no unnecessary keystrokes.]*

—Corrupted from William Strunk, Jr. (1869–1946)

2.1 THE ALGORITHM METHOD

The preceding chapter contains techniques used to translate arithmetic or algebraic formulas into RPN. For some problems these translations can be done on sight (i.e., without writing down the RPN keystrokes). For more complex problems one may write out the RPN keystrokes, along with any other instructions or notes necessary to do the problem. Such a keystroke

*From Strunk and White (1972), p. 17.

procedure or recipe for a problem is called an algorithm. The word algorithm probably came from Al-Khowârizmî, author of a famous ninth-century book on mathematics. For programmable calculators, in which most or all of the algorithm is stored in the calculator's memory, such an algorithm should properly be called a program, and such a calculator should be called a computer. In this book at least, these semantic distinctions can be ignored and all keystroke procedures can be called algorithms.

Even for complex problems done only once (one set of numerical values for the parameters), one may be able to do the translations into RPN on sight. More typically, however, one needs to do a problem repeatedly with different sets of numerical values for the parameters; thus the need to write down an RPN algorithm. Books containing selections of algorithms are available from Hewlett-Packard for their calculators (see references in Appendix C), and Appendix A in this book contains a selection of algorithms for a variety of problems.

One need not derive an algebraic expression to be able to put numbers into it and get numerical answers; so also one need not understand how or why an algorithm works to be able to use it. Furthermore, "blindly" using an algorithm is not necessarily undesirable; one cannot know *everything*. But one *should* be able to write one's own algorithms; problems do come up without algorithms "in the book." This chapter is intended to help.

Many possible notational schemes exist for RPN algorithms. I use solid boxes enclosing keystroke symbols (e.g., [÷]), but usually no boxes around numbers (e.g., 3). Key symbols following a shift key are on the land area above or below the key on the calculator, or on the side rather than the top of the key, and are indicated herein by dashed boxes (e.g., [y^x]). A symbol for a parameter (e.g., A) occurs in an algorithm where the numerical value should be keyed. Explanations and notes within the algorithm, and parameters requiring units or other explanation, are usually enclosed in parentheses. More compact notations are obviously possible, and Section 2.6 contains a different notation for algebraic manipulations in RPN. McKelvey (1975) and Ball (1976a) used an intermediate notation with commas as separators.

2.2 WHAT IS A GOOD ALGORITHM?

Several different algorithms are usually possible, even for the simplest problems. As an absolute figure of merit for an algorithm, I recommend *the total number of keystrokes*—the fewer the better. This statement needs to be qualified only by saying that each numerical parameter should be keyed only once, because keying a parameter can cost up to 16 keystrokes, in principle if rarely in practice. And this statement needs only the following qualification: that some parameters known to be small integers (e.g., n in $J_n(x)$) can be rekeyed if necessary. The number of keystrokes is an appropriate figure of merit to choose among algorithms because keystrokes cost time and effort, and also because each keystroke is a potential error. The same criterion applies to programmable calculators, although one might argue that it is somewhat less important; if the algorithm *fits* and *works*, then in some sense, it is good enough. With cyclic or iterative algorithms, the total number of keystrokes as written is less than the number of keystrokes actually pressed on a nonprogrammable calculator; one must multiply by the number of cycles. On a programmable calculator, the number of keystrokes as written determines whether the algorithm will fit; the total number of operations determines how long the algorithm will run—usually a consideration of lesser importance.

Only if two or more alternative algorithms exist for the same problem with the same number of keystrokes, are subsidiary criteria required. I feel less strongly about these, but I recommend algorithms with repeated keystrokes (e.g., [↑] [↑] [↑]) or with cyclic patterns of keystrokes, since one usually makes fewer errors with such patterns.

Although some rough rules of thumb aid in writing more concise algorithms, the process is mostly dependent on cleverness, which has, by definition, no rules. This section contains some of these rough rules that I find useful, and some examples of problems with algorithms.

Although all RPN calculators have at least one [STO] register, to [STO] and [RCL] a number frequently costs more keystrokes than to leave it on the stack for subsequent use. Try to rearrange the order of a problem to have the numbers on the stack in the correct order for subsequent steps with few stack manipulations ([R↓], [$x \gtrless y$], etc.), and without [STO] and [RCL].

Try putting a number needed repeatedly into T (e.g., by [↑] [↑] [↑]) rather than [STO], thus avoiding all [RCL]s. The number from T will turn up in Y when needed, usually to no disadvantage if one remembers to negate the number ([CHS]) or invert it ([$1/x$]) if necessary before putting it into T. Section 2.4 contains examples of the usefulness of this idea. This suggestion does not apply to National Semiconductor or Novus calculators.

To double a number (e.g., $2A$) use [↑] [+] rather than [↑] 2 [×]. The corresponding rule for squaring ([↑] [×] instead of [x^2]) saves keystrokes only with calculators on which [x^2] needs *two* prefix keys. If A is already in the stack twice, maybe even the [↑] can be omitted. To key a large power of 10 (e.g., 1,000,000 or 0.0001), use just [EEX], [CHS] if necessary, and an integer for the power. For example, [EEX] 6 gets 1,000,000 in two keystrokes, and [EEX] [CHS] 4 gets 0.0001 in three keystrokes. This suggestion does not apply to Novus calculators.

Often the most useful rule of thumb is the least clear-cut: rewrite the expression algebraically. The length or elegance of the algebraic expression is poorly correlated with the terseness of the resulting RPN algorithm, but sometimes rewriting can save dozens of keystrokes, even if the resulting algebraic expression is longer! Section 2.6 discusses the approach of rewriting, instead, the RPN algorithm.

2.3 EXAMPLES OF WRITING ALGORITHMS

2.3.1 Normal or Gaussian Probability Function

Consider first a simple example—the normal or Gaussian probability function (HMF* 26.2),

$$Z(x) = \frac{1}{\sqrt{2\pi}} \exp\left(\frac{-x^2}{2}\right). \tag{2.3.1}$$

A naïve algorithm for the HP-35 might be

2 [π] [×] [√x] [1/x] x [↑] [×] 2 [÷] [CHS] [e^x] [×] (see $Z(x)$). (2.3.2)

The first 2 and [π] can be in either order, and either way an [↑] is not needed. The [CHS] can also be moved around. If we calculate the exponential first, so that $\sqrt{2\pi}$ will be in X for a [÷], we save a keystroke ([1/x]):

x [↑] [×] 2 [÷] [CHS] [e^x] 2 [π] [×] [√x] [÷] (see $Z(x)$). (2.3.3)

To shorten this algorithm any further, rewrite the expression algebraically, either mentally or on paper. Since the /2 in the exponential corresponds to a square root, and the minus sign to a reciprocal, we can write

$$Z(x) = \frac{1}{\sqrt{2\pi \exp(x^2)}}, \tag{2.3.4}$$

and so write

x [↑] [×] [e^x] 2 [×] [π] [×] [√x] [1/x] (see $Z(x)$). (2.3.5)

*References are in Appendix C.

Or, for a calculator that requires a prefix keystroke for [1/x], we could trade a [CHS] for a [1/x] by

$$x\,[\uparrow]\,[\times]\,[\text{CHS}]\,[e^x]\,2\,[\div]\,[\pi]\,[\div]\,[\sqrt{x}] \qquad (\text{see } Z(x)). \tag{2.3.6}$$

2.3.2 Coordinate Translation and Rotation

As another example, consider a point P with coordinates x and y in a standard two-dimensional Cartesian reference frame. Suppose we want to find the coordinates x' and y' of P in a new reference frame defined by the coordinates x_0 and y_0 of the new origin in the old frame, and by α, the angle of rotation (positive counterclockwise) of the new frame with respect to the old. The situation is sketched in Figure 2.3.1, from which the transformation equations are

$$x' = (x - x_0)\cos\alpha + (y - y_0)\sin\alpha,$$

$$y' = (y - y_0)\cos\alpha - (x - x_0)\sin\alpha. \tag{2.3.7}$$

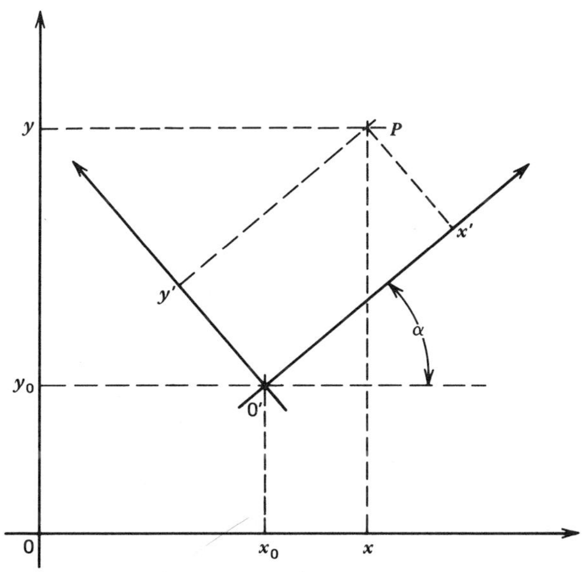

Figure 2.3.1 Translation and rotation of Cartesian coordinates.

In these equations, α appears four times, and $x-x_0$ and $y-y_0$ each appear twice; therefore these three quantities must all be saved. A naïve solution might require three STO registers,* but we can also save some quantities in the four-register stack.

First consider using a calculator that has a polar-to-rectangular coordinate converter (→R), such as the HP-21. This feature is useful because we can easily calculate $A\cos\alpha$ and $A\sin\alpha$ simultaneously. Suppose we begin by

$$x \boxed{\uparrow} x_0 \boxed{-} \alpha \boxed{x \gtrless y} \boxed{B} \rightarrow R. \tag{2.3.8}$$

This gives $(x-x_0)\cos\alpha$ in X and $(x-x_0)\sin\alpha$ in Y, but α is not available for the next step. Also the $\boxed{x \gtrless y}$ after α seems to be a wasted keystroke. So try instead

$$\alpha \boxed{\text{STO}} x \boxed{\uparrow} x_0 \boxed{-} \boxed{B} \rightarrow R. \tag{2.3.9}$$

No $\boxed{\uparrow}$ is needed after α; STO serves as a number separator and enables auto-enter on the HP-21. Then

$$y \boxed{\uparrow} y_0 \boxed{-} \boxed{\text{RCL}} \boxed{x \gtrless y} \boxed{B} \rightarrow R. \tag{2.3.10}$$

At this point the stack has

$$\begin{aligned} T&: \quad (x-x_0)\sin\alpha, \\ Z&: \quad (x-x_0)\cos\alpha, \\ Y&: \quad (y-y_0)\sin\alpha, \\ X&: \quad (y-y_0)\cos\alpha. \end{aligned} \tag{2.3.11}$$

The RCL cannot be moved in front of the y to eliminate the $\boxed{x \gtrless y}$ because then a five-register stack would be needed to hold y_0, y, α, $(x-x_0)\cos\alpha$, and $(x-x_0)\sin\alpha$ when y_0 is keyed. Then

$$\boxed{R\downarrow} \boxed{+} \qquad \text{(see } x'\text{)} \tag{2.3.12}$$

puts $(x-x_0)\cos\alpha+(y-y_0)\sin\alpha=x'$ into X, and then

$$\boxed{R\downarrow} \boxed{-} \qquad \text{(see } y'\text{)} \tag{2.3.13}$$

puts $(y-y_0)\cos\alpha-(x-x_0)\sin\alpha=y'$ into X.

*The *HP-45 Applications Book*, pp. 70 and 71, has an algorithm that requires three STO registers and 31 keystrokes, plus data.

The same algorithm works on a calculator that has several storage registers, such as the HP-45; change [B] to [G] and add a storage register number after [STO] and [RCL]. Since this adds two keystrokes, however, we might try to write an algorithm for this problem without using [STO] and [RCL]. Consider the problem rather in reverse order. Suppose, somehow, that the stack contained

$$\begin{aligned} T&:\ \alpha,\\ Z&:\ x-x_0,\\ Y&:\ \alpha,\\ X&:\ y-y_0; \end{aligned} \tag{2.3.14}$$

then we could finish up the algorithm nicely by

[B] [→R] [R↓] [R↓] [B] [→R], (2.3.15)

which gives

$$\begin{aligned} T&:\ (y-y_0)\sin\alpha,\\ Z&:\ (y-y_0)\cos\alpha,\\ Y&:\ (x-x_0)\sin\alpha,\\ X&:\ (x-x_0)\cos\alpha, \end{aligned} \tag{2.3.16}$$

and then

[R↓] [−] (see y'),

[R↓] [+] (see x'). (2.3.17)

To get the initial configuration assumed earlier, α must be keyed either first (so that the pop on [−] will duplicate α in T and Z) or last (so that an [↑] can be used to duplicate α). For example,

x [↑] x_0 [−] y [↑] y_0 [−] α [↑] (2.3.18)

gives

$$\begin{aligned} T&: \quad x-x_0, \\ Z&: \quad y-y_0, \\ Y&: \quad \alpha, \\ X&: \quad \alpha, \end{aligned} \tag{2.3.19}$$

and then

$$\boxed{\mathrm{R}\!\downarrow}\ \boxed{x \gtrless y} \tag{2.3.20}$$

gives the initial configuration previously assumed.

Starting at the end of an algorithm and working backward or starting in the middle and working both ways is often useful for more complex problems.

Or α can be entered first, for example,

$$\alpha \boxed{\uparrow} x \boxed{\uparrow} x_0 \boxed{-}\ y \boxed{\uparrow} y_0 \boxed{-} \tag{2.3.21}$$

gives

$$\begin{aligned} T&: \quad \alpha, \\ Z&: \quad \alpha, \\ Y&: \quad x-x_0, \\ X&: \quad y-y_0, \end{aligned} \tag{2.3.22}$$

and we can finish this algorithm by

$$\boxed{\mathrm{R}\!\downarrow}\ \boxed{\mathrm{B}}\ \boxed{\to\mathrm{R}}\ \boxed{\mathrm{R}\!\downarrow}\ \boxed{\mathrm{R}\!\downarrow}\ \boxed{x \gtrless y}\ \boxed{\mathrm{B}}\ \boxed{\to\mathrm{R}}$$

$$\boxed{\mathrm{R}\!\downarrow}\ \boxed{+} \qquad \text{(see } x'\text{)}$$

$$\boxed{\mathrm{R}\!\downarrow}\ \boxed{-} \qquad \text{(see } y'\text{).} \tag{2.3.23}$$

On the HP-21 these last two algorithms each have 17 keystrokes (plus data) compared with 15 keystrokes for the first algorithm. All three algorithms have the same number of keystrokes (17) on the HP-45.

To write a shorter algorithm for this problem, we reconsider Figure 2.3.1. Clearly the translation and rotation can be done separately; the translation alone is just

$$x'' = x - x_0,$$

$$y'' = y - y_0. \qquad (2.3.24)$$

Now if x'' and y'' were expressed in polar coordinates, the rotation would be trivial—just subtract α from the polar angle. Thus the rotation can be done by converting to polar coordinates, rotating, and converting back to rectangular coordinates. Then the whole algorithm would be

y' [↑] y_0 [−] x [↑] x_0 [ ] [B] [→P] [$x \gtrless y$] α [−] [$x \gtrless y$] [B] [→R] (see x')

[$x \gtrless y$] (see y'). (2.3.25)

The two [$x \gtrless y$]s are necessary because [→P] leaves, and [→R] needs θ in Y. This algorithm for the HP-21 requires only 12 keystrokes plus data; the HP-45 version requires only 11 keystrokes (because [→P] does not need a prefix keystroke).

Now suppose we want to do the same problem on a calculator such as the HP-35 or the Novus Scientist, without [→P] or [→R]. Equation 2.3.7 shows that this will be a little more difficult because there are not enough registers to save everything. But our success with the polar-coordinate representation suggests another approach: we could write algorithms to duplicate the effect of [→P] and [→R] for a calculator without them, then combine these algorithms into a composite algorithm to do the coordinate rotation.

The formulas for polar-to-rectangular coordinate conversion are just

$$x = R\cos\theta,$$

$$y = R\sin\theta. \qquad (2.3.26)$$

Since both θ and R are used twice in these formulas, they must be saved, but [STO] is not needed. One approach is to begin by

θ [↑] [COS] R [↑] [R↓]. (2.3.27)

At this point the stack has

$$\begin{aligned} T&: \quad R,\\ Z&: \quad \theta,\\ Y&: \quad \cos\theta,\\ X&: \quad R. \end{aligned} \tag{2.3.28}$$

The [↑] after both θ and R serves as a duplicator in this algorithm, and the [COS] enables auto-enter. The alternatives of beginning by θ [↑] [↑] $R \cdots$ or R [↑] [↑] $\theta \cdots$ cost more keystrokes. We can finish the foregoing algorithm by

[×] (see x),

[R↓] [SIN] [×] (see y). (2.3.29)

But the [SIN] in the last line loses x from T on the HP-35 or the Novus Scientist.

A shorter algorithm for polar-to-rectangular coordinate conversion results from using

$$y = x\tan\theta. \tag{2.3.30}$$

That is,

θ [↑] [COS] R [×] (see x)

[$x \gtrless y$] [TAN] [×] (see y). (2.3.31)

This algorithm, however, does not work around $\theta = \pm 90°$ because the formula for y degenerates into the form $0 \times \infty$. Alternatively,

θ [↑] [SIN] R [×] (see y)

[$x \gtrless y$] [TAN] [×] (see x) (2.3.32)

works except around $\theta = 0$ or $180°$.

The formulas for rectangular-to-polar coordinate conversion are

$$\begin{aligned} R &= \sqrt{x^2 + y^2}\,,\\ \theta &= \tan^{-1}\left(\frac{y}{x}\right), \end{aligned} \tag{2.3.33}$$

with θ placed in the correct quadrant. We can also use, for example,

$$\theta = \cos^{-1}\left(\frac{x}{R}\right), \tag{2.3.34}$$

if it saves keystrokes. Some experimentation leads to

y [↑] [×] x [↑] [↑] [R↓] [×] [+] [$\sqrt{x}$] (see R) [÷] [ARC]

[COS] (if $y<0$: [CHS]; see θ). (2.3.35)

Note that R is used in calculating θ and is not saved anywhere. We can save keystrokes by using

$$R = \frac{y}{\sin\theta}, \tag{2.3.36}$$

rather than the square root relation, provided θ is not too near 0 or 180°. Then

y [↑] [↑] x [÷] [ARC] [TAN] (if $x<0$: 180 [+]; see θ)

[SIN] [÷] (see R). (2.3.37)

These algorithms are, of course, useful in themselves, but the goal is to combine them into an algorithm for coordinate rotation (and translation). First suppose we try for a version with no restrictions against particular values of the parameters. Begin by

y [↑] y_0 [−]. (2.3.38)

We need to remember the sign of $y-y_0$ in the display at this point. Then

[↑] [×] x [↑] x_0 [−] [↑] [↑] [R↓] [×] [+] [$\sqrt{x}$]. (2.3.39)

Now R' is in X and we need to save it. So

[STO] [÷] [ARC] [COS] (if $y-y_0<0$: [CHS] [↑]) (2.3.40)

The added [↑] is to preserve the negation; refer to the peculiarities of [CHS] on the HP-35 in Section 1.3. At this point, since θ is in X and R' is in the [STO] register, subtract α from θ and convert back to rectangular coordinates. That is,

$$\alpha \boxed{-}\boxed{\uparrow}\boxed{\text{COS}}\boxed{\text{RCL}}\boxed{\times} \qquad \text{(see } x'\text{)}$$

$$\boxed{x \gtrless y}\boxed{\text{SIN}}\boxed{\text{RCL}}\boxed{\times} \qquad \text{(see } y'\text{).} \qquad (2.3.41)$$

This algorithm has 25 keystrokes if $y - y_0 > 0$ or 27 keystrokes if $y - y_0 < 0$, and it works for all values of the parameters, although it gives an error indication (flashing zero) in case x' and y' are both zero.

We can write a slightly shorter algorithm if we avoid certain values of the parameters. Try

$$y\boxed{\uparrow}y_0\boxed{-}\boxed{\uparrow}\boxed{\uparrow}x\boxed{\uparrow}x_0\boxed{-} \qquad \text{(see } x - x_0\text{; note sign)}$$

$$\boxed{\div}\boxed{\text{ARC}}\boxed{\text{TAN}} \qquad (\text{if } x - x_0 < 0\text{: } 180\boxed{+}) \qquad \boxed{\text{STO}}\boxed{\text{SIN}}\boxed{\div}. \qquad (2.3.42)$$

At this point R' is in X and θ is in $\boxed{\text{STO}}$; we must save θ somewhere. So now a

$$\boxed{\text{RCL}}\alpha\boxed{-} \qquad (2.3.43)$$

puts R' into Y and $\theta' = \theta - \alpha$ into X, ready to be converted back into rectangular coordinates. Unfortunately the polar-to-rectangular coordinate converter needs to be modified a little to accept arguments in this order. In particular, we should save θ' in $\boxed{\text{STO}}$, that is,

$$\boxed{\text{STO}}\boxed{\text{COS}}\boxed{\times} \qquad \text{(see } x'\text{)}$$

$$\boxed{\text{RCL}}\boxed{\text{TAN}}\boxed{\times} \qquad \text{(see } y'\text{).} \qquad (2.3.44)$$

Note that x must not equal x_0, y must not equal y_0, and θ' must not equal $\pm 90°$; that is, x' must not be zero. This algorithm has 20 keystrokes if $x - x_0 > 0$, or 24 keystrokes if $x - x_0 < 0$. Several other similar algorithms are possible with different forbidden values.

Like a good fable, this rather belabored example has more than one moral, including one not yet stated: don't give up too soon, but do stop.

2.4 POLYNOMIALS, POWER SERIES, AND CONTINUED FRACTIONS

Efficient translation of polynomials or power series into RPN usually requires that they be written in one of several possible parenthetical forms.

A general polynomial or finite power series

$$\sum_{k=0}^{n} a_k x^k = a_0 + a_1 x + a_2 x^2 + \cdots + a_n x^n \tag{2.4.1}$$

can be written as

$$a_0 + x(a_1 + x(a_2 + x(\cdots + x a_n)\cdots) \tag{2.4.2}$$

and evaluated by

$$x \boxed{\uparrow}\boxed{\uparrow}\boxed{\uparrow} a_n \boxed{\times} a_{n-1} \boxed{+}$$
$$\boxed{\times} a_{n-2} \boxed{+}\boxed{\times} \cdots a_1 \boxed{+}\boxed{\times} a_0 \boxed{+}. \tag{2.4.3}$$

This method is possible because the RPN stack is only four high and T does not clear on pop (on most RPN calculators); thus x remains available. If any of the a_k other than a_n are negative, enter them as positive but replace the following $\boxed{+}$ by $\boxed{-}$.

As an example, the polynomial

$$32x^6 - 48x^4 + 18x^2 - 1 \tag{2.4.4}$$

could be evaluated by

$$x \boxed{\uparrow}\boxed{\uparrow}\boxed{\uparrow} 32 \boxed{\times}\boxed{\times} 48 \boxed{-}\boxed{\times}\boxed{\times} 18 \boxed{+}\boxed{\times}\boxed{\times} 1 \boxed{-}, \tag{2.4.5}$$

or perhaps by

$$x \boxed{\uparrow}\boxed{\times}\boxed{\uparrow}\boxed{\uparrow}\boxed{\uparrow} 32 \boxed{\times} 48 \boxed{-}\boxed{\times} 18 \boxed{+}\boxed{\times} 1 \boxed{-}. \tag{2.4.6}$$

For $x = 0.5$, get 1. But this is Chebyshev polynomial $T_6(x)$, and Section A.5.6 contains a shorter algorithm for such polynomials, based on their relationship with cosines.

As an application of this procedure, consider the problem of converting octal numbers to decimal. An octal number

$$d_n d_{n-1} \cdots d_1 d_0, \tag{2.4.7}$$

where each d represents a digit, is just

$$8^n d_n + 8^{n-1} d_{n-1} + \cdots + 8 d_1 + d_0. \tag{2.4.8}$$

Therefore this is a polynomial and can be evaluated by

$$8\boxed{\uparrow}\boxed{\uparrow}\boxed{\uparrow}d_n\boxed{\times}d_{n-1}\boxed{+}\boxed{\times}\cdots d_1\boxed{+}\boxed{\times}d_0\boxed{+}. \tag{2.4.9}$$

An octal fraction such as

$$d_0.d_{-1}d_{-2}\cdots d_{-n+1}d_{-n} \tag{2.4.10}$$

can be handled by

$$8\boxed{1/x}\boxed{\uparrow}\boxed{\uparrow}\boxed{\uparrow}d_{-n}\boxed{\times}d_{-n+1}\boxed{+}\boxed{\times}\cdots d_{-1}\boxed{+}\boxed{\times}d_0\boxed{+}. \tag{2.4.11}$$

As an example, $3721_{(8)}$ would be

$$8\boxed{\uparrow}\boxed{\uparrow}\boxed{\uparrow}3\boxed{\times}7\boxed{+}\boxed{\times}2\boxed{+}\boxed{\times}1\boxed{+} \qquad (\text{see } 2001_{(10)}). \tag{2.4.12}$$

Hexadecimal (or sedenary) numbers can be handled in the same way with 16 substituted for 8 and with the letters entered as their two-digit equivalents.

The preceding formulation is usually best for ordinary polynomials; however an alternative scheme sometimes results in a shorter algorithm for power series if the coefficients have a special relationship to each other. The general polynomial can also be written as

$$\sum_{k=0}^{n} a_k x^k = a_0 + a_1 x\left(1 + \frac{a_2}{a_1}x\left(1 + \frac{a_3}{a_2}x(\cdots \frac{a_n}{a_{n-1}}x\right)\cdots\right). \tag{2.4.13}$$

This form is preferable whenever a_k/a_{k-1} is simpler than a_k alone. A corresponding RPN algorithm is

$$x\boxed{\uparrow}\boxed{\uparrow}\boxed{\uparrow}\left(\frac{a_n}{a_{n-1}}\right)\boxed{\times}1\boxed{+}\boxed{\times}\left(\frac{a_{n-1}}{a_{n-2}}\right)\boxed{\times}1\boxed{+}\boxed{\times}\cdots\left(\frac{a_2}{a_1}\right) \tag{2.4.14}$$

$$\boxed{\times}1\boxed{+}\boxed{\times}a_1\boxed{\times}a_0\boxed{+}.$$

As an example, consider the ascending series expansion for the Bessel function $J_0(x)$ (HMF 9.1.10),

$$J_0(x) = \sum_{k=0}^{\infty} \frac{(-x^2/4)^k}{(k!)^2}. \tag{2.4.15}$$

This series, truncated after five terms, can be written as

$$J_0(x) \cong 1 + y + \frac{y^2}{(2!)^2} + \frac{y^3}{(3!)^2} + \frac{y^4}{(4!)^2} + \frac{y^5}{(5!)^2}, \tag{2.4.16}$$

where $y = -x^2/4$, or as

$$J_0(x) \cong 1 + y\left(1 + \frac{y}{2^2}\left(1 + \frac{y}{3^2}\left(1 + \frac{y}{4^2}\left(1 + \frac{y}{5^2}\right)\right)\right)\right). \tag{2.4.17}$$

This form is preferable because the factorials are eliminated. A corresponding RPN algorithm is

$$x \boxed{x^2} 4 \boxed{\div} \boxed{\text{CHS}} \boxed{\uparrow} \boxed{\uparrow} \boxed{\uparrow} 25 \boxed{\div} 1 \boxed{+} \boxed{\times} 16 \boxed{\div} 1 \boxed{+} \boxed{\times} 9 \boxed{\div} 1 \boxed{+}$$
$$\boxed{\times} 4 \boxed{\div} 1 \boxed{+} \boxed{\times} 1 \boxed{+}. \tag{2.4.18}$$

This ascending series of five terms gives $J_0(x)$ with an absolute error $< 10^{-5}$ for $0 < x \lesssim 2.3$.

As another example, consider the general Taylor series expansion of any differentiable function (HMF 3.6.1),

$$f(x+h) = \sum_{k=0}^{\infty} \frac{h^k}{k!} f^{(k)}(x), \tag{2.4.19}$$

which can be written as

$$f(x+h) = f + h\left(f' + \frac{h}{2}\left(f'' + \frac{h}{3}(f''' + \cdots)\cdots\right)\right), \tag{2.4.20}$$

where the x dependence after each f was dropped. This form comes from a combination of the two schemes given earlier. An RPN algorithm for this series truncated after the nth term is

$$h \boxed{\uparrow} \boxed{\uparrow} \boxed{\uparrow} f^{(n)} \boxed{\times} (n) \boxed{\div} f^{(n-1)} \boxed{+} \boxed{\times} (n-1)$$
$$\boxed{\div} f^{(n-2)} \boxed{+} \boxed{\times} (n-2) \boxed{\div} \cdots f'' \boxed{+} \boxed{\times} \tag{2.4.21}$$
$$2 \boxed{\div} f' \boxed{+} \boxed{\times} f \boxed{+}.$$

Minimax, near minimax, or best polynomial approximations to transcendental functions (see e.g., Fike, 1968) can, of course, be evaluated in the same way. However, such polynomials usually have irrational coefficients that must be entered as up to 10-digit numbers. If the calculator has enough storage registers to hold these coefficients, and if the function is needed at a number of points, such polynomials may be desirable. More typically, however, a power series with coefficients that are small integers or ratios of small integers results in a shorter algorithm even though more terms in the series must be taken to achieve the same precision.

But adjusting the last coefficient in the series (i.e., the first number in the algorithm) in an informal way, may achieve a specified precision over a wider range of x. This can be done without increasing the number of keystrokes if a different integer, rather than a multidigit number, is chosen for the last coefficient.

As an example, in algorithm 2.4.18 for $J_0(x)$, if we change the coefficient 25 to 26, the range over which the absolute error is $<10^{-5}$ extends from $0<x\lesssim 2.3$ to $0<x\lesssim 2.65$, as shown in Figure A.5.1. The change to 26 increases the absolute error by an amount that may be negligible, for $x\lesssim 1.9$. The number 26 was determined by trial and error.

We can also write algorithms for power series in ascending order. Examples using the $\boxed{\Sigma+}$ feature of the HP-45 appear in Section A.5.9 for Bessel functions and in Section A.5.10 for the error function. Other examples are in the *HP-35 Math Pac*, p. 90; the *HP-45 Applications Book*, p. 27; and the *HP-55 Mathematics Programs*, pp. 100 ff. For a given number of terms, these ascending-order algorithms are almost always longer than the parenthetical forms already given. But an ascending-order algorithm allows one to add terms until they become less than some limit, thus achieving a specified precision with a minimum number of terms. Ascending-order algorithms for power series are often desirable for programmable RPN calculators because the looping feature makes such cyclic procedures easy.

A *continued fraction* can be thought of as a special form of a rational function (i.e., the ratio of two polynomials) and formulas exist (Wall, 1948, Chapter 9; Demidovich and Maron, 1973, Section 2.5) for converting a rational function into a continued fraction or a continued fraction into a rational function. A finite continued fraction such as

$$b_0+\cfrac{a_1}{b_1+\cfrac{a_2}{b_2+\cfrac{a_3}{b_3+\cfrac{a_4}{b_4}}}} \tag{2.4.22}$$

can be evaluated by

$$a_4\boxed{\uparrow}b_4\boxed{\div}b_3\boxed{+}a_3\boxed{x\gtrless y}\boxed{\div}b_2\boxed{+}a_2\boxed{x\gtrless y}\boxed{\div}b_1\boxed{+}a_1\boxed{x\gtrless y}\boxed{\div}b_0\boxed{+}. \tag{2.4.23}$$

For example, the error function is (HMF 7.1.14)

$$\mathrm{erf}(x) = 1 - \frac{e^{-x^2}}{\sqrt{\pi}} \times \cfrac{1}{x + \cfrac{1}{2x + \cfrac{2}{x + \cfrac{3}{2x + \cfrac{4}{x + \cfrac{5}{2x + \cfrac{6}{x + \cfrac{7}{2x + \cdots}}}}}}}} \tag{2.4.24}$$

A corresponding HP-45 algorithm is

$$x \boxed{\uparrow} \boxed{\uparrow} \boxed{\uparrow} \boxed{+} 7 \boxed{x \gtrless y} \boxed{\div} \boxed{+} 6 \boxed{x \gtrless y} \boxed{\div} \boxed{+} \boxed{+} \tag{2.4.25}$$

$$5 \boxed{x \gtrless y} \boxed{\div} \boxed{+} 4 \boxed{x \gtrless y} \boxed{\div} \boxed{+} \boxed{+}$$

$$3 \boxed{x \gtrless y} \boxed{\div} \boxed{+} 2 \boxed{x \gtrless y} \boxed{\div} \boxed{+} \boxed{+}$$

$$\boxed{1/x} \boxed{+} \boxed{x \gtrless y} \boxed{x^2} \boxed{e^x} \boxed{\times} \boxed{G} \boxed{\pi}$$

$$\boxed{G} \boxed{\sqrt{x}} \boxed{\times} \boxed{1/x} 1 \boxed{x \gtrless y} \boxed{-} \qquad (\text{see} \cong \mathrm{erf}\,(x)).$$

This continued fraction of seven terms gives $\mathrm{erf}(x)$ to four or more significant figures for $1.3 \lesssim x$. If we change the last numerator (i.e., the first number in the algorithm) from 7 to 3, the range over which this algorithm gives four or more significant figures extends to $1 \lesssim x$, as shown in Figure A.5.3.

In the "simple form" for a continued fraction, all the a_j are unity (see, e.g., Knuth, 1969, Vol. II, Section 4.5.3) and

$$/b_1, b_2, b_3, \cdots, b_n/ \equiv \frac{1}{b_1 + 1/(b_2 + 1/(b_3 + \cdots + 1/b_n)\cdots)}. \tag{2.4.26}$$

This expression can be evaluated by

$$b_n \boxed{1/x} b_{n-1} \boxed{+} \boxed{1/x} b_{n-2} \boxed{+} \boxed{1/x} \cdots b_2 \boxed{+} \boxed{1/x} b_1 \boxed{+} \boxed{1/x}. \tag{2.4.27}$$

Every real number x between 0 and 1 has a unique "regular continued fraction" expansion in this simple form with the b_j all positive integers. But n is infinite if x is irrational. This continued fraction expansion of a (positive) number x can be determined by

SETUP: x [↑] (mentally set $j=0$)

LOOP: (see b_j as the *integer part* of the display) b_j (i.e., key it back in) [−] [1/x] (mentally add 1 to j) :|. (2.4.28)

The :| symbol in this algorithm means loop back to the last preceding colon (:), in this case after the word "loop." For a rational x, this process terminates with zero, except possibly for round-off error. For an irrational x, the process would go on forever, but on a calculator with a finite number of digits, the algorithm fails to give the correct b_j after a finite number of loops.

If instead of all being unity, the a_j are prescribed numbers, we have

SETUP: x [↑] (mentally set $j=0$)

LOOP: (see b_j as the *integer part* of the display) b_j (i.e., key it back in) [−] (mentally add 1 to j) a_j [$x \gtrless y$] [÷] :|. (2.4.29)

Or if the b_j are prescribed and the a_j are integers to be found, then

SETUP: x [↑] b_0 [−] (mentally set $j=1$)

LOOP: b_j [↑] [↑] [R↓] [R↓] [×] (note *integer part* of display; key it back in) [↑] 1 [+] (see a_j) [$x \gtrless y$] [÷] 1 [−] [×] (mentally add 1 to j) :|. (2.4.30)

These algorithms are based on the assumption that each of the partial fractions is $\leqslant 1$ and that all the a_j and b_j are positive. These are not necessary assumptions, of course, but they are conventional.

Fewer arithmetic operations are usually required to evaluate a rational function if it is converted into a continued fraction; however the total number of keystrokes may be either greater or smaller depending on the numerical values of the coefficients. In references such as HMF, continued fractions usually appear when the coefficients in this form are simpler than in some other form.

2.5 STACK REARRANGEMENTS (RPN CALCULATORS)

As discussed in Section 1.3.3, for any but the simplest problems, numbers in the stack sometimes turn up in the wrong order. The [$x \gtrless y$] and [R↓] keys are provided just to alleviate such difficulties. But one might need any of a large number of possible stack rearrangements. When working through a complex algorithm, I find a table of such rearrangements to be helpful. In computer jargon, they are macros to be inserted where needed into a main algorithm.

Consider four numbers—A, B, C, and D—in order in the stack; that is, A in X, B in Y, C in Z, and D in T. And consider rearranging these numbers in an arbitrary way and also zeroing any registers. Then the $5^4-1=624$ possible rearrangements can all be done with some combination of [R↓], [$x \gtrless y$], [↑], [CLX], [+], [−], and [0]. A few of these rearrangements are in the *HP-35 Math Pac*, pp. 122–127, and in the *HP-45 Applications Book*, pp. 168–174. Table 2.5.1 contains most of these rearrangements in the following format. The first four columns are the rearranged register contents from left to right; the first column is the new contents of the X register, the second column the new contents of the Y register, and so on. An algorithm to achieve this rearrangement follows. A few blank lines are left in the table as exercises for the reader.

Auto-enter is assumed to be enabled at the start of the algorithms, but this is necessary only for algorithms beginning with [0]. The status of auto-enter at the end is ignored; if auto-enter needs to be enabled, append [$x \gtrless y$] to algorithms ending in [↑], or append [$x \gtrless y$] [$x \gtrless y$] to algorithms ending in [CLX] or [0]. I assume that T does not clear on pop, but this is necessary only for algorithms containing [+] or [−]; thus most such algorithms will not work on National Semiconductor or Novus calculators.

Consider an example from the *HP-35 Math Pac*, p. 125, and the *HP-45 Applications Book*, p. 171. To do the rearrangement

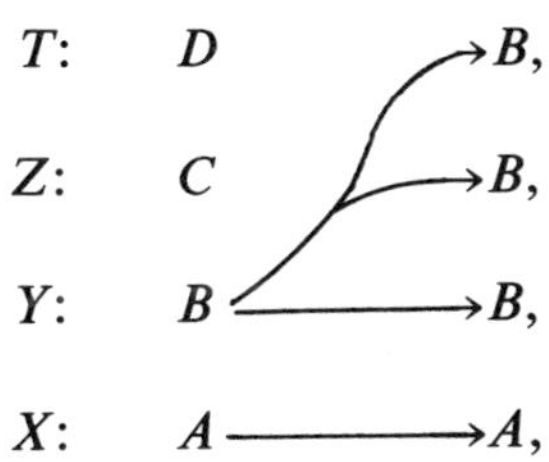

(2.5.1)

which HP calls "copy y into Z and T," HP recommends

[$x \gtrless y$] [↑] [↑] [R↓] [R↓] [R↓]. (2.5.2)

In my notation this rearrangement is $ABBB$, and Table 2.5.1 contains the rather shorter

[↑] [↑] [−] [−]. (2.5.3)

Table 2.5.1 Stack Rearrangements

X Y Z T	Keystrokes
0 0 0 0	[CLX] [↑] [↑] [↑]
0 0 0 *A*	[0] [↑] [↑]
0 0 0 *B*	[CLX] [↑] [↑]
0 0 0 *C*	[CLX] [x≷y] [CLX] [↑]
0 0 0 *D*	[R↓] [CLX] [x≷y] [CLX] [↑]
0 0 *A* 0	[0] [↑] [↑] [R↓]
0 0 *A A*	[0] [↑] [↑] [−]
0 0 *A B*	[0] [↑]
0 0 *A C*	[x≷y] [CLX] [↑]
0 0 *A D*	[x≷y] [R↓] [x≷y] [CLX] [↑]
0 0 *B* 0	[CLX] [↑] [↑] [R↓]
0 0 *B A*	[x≷y] [0] [↑]
0 0 *B B*	[0] [x≷y] [↑] [−]
0 0 *B C*	[CLX] [↑]
0 0 *B D*	[R↓] [x≷y] [CLX] [↑]
0 0 *C* 0	[CLX] [x≷y] [CLX] [↑] [R↓]
0 0 *C A*	[0] [x≷y] [R↓] [x≷y] [CLX]
0 0 *C B*	[R↓] [x≷y] [0] [↑]
0 0 *C C*	[CLX] [x≷y] [↑] [−]
0 0 *C D*	[CLX] [x≷y] [CLX]
0 0 *D* 0	[CLX] [R↓] [CLX] [x≷y] [CLX]
0 0 *D A*	[R↓] [CLX] [x≷y] [CLX]
0 0 *D B*	[CLX] [x≷y] [R↓] [x≷y] [CLX]
0 0 *D C*	[R↓] [R↓] [x≷y] [0] [↑]
0 0 *D D*	[−] [CLX] [x≷y] [CLX]

Table 2.5.1 (Continued)

$X\ Y\ Z\ T$	Keystrokes
$0\ A\ 0\ 0$	[0] [↑] [↑] [R↓] [R↓]
$0\ A\ 0\ A$	[↑] [↑] [CLX] [$x \gtrless y$] [0]
$0\ A\ 0\ B$	[0] [$x \gtrless y$] [0]
$0\ A\ 0\ C$	[$x \gtrless y$] [CLX] [$x \gtrless y$] [0]
$0\ A\ 0\ D$	[R↓] [CLX] [R↓] [CLX] [$x \gtrless y$] [R↓]
$0\ A\ A\ 0$	[0] [↑] [↑] [−] [R↓]
$0\ A\ A\ A$	[↑] [↑] [↑] [−]
$0\ A\ A\ B$	[↑] [↑] [CLX]
$0\ A\ A\ C$	[R↓] [CLX] [−] [R↓] [CLX]
$0\ A\ A\ D$	[R↓] [−] [CLX] [$x \gtrless y$] [R↓]
$0\ A\ B\ 0$	[0] [↑] [R↓]
$0\ A\ B\ A$	[↑] [↑] [R↓] [CLX]
$0\ A\ B\ B$	[↑] [↑] [−]
$0\ A\ B\ C$	[0]
$0\ A\ B\ D$	[R↓] [R↓] [CLX] [$x \gtrless y$] [R↓]
$0\ A\ C\ 0$	[$x \gtrless y$] [CLX] [↑] [R↓]
$0\ A\ C\ A$	[↑] [R↓] [$x \gtrless y$] [CLX]
$0\ A\ C\ B$	[$x \gtrless y$] [0] [$x \gtrless y$] [R↓]
$0\ A\ C\ C$	[$x \gtrless y$] [↑] [−]
$0\ A\ C\ D$	[$x \gtrless y$] [CLX]
$0\ A\ D\ 0$	[$x \gtrless y$] [CLX] [R↓] [$x \gtrless y$] [CLX]
$0\ A\ D\ A$	[R↓] [−] [R↓] [$x \gtrless y$] [0]
$0\ A\ D\ B$	[$x \gtrless y$] [R↓] [$x \gtrless y$] [CLX]
$0\ A\ D\ C$	[$x \gtrless y$] [R↓] [$x \gtrless y$] [0] [$x \gtrless y$] [R↓]
$0\ A\ D\ D$	[$x \gtrless y$] [R↓] [$x \gtrless y$] [↑] [−]

Table 2.5.1 (Continued)

$X\ Y\ Z\ T$	Keystrokes
$0\ B\ 0\ 0$	[CLX] [↑] [↑] [R↓] [R↓]
$0\ B\ 0\ A$	[$x \gtrless y$] [0] [$x \gtrless y$] [0]
$0\ B\ 0\ B$	[CLX] [$x \gtrless y$] [↑] [↑] [R↓] [CLX]
$0\ B\ 0\ C$	[CLX] [$x \gtrless y$] [0]
$0\ B\ 0\ D$	[R↓] [$x \gtrless y$] [CLX] [$x \gtrless y$] [0]
$0\ B\ A\ 0$	[$x \gtrless y$] [0] [↑] [R↓]
$0\ B\ A\ A$	[↑] [↑] [R↓] [−]
$0\ B\ A\ B$	[$x \gtrless y$] [↑] [↑] [R↓] [CLX]
$0\ B\ A\ C$	[$x \gtrless y$] [0]
$0\ B\ A\ D$	[$x \gtrless y$] [R↓] [R↓] [CLX] [$x \gtrless y$] [R↓]
$0\ B\ B\ 0$	[0] [$x \gtrless y$] [↑] [−] [R↓]
$0\ B\ B\ A$	[$x \gtrless y$] [↑] [↑] [CLX]
$0\ B\ B\ B$	[↑] [↑] [CLX] [−] [−]
$0\ B\ B\ C$	[R↓] [↑] [↑] [CLX]
$0\ B\ B\ D$	[$x \gtrless y$] [R↓] [−] [CLX] [$x \gtrless y$] [R↓]
$0\ B\ C\ 0$	[CLX] [↑] [R↓]
$0\ B\ C\ A$	[0] [$x \gtrless y$] [R↓]
$0\ B\ C\ B$	[CLX] [$x \gtrless y$] [↑] [R↓] [$x \gtrless y$]
$0\ B\ C\ C$	[↑] [−]
$0\ B\ C\ D$	[CLX]
$0\ B\ D\ 0$	[CLX] [R↓] [$x \gtrless y$] [CLX]
$0\ B\ D\ A$	[R↓] [$x \gtrless y$] [CLX]
$0\ B\ D\ B$	[R↓] [↑] [R↓] [$x \gtrless y$] [CLX]
$0\ B\ D\ C$	[R↓] [$x \gtrless y$] [0] [$x \gtrless y$] [R↓]
$0\ B\ D\ D$	[R↓] [$x \gtrless y$] [↑] [−]

Table 2.5.1 (Continued)

X Y Z T	Keystrokes
0 C 0 0	[0] [R↓] [−] [CLX]
0 C 0 A	[0] [R↓] [R↓] [CLX]
0 C 0 B	[CLX] [x≷y] [0] [R↓] [R↓]
0 C 0 C	[↑] [−] [R↓] [CLX] [x≷y] [R↓]
0 C 0 D	[R↓] [CLX] [x≷y] [0]
0 C A 0	[0] [x≷y] [R↓] [R↓] [CLX]
0 C A A	[↑] [R↓] [R↓] [CLX]
0 C A B	[R↓] [R↓] [x≷y] [CLX]
0 C A C	[0] [−] [R↓] [CLX] [x≷y] [R↓]
0 C A D	[x≷y] [R↓] [x≷y] [0]
0 C B 0	[R↓] [x≷y] [0] [↑] [R↓]
0 C B A	[x≷y] [R↓] [R↓] [x≷y] [CLX]
0 C B B	[CLX] [x≷y] [↑] [R↓] [R↓]
0 C B C	[↑] [−] [R↓] [x≷y] [0]
0 C B D	[R↓] [x≷y] [0]
0 C C 0	[↑] [−] [R↓] [CLX]
0 C C A	[0] [−] [R↓] [CLX]
0 C C B	[↑] [−] [x≷y] [R↓]
0 C C C	[↑] [−] [−] [CLX]
0 C C D	[R↓] [R↓] [↑] [↑] [CLX]
0 C D 0	[CLX] [R↓] [CLX]
0 C D A	[R↓] [CLX]
0 C D B	[CLX] [x≷y] [R↓]
0 C D C	[R↓] [CLX] [x≷y] [↑] [R↓] [x≷y]
0 C D D	[−] [CLX]

Table 2.5.1 (Continued)

$X\ Y\ Z\ T$	Keystrokes
$0\ D\ 0\ 0$	CLX R↓ − CLX
$0\ D\ 0\ A$	$x \gtrless y$ CLX R↓ R↓ CLX
$0\ D\ 0\ B$	CLX R↓ R↓ CLX
$0\ D\ 0\ C$	CLX R↓ CLX $x \gtrless y$ R↓
$0\ D\ 0\ D$	− R↓ CLX $x \gtrless y$ 0
$0\ D\ A\ 0$	R↓ CLX R↓ CLX
$0\ D\ A\ A$	R↓ − CLX
$0\ D\ A\ B$	R↓ R↓ CLX
$0\ D\ A\ C$	R↓ CLX $x \gtrless y$ R↓
$0\ D\ A\ D$	$x \gtrless y$ CLX − R↓ CLX $x \gtrless y$ R↓
$0\ D\ B\ 0$	CLX $x \gtrless y$ R↓ R↓ CLX
$0\ D\ B\ A$	$x \gtrless y$ R↓ R↓ CLX
$0\ D\ B\ B$	$x \gtrless y$ R↓ − CLX
$0\ D\ B\ C$	CLX $x \gtrless y$ R↓ $x \gtrless y$ R↓
$0\ D\ B\ D$	CLX − R↓ CLX $x \gtrless y$ R↓
$0\ D\ C\ 0$	CLX R↓ R↓ $x \gtrless y$ 0
$0\ D\ C\ A$	R↓ R↓ $x \gtrless y$ 0
$0\ D\ C\ B$	$x \gtrless y$ R↓ R↓ $x \gtrless y$ 0
$0\ D\ C\ C$	− CLX − R↓ −
$0\ D\ C\ D$	− R↓ $x \gtrless y$ 0
$0\ D\ D\ 0$	− CLX R↓ CLX
$0\ D\ D\ A$	$x \gtrless y$ CLX − R↓ CLX
$0\ D\ D\ B$	CLX − R↓ CLX
$0\ D\ D\ C$	− CLX $x \gtrless y$ R↓
$0\ D\ D\ D$	− − CLX

Table 2.5.1 (Continued)

$XYZT$	Keystrokes
$A\,0\,0\,0$	[0] [↑] [↑] [R↓] [R↓] [R↓]
$A\,0\,0\,A$	[0] [$x \gtrless y$] [0] [$x \gtrless y$] [↑] [R↓]
$A\,0\,0\,B$	[0] [$x \gtrless y$] [0] [$x \gtrless y$]
$A\,0\,0\,C$	[$x \gtrless y$] [CLX] [$x \gtrless y$] [0] [$x \gtrless y$]
$A\,0\,0\,D$	[R↓] [CLX] [R↓] [CLX] [R↓] [R↓]
$A\,0\,A\,0$	[↑] [↑] [CLX] [$x \gtrless y$] [0] [R↓]
$A\,0\,A\,A$	[↑] [↑] [↑] [−] [$x \gtrless y$]
$A\,0\,A\,B$	[↑] [↑] [CLX] [$x \gtrless y$]
$A\,0\,A\,C$	[R↓] [CLX] [−] [R↓] [CLX] [$x \gtrless y$]
$A\,0\,A\,D$	[R↓] [−] [CLX] [$x \gtrless y$] [R↓] [$x \gtrless y$]
$A\,0\,B\,0$	[0] [$x \gtrless y$] [0] [R↓]
$A\,0\,B\,A$	[0] [$x \gtrless y$] [↑] [R↓]
$A\,0\,B\,B$	[↑] [↑] [−] [$x \gtrless y$]
$A\,0\,B\,C$	[0] [$x \gtrless y$]
$A\,0\,B\,D$	[R↓] [$x \gtrless y$] [CLX] [R↓] [R↓] [R↓]
$A\,0\,C\,0$	[$x \gtrless y$] [CLX] [$x \gtrless y$] [0] [R↓]
$A\,0\,C\,A$	[$x \gtrless y$] [CLX] [$x \gtrless y$] [↑] [R↓]
$A\,0\,C\,B$	[$x \gtrless y$] [0] [$x \gtrless y$] [R↓] [$x \gtrless y$]
$A\,0\,C\,C$	[$x \gtrless y$] [↑] [−] [$x \gtrless y$]
$A\,0\,C\,D$	[$x \gtrless y$] [CLX] [$x \gtrless y$]
$A\,0\,D\,0$	[$x \gtrless y$] [CLX] [R↓] [$x \gtrless y$] [CLX] [$x \gtrless y$]
$A\,0\,D\,A$	[R↓] [−] [CLX] [R↓] [R↓] [R↓]
$A\,0\,D\,B$	[$x \gtrless y$] [R↓] [$x \gtrless y$] [CLX] [$x \gtrless y$]
$A\,0\,D\,C$	[R↓] [CLX] [R↓] [$x \gtrless y$] [R↓] [R↓]
$A\,0\,D\,D$	[$x \gtrless y$] [R↓] [$x \gtrless y$] [↑] [−] [$x \gtrless y$]

Table 2.5.1 (Continued)

X Y Z T	Keystrokes
$A\ A\ 0\ 0$	[0] [$x \gtrless y$] [0] [$x \gtrless y$] [↑]
$A\ A\ 0\ A$	[0] [$x \gtrless y$] [↑] [↑] [R↓]
$A\ A\ 0\ B$	[0] [$x \gtrless y$] [↑]
$A\ A\ 0\ C$	[$x \gtrless y$] [CLX] [$x \gtrless y$] [↑]
$A\ A\ 0\ D$	[R↓] [−] [CLX] [R↓] [R↓]
$A\ A\ A\ 0$	[0] [$x \gtrless y$] [↑] [↑]
$A\ A\ A\ A$	[↑] [↑] [↑]
$A\ A\ A\ B$	[↑] [↑]
$A\ A\ A\ C$	[$x \gtrless y$] [R↓] [↑] [↑]
$A\ A\ A\ D$	[R↓] [−] [CLX] [−] [R↓]
$A\ A\ B\ 0$	[↑] [↑] [CLX] [R↓]
$A\ A\ B\ A$	[↑] [↑] [R↓]
$A\ A\ B\ B$	[↑] [↑] [CLX] [−]
$A\ A\ B\ C$	[↑]
$A\ A\ B\ D$	[R↓] [$x \gtrless y$] [CLX] [−] [R↓] [R↓]
$A\ A\ C\ 0$	[$x \gtrless y$] [0] [R↓] [R↓] [↑]
$A\ A\ C\ A$	[$x \gtrless y$] [R↓] [↑] [↑] [R↓]
$A\ A\ C\ B$	[$x \gtrless y$] [↑] [R↓] [R↓] [↑]
$A\ A\ C\ C$	[0] [−] [$x \gtrless y$] [R↓] [↑]
$A\ A\ C\ D$	[$x \gtrless y$] [R↓] [↑]
$A\ A\ D\ 0$	[R↓] [−] [CLX] [$x \gtrless y$] [R↓] [R↓]
$A\ A\ D\ A$	[R↓] [−] [R↓] [$x \gtrless y$] [↑]
$A\ A\ D\ B$	[$x \gtrless y$] [R↓] [$x \gtrless y$] [R↓] [↑]
$A\ A\ D\ C$	[R↓] [CLX] [−] [$x \gtrless y$] [R↓] [R↓]
$A\ A\ D\ D$	[R↓] [−] [R↓] [↑] [R↓] [R↓]

Table 2.5.1 (Continued)

X Y Z T	Keystrokes
A B 0 0	[0] [↑] [R↓] [R↓]
A B 0 *A*	[↑] [↑] [CL*X*] [R↓] [R↓]
A B 0 *B*	[↑] [↑] [−] [R↓] [$x \gtrless y$] [R↓]
A B 0 *C*	[R↓] [R↓] [$x \gtrless y$] [CL*X*] [R↓] [R↓]
A B 0 *D*	[R↓] [R↓] [CL*X*] [R↓] [R↓]
A B A 0	[↑] [↑] [R↓] [CL*X*] [R↓]
A B A A	[↑] [↑] [R↓] [R↓]
A B A B	[↑] [↑] [CL*X*] [−] [R↓] [$x \gtrless y$] [R↓]
A B A C	[↑] [R↓] [$x \gtrless y$] [R↓] [R↓] [R↓]
A B A D	[R↓] [$x \gtrless y$] [CL*X*] [−] [$x \gtrless y$] [R↓] [$x \gtrless y$]
A B B 0	[↑] [↑] [−] [R↓]
A B B A	[↑] [↑] [CL*X*] [−] [R↓]
A B B B	[↑] [↑] [−] [−]
A B B C	[$x \gtrless y$] [↑] [R↓] [R↓] [$x \gtrless y$] [R↓]
A B B D	[R↓] [R↓] [CL*X*] [−] [R↓]
A B C 0	[0] [R↓]
A B C A	[↑] [R↓]
A B C B	[$x \gtrless y$] [↑] [R↓] [$x \gtrless y$]
A B C C	[0] [−]
A B D 0	[R↓] [R↓] [CL*X*] [$x \gtrless y$] [R↓] [R↓]
A B D A	[R↓] [$x \gtrless y$] [CL*X*] [−] [R↓] [R↓] [R↓]
A B D B	[R↓] [R↓] [CL*X*] [−] [R↓] [$x \gtrless y$] [R↓]
A B D C	[R↓] [R↓] [$x \gtrless y$] [R↓] [R↓]
A B D D	[R↓] [R↓] [R↓] [↑] [R↓] [R↓]

Table 2.5.1 (Continued)

$XYZT$	Keystrokes
$A\ C\ 0\ 0$	[$x \gtrless y$] [CLX] [↑] [R↓] [R↓]
$A\ C\ 0\ A$	[$x \gtrless y$] [0] [R↓] [R↓] [↑] [R↓]
$A\ C\ 0\ B$	[$x \gtrless y$] [0] [R↓] [R↓]
$A\ C\ 0\ C$	[$x \gtrless y$] [↑] [−] [R↓] [$x \gtrless y$] [R↓]
$A\ C\ 0\ D$	[R↓] [CLX] [$x \gtrless y$] [R↓] [R↓] [R↓]
$A\ C\ A\ 0$	[↑] [R↓] [$x \gtrless y$] [CLX] [R↓]
$A\ C\ A\ A$	[↑] [R↓] [$x \gtrless y$] [CLX] [−]
$A\ C\ A\ B$	[↑] [R↓] [$x \gtrless y$] [R↓]
$A\ C\ A\ C$	[0] [−] [R↓] [CLX] [−] [R↓] [$x \gtrless y$]
$A\ C\ A\ D$	[R↓] [CLX] [−] [$x \gtrless y$] [R↓] [$x \gtrless y$]
$A\ C\ B\ 0$	[$x \gtrless y$] [0] [$x \gtrless y$] [R↓] [R↓]
$A\ C\ B\ A$	[$x \gtrless y$] [↑] [R↓] [R↓] [↑] [R↓]
$A\ C\ B\ B$	[$x \gtrless y$] [↑] [R↓] [R↓]
$A\ C\ B\ C$	[0] [−] [$x \gtrless y$] [R↓] [$x \gtrless y$] [R↓]
$A\ C\ B\ D$	[R↓] [$x \gtrless y$] [R↓] [R↓] [R↓]
$A\ C\ C\ 0$	[$x \gtrless y$] [↑] [−] [R↓]
$A\ C\ C\ A$	[0] [−] [$x \gtrless y$] [R↓] [↑] [R↓]
$A\ C\ C\ B$	[0] [−] [$x \gtrless y$] [R↓]
$A\ C\ C\ C$	[$x \gtrless y$] [↑] [−] [−]
$A\ C\ C\ D$	[R↓] [R↓] [↑] [R↓] [R↓] [R↓]
$A\ C\ D\ 0$	[$x \gtrless y$] [CLX] [R↓]
$A\ C\ D\ A$	[$x \gtrless y$] [R↓] [↑] [R↓]
$A\ C\ D\ B$	[$x \gtrless y$] [R↓]
$A\ C\ D\ C$	[$x \gtrless y$] [R↓] [$x \gtrless y$] [↑] [R↓] [$x \gtrless y$]
$A\ C\ D\ D$	[$x \gtrless y$] [CLX] [−]

Table 2.5.1 (Continued)

X Y Z T	Keystrokes
A D 0 0	[x ≷ y] [CLX] [R↓] [x ≷ y] [CLX] [R↓]
A D 0 *A*	[R↓] [−] [CLX] [x ≷ y] [R↓] [R↓] [R↓]
A D 0 *B*	[R↓] [x ≷ y] [CLX] [R↓] [R↓] [x ≷ y]
A D 0 *C*	[x ≷ y] [CLX] [R↓] [x ≷ y] [R↓]
A D 0 *D*	[R↓] [CLX] [R↓] [R↓] [↑] [R↓] [x ≷ y]
A D A 0	[R↓] [−] [CLX] [R↓] [x ≷ y]
A D A A	[R↓] [−] [CLX] [−] [x ≷ y]
A D A B	[R↓] [x ≷ y] [CLX] [−] [R↓] [x ≷ y]
A D A C	[R↓] [CLX] [−] [R↓] [x ≷ y]
A D A D	[R↓] [−] [R↓] [↑] [R↓] [x ≷ y]
A D B 0	[x ≷ y] [R↓] [x ≷ y] [CLX] [R↓]
A D B A	[x ≷ y] [R↓] [x ≷ y] [R↓] [↑] [R↓]
A D B B	[R↓] [R↓] [CLX] [−] [x ≷ y]
A D B C	[x ≷ y] [R↓] [x ≷ y] [R↓]
A D B D	[R↓] [R↓] [R↓] [↑] [R↓] [x ≷ y]
A D C 0	[R↓] [CLX] [x ≷ y] [R↓] [R↓] [x ≷ y]
A D C A	[R↓] [CLX] [−] [x ≷ y] [R↓] [R↓] [R↓]
A D C B	[R↓] [x ≷ y] [R↓] [R↓] [x ≷ y]
A D C C	[x ≷ y] [R↓] [x ≷ y] [↑] [R↓] [R↓]
A D C D	[x ≷ y] [CLX] [−] [x ≷ y] [R↓] [x ≷ y] [R↓]
A D D 0	[x ≷ y] [R↓] [x ≷ y] [↑] [−] [R↓]
A D D A	[R↓] [−] [R↓] [↑] [R↓] [R↓] [R↓]
A D D C	[x ≷ y] [CLX] [−] [x ≷ y] [R↓]
A D D D	[x ≷ y] [R↓] [x ≷ y] [↑] [−] [−]

Table 2.5.1 (Continued)

$X\ Y\ Z\ T$	Keystrokes
B 0 0 0	[CLX] [↑] [↑] [R↓] [R↓] [R↓]
B 0 0 A	[0] [↑] [R↓] [R↓] [R↓]
B 0 0 B	[CLX] [$x \gtrless y$] [0] [$x \gtrless y$] [↑] [R↓]
B 0 0 C	[CLX] [$x \gtrless y$] [0] [$x \gtrless y$]
B 0 0 D	[R↓] [$x \gtrless y$] [CLX] [$x \gtrless y$] [0] [$x \gtrless y$]
B 0 A 0	[$x \gtrless y$] [0] [$x \gtrless y$] [0] [R↓]
B 0 A A	[↑] [↑] [R↓] [−] [$x \gtrless y$]
B 0 A B	[$x \gtrless y$] [0] [$x \gtrless y$] [↑] [R↓]
B 0 A C	[$x \gtrless y$] [0] [$x \gtrless y$]
B 0 A D	[$x \gtrless y$] [R↓] [$x \gtrless y$] [CLX] [R↓] [R↓] [R↓]
B 0 B 0	[0] [$x \gtrless y$] [↑] [−] [R↓] [$x \gtrless y$]
B 0 B A	[$x \gtrless y$] [↑] [↑] [CLX] [$x \gtrless y$]
B 0 B B	[CLX] [$x \gtrless y$] [↑] [↑] [R↓] [R↓]
B 0 B C	[R↓] [↑] [↑] [CLX] [$x \gtrless y$]
B 0 B D	[$x \gtrless y$] [R↓] [−] [CLX] [$x \gtrless y$] [R↓] [$x \gtrless y$]
B 0 C 0	[CLX] [$x \gtrless y$] [0] [R↓]
B 0 C A	[0] [$x \gtrless y$] [R↓] [$x \gtrless y$]
B 0 C B	[CLX] [$x \gtrless y$] [↑] [R↓]
B 0 C C	[↑] [−] [$x \gtrless y$]
B 0 C D	[CLX] [$x \gtrless y$]
B 0 D 0	[CLX] [R↓] [$x \gtrless y$] [CLX] [$x \gtrless y$]
B 0 D A	[R↓] [$x \gtrless y$] [CLX] [$x \gtrless y$]
B 0 D B	[R↓] [$x \gtrless y$] [CLX] [$x \gtrless y$] [↑] [R↓]
B 0 D C	[R↓] [$x \gtrless y$] [0] [$x \gtrless y$] [R↓] [$x \gtrless y$]
B 0 D D	[R↓] [$x \gtrless y$] [↑] [−] [$x \gtrless y$]

Table 2.5.1 (Continued)

$X\ Y\ Z\ T$	Keystrokes
$B\ A\ 0\ 0$	[x≷y] [0] [↑] [R↓] [R↓]
$B\ A\ 0\ A$	[↑] [↑] [R↓] [CLX] [R↓] [R↓]
$B\ A\ 0\ B$	[x≷y] [↑] [↑] [CLX] [R↓] [R↓]
$B\ A\ 0\ C$	[0] [x≷y] [R↓] [R↓] [x≷y] [R↓]
$B\ A\ 0\ D$	[x≷y] [R↓] [R↓] [CLX] [R↓] [R↓]
$B\ A\ A\ 0$	[↑] [↑] [R↓] [−] [R↓]
$B\ A\ A\ A$	[↑] [↑] [R↓] [R↓] [R↓]
$B\ A\ A\ B$	[x≷y] [↑] [↑] [CLX] [−] [R↓]
$B\ A\ A\ C$	[↑] [R↓] [R↓] [x≷y] [R↓]
$B\ A\ A\ D$	[R↓] [x≷y] [CLX] [−] [x≷y] [R↓]
$B\ A\ B\ 0$	[↑] [↑] [−] [R↓] [x≷y]
$B\ A\ B\ A$	[↑] [↑] [CLX] [−] [R↓] [x≷y]
$B\ A\ B\ B$	[x≷y] [↑] [↑] [R↓] [R↓]
$B\ A\ B\ C$	[x≷y] [↑] [R↓] [x≷y] [R↓] [R↓] [R↓]
$B\ A\ B\ D$	[R↓] [R↓] [CLX] [−] [R↓] [x≷y]
$B\ A\ C\ 0$	[x≷y] [0] [R↓]
$B\ A\ C\ A$	[↑] [R↓] [x≷y]
$B\ A\ C\ B$	[x≷y] [↑] [R↓]
$B\ A\ C\ C$	[0] [−] [x≷y]
$B\ A\ C\ D$	[x≷y]
$B\ A\ D\ 0$	[x≷y] [R↓] [R↓] [CLX] [x≷y] [R↓] [R↓]
$B\ A\ D\ C$	[x≷y] [R↓] [R↓] [x≷y] [R↓] [R↓]
$B\ A\ D\ D$	[x≷y] [R↓] [R↓] [R↓] [↑] [R↓] [R↓]

Table 2.5.1 (Continued)

$X\ Y\ Z\ T$	Keystrokes
$B\ B\ 0\ 0$	[CLX] [$x \gtrless y$] [0] [$x \gtrless y$] [↑]
$B\ B\ 0\ A$	[$x \gtrless y$] [0] [$x \gtrless y$] [↑]
$B\ B\ 0\ B$	[CLX] [$x \gtrless y$] [↑] [↑] [R↓]
$B\ B\ 0\ C$	[CLX] [$x \gtrless y$] [↑]
$B\ B\ 0\ D$	[R↓] [$x \gtrless y$] [CLX] [$x \gtrless y$] [↑]
$B\ B\ A\ 0$	[$x \gtrless y$] [↑] [↑] [CLX] [R↓]
$B\ B\ A\ A$	[$x \gtrless y$] [↑] [↑] [CLX] [−]
$B\ B\ A\ B$	[$x \gtrless y$] [↑] [↑] [R↓]
$B\ B\ A\ C$	[$x \gtrless y$] [↑]
$B\ B\ A\ D$	[R↓] [R↓] [CLX] [−] [$x \gtrless y$] [R↓] [R↓]
$B\ B\ B\ 0$	[CLX] [$x \gtrless y$] [↑] [↑]
$B\ B\ B\ A$	[$x \gtrless y$] [↑] [↑]
$B\ B\ B\ B$	[R↓] [↑] [↑] [↑]
$B\ B\ B\ C$	[R↓] [↑] [↑]
$B\ B\ B\ D$	[R↓] [$x \gtrless y$] [R↓] [↑] [↑]
$B\ B\ C\ 0$	[0] [R↓] [R↓] [↑]
$B\ B\ C\ A$	[↑] [R↓] [R↓] [↑]
$B\ B\ C\ B$	[R↓] [↑] [↑] [R↓]
$B\ B\ C\ C$	[↑] [−] [R↓] [↑]
$B\ B\ C\ D$	[R↓] [↑]
$B\ B\ D\ 0$	[CLX] [R↓] [$x \gtrless y$] [R↓] [↑]
$B\ B\ D\ A$	[R↓] [$x \gtrless y$] [R↓] [↑]
$B\ B\ D\ B$	[R↓] [$x \gtrless y$] [R↓] [↑] [↑] [R↓]
$B\ B\ D\ C$	[R↓] [$x \gtrless y$] [↑] [R↓] [R↓] [↑]
$B\ B\ D\ D$	[CLX] [−] [$x \gtrless y$] [R↓] [↑]

Table 2.5.1 (Continued)

X Y Z T	Keystrokes
B C 0 0	[CLX] [↑] [R↓] [R↓]
B C 0 *A*	[0] [R↓] [R↓]
B C 0 *B*	[0] [R↓] [R↓] [↑] [R↓]
B C 0 *C*	[↑] [−] [R↓] [$x \gtrless y$] [R↓]
B C 0 *D*	[CLX] [$x \gtrless y$] [R↓] [$x \gtrless y$] [R↓] [R↓] [R↓]
B C A 0	[0] [$x \gtrless y$] [R↓] [R↓]
B C A A	[↑] [R↓] [R↓]
B C A B	[↑] [R↓] [R↓] [↑] [R↓]
B C A C	[0] [−] [R↓] [$x \gtrless y$] [R↓]
B C A D	[$x \gtrless y$] [R↓] [$x \gtrless y$] [R↓] [R↓] [R↓]
B C B 0	[CLX] [$x \gtrless y$] [↑] [R↓] [$x \gtrless y$] [R↓]
B C B A	[$x \gtrless y$] [↑] [R↓] [$x \gtrless y$] [R↓]
B C B B	[R↓] [↑] [↑] [R↓] [R↓]
B C B C	[↑] [−] [R↓] [↑] [R↓] [$x \gtrless y$] [R↓]
B C B D	[R↓] [↑] [R↓] [$x \gtrless y$] [R↓] [R↓] [R↓]
B C C 0	[↑] [−] [R↓]
B C C A	[0] [−] [R↓]
B C C B	[↑] [−] [R↓] [↑] [R↓]
B C C C	[↑] [−] [−]
B C C D	[CLX] [R↓] [R↓] [R↓] [+] [R↓]
B C D 0	[CLX] [R↓]
B C D A	[R↓]
B C D B	[R↓] [↑] [R↓]
B C D C	[R↓] [$x \gtrless y$] [↑] [R↓] [$x \gtrless y$]
B C D D	[CLX] [−]

Table 2.5.1 (Continued)

X Y Z T	Keystrokes
B D 0 0	CLX R↓ x≷y CLX R↓
B D 0 A	R↓ R↓ CLX x≷y R↓ R↓ R↓
B D 0 B	CLX R↓ x≷y R↓ ↑ R↓
B D 0 C	CLX R↓ x≷y R↓
B D 0 D	R↓ x≷y ↑ − R↓ x≷y R↓
B D A 0	R↓ x≷y CLX R↓
B D A A	R↓ x≷y CLX −
B D A B	R↓ x≷y R↓ ↑ R↓
B D A C	R↓ x≷y R↓
B D A D	x≷y R↓ R↓ R↓ ↑ R↓ x≷y
B D B 0	R↓ ↑ R↓ x≷y CLX R↓
B D B A	R↓ R↓ CLX − x≷y R↓ x≷y
B D B B	x≷y R↓ − CLX − x≷y
B D B C	R↓ ↑ R↓ x≷y R↓
B D B D	CLX − R↓ CLX − R↓ x≷y
B D C 0	R↓ x≷y 0 x≷y R↓ R↓
B D C A	R↓ R↓ x≷y R↓ R↓ R↓
B D C B	R↓ x≷y ↑ R↓ R↓ ↑ R↓
B D C C	R↓ x≷y ↑ R↓ R↓
B D C D	CLX − x≷y R↓ x≷y R↓
B D D 0	R↓ x≷y ↑ − R↓
B D D A	R↓ R↓ R↓ ↑ R↓ R↓ R↓
B D D B	CLX − x≷y R↓ ↑ R↓
B D D C	CLX − x≷y R↓
B D D D	R↓ x≷y ↑ − −

Table 2.5.1 (Continued)

$X\ Y\ Z\ T$	Keystrokes
$C\ 0\ 0\ 0$	[0] [R↓] [−] [CLX] [R↓]
$C\ 0\ 0\ A$	[0] [R↓] [R↓] [CLX] [$x \gtrless y$]
$C\ 0\ 0\ B$	[CLX] [↑] [R↓] [R↓] [R↓]
$C\ 0\ 0\ C$	[↑] [−] [R↓] [CLX] [R↓] [R↓]
$C\ 0\ 0\ D$	[R↓] [CLX] [$x \gtrless y$] [0] [$x \gtrless y$]
$C\ 0\ A\ 0$	[0] [R↓] [R↓] [CLX] [R↓]
$C\ 0\ A\ A$	[↑] [R↓] [R↓] [CLX] [$x \gtrless y$]
$C\ 0\ A\ B$	[0] [R↓] [R↓] [R↓]
$C\ 0\ A\ C$	[$x \gtrless y$] [↑] [−] [R↓] [R↓] [R↓]
$C\ 0\ A\ D$	[$x \gtrless y$] [R↓] [$x \gtrless y$] [0] [$x \gtrless y$]
$C\ 0\ B\ 0$	[CLX] [$x \gtrless y$] [0] [R↓] [R↓] [R↓]
$C\ 0\ B\ A$	[$x \gtrless y$] [0] [R↓] [R↓] [R↓]
$C\ 0\ B\ B$	[0] [R↓] [R↓] [↑] [R↓] [R↓]
$C\ 0\ B\ C$	[↑] [−] [R↓] [R↓] [R↓]
$C\ 0\ B\ D$	[R↓] [$x \gtrless y$] [0] [$x \gtrless y$]
$C\ 0\ C\ 0$	[↑] [−] [R↓] [CLX] [$x \gtrless y$]
$C\ 0\ C\ A$	[0] [−] [R↓] [CLX] [$x \gtrless y$]
$C\ 0\ C\ B$	[↑] [−] [$x \gtrless y$] [R↓] [$x \gtrless y$]
$C\ 0\ C\ C$	[↑] [−] [−] [CLX] [$x \gtrless y$]
$C\ 0\ C\ D$	[R↓] [R↓] [↑] [↑] [CLX] [$x \gtrless y$]
$C\ 0\ D\ 0$	[CLX] [R↓] [CLX] [$x \gtrless y$]
$C\ 0\ D\ A$	[R↓] [CLX] [$x \gtrless y$]
$C\ 0\ D\ B$	[CLX] [$x \gtrless y$] [R↓] [$x \gtrless y$]
$C\ 0\ D\ C$	[R↓] [CLX] [$x \gtrless y$] [↑] [R↓]
$C\ 0\ D\ D$	[−] [CLX] [$x \gtrless y$]

Table 2.5.1 (Continued)

$X\ Y\ Z\ T$	Keystrokes
$C\ A\ 0\ 0$	[0] [$x \gtrless y$] [R↓] [R↓] [CLX] [R↓]
$C\ A\ 0\ A$	[↑] [R↓] [$x \gtrless y$] [CLX] [R↓] [R↓]
$C\ A\ 0\ B$	[0] [$x \gtrless y$] [R↓] [R↓] [R↓]
$C\ A\ 0\ C$	[0] [−] [R↓] [CLX] [R↓] [R↓]
$C\ A\ 0\ D$	[R↓] [CLX] [R↓] [$x \gtrless y$] [R↓]
$C\ A\ A\ 0$	[↑] [R↓] [R↓] [CLX] [R↓]
$C\ A\ A\ A$	[↑] [R↓] [R↓] [CLX] [−]
$C\ A\ A\ B$	[↑] [R↓] [R↓] [R↓]
$C\ A\ A\ C$	[0] [−] [R↓] [CLX] [−] [R↓]
$C\ A\ A\ D$	[R↓] [CLX] [−] [$x \gtrless y$] [R↓]
$C\ A\ B\ 0$	[R↓] [R↓] [$x \gtrless y$] [CLX] [R↓]
$C\ A\ B\ A$	[↑] [R↓] [$x \gtrless y$] [R↓] [R↓]
$C\ A\ B\ B$	[$x \gtrless y$] [↑] [R↓] [R↓] [$x \gtrless y$]
$C\ A\ B\ C$	[0] [−] [R↓] [R↓] [R↓]
$C\ A\ B\ D$	[R↓] [R↓] [$x \gtrless y$] [R↓]
$C\ A\ C\ 0$	[$x \gtrless y$] [↑] [−] [R↓] [$x \gtrless y$]
$C\ A\ C\ A$	[0] [−] [$x \gtrless y$] [R↓] [↑] [R↓] [$x \gtrless y$]
$C\ A\ C\ B$	[0] [−] [$x \gtrless y$] [R↓] [$x \gtrless y$]
$C\ A\ C\ C$	[$x \gtrless y$] [↑] [−] [−] [$x \gtrless y$]
$C\ A\ C\ D$	[R↓] [R↓] [↑] [R↓] [$x \gtrless y$] [R↓]
$C\ A\ D\ 0$	[$x \gtrless y$] [CLX] [R↓] [$x \gtrless y$]
$C\ A\ D\ A$	[$x \gtrless y$] [R↓] [↑] [R↓] [$x \gtrless y$]
$C\ A\ D\ B$	[$x \gtrless y$] [R↓] [$x \gtrless y$]
$C\ A\ D\ C$	[$x \gtrless y$] [R↓] [$x \gtrless y$] [↑] [R↓]
$C\ A\ D\ D$	[$x \gtrless y$] [CLX] [−] [$x \gtrless y$]

Table 2.5.1 (Continued)

$XYZT$	Keystrokes
$C\,B\,0\,0$	CLX ↑ R↓ R↓ $x \gtrless y$
$C\,B\,0\,A$	0 R↓ R↓ $x \gtrless y$
$C\,B\,0\,B$	0 R↓ R↓ ↑ R↓ $x \gtrless y$
$C\,B\,0\,C$	↑ − $x \gtrless y$ R↓ R↓ R↓
$C\,B\,0\,D$	CLX $x \gtrless y$ R↓ R↓ $x \gtrless y$ R↓
$C\,B\,A\,0$	0 $x \gtrless y$ R↓ R↓ $x \gtrless y$
$C\,B\,A\,A$	↑ R↓ R↓ $x \gtrless y$
$C\,B\,A\,B$	$x \gtrless y$ ↑ R↓ $x \gtrless y$ R↓ R↓
$C\,B\,A\,C$	0 − $x \gtrless y$ R↓ R↓ R↓
$C\,B\,A\,D$	$x \gtrless y$ R↓ R↓ $x \gtrless y$ R↓
$C\,B\,B\,0$	CLX $x \gtrless y$ ↑ R↓ R↓ R↓
$C\,B\,B\,A$	$x \gtrless y$ ↑ R↓ R↓ R↓
$C\,B\,B\,B$	R↓ ↑ ↑ R↓ R↓ R↓
$C\,B\,B\,C$	↑ − $x \gtrless y$ R↓ − R↓
$C\,B\,B\,D$	R↓ ↑ R↓ R↓ $x \gtrless y$ R↓
$C\,B\,C\,0$	↑ − R↓ $x \gtrless y$
$C\,B\,C\,A$	0 − R↓ $x \gtrless y$
$C\,B\,C\,B$	↑ − R↓ ↑ R↓ $x \gtrless y$
$C\,B\,C\,C$	↑ − − $x \gtrless y$
$C\,B\,C\,D$	$x \gtrless y$ R↓ R↓ ↑ R↓ $x \gtrless y$ R↓
$C\,B\,D\,0$	CLX R↓ $x \gtrless y$
$C\,B\,D\,A$	R↓ $x \gtrless y$
$C\,B\,D\,B$	R↓ ↑ R↓ $x \gtrless y$
$C\,B\,D\,C$	R↓ $x \gtrless y$ ↑ R↓
$C\,B\,D\,D$	CLX − $x \gtrless y$

Table 2.5.1 (Continued)

X Y Z T	Keystrokes
C C 0 0	[↑] [−] [R↓] [CLX] [R↓]
C C 0 A	[x≷y] [↑] [−] [R↓] [R↓]
C C 0 B	[↑] [−] [R↓] [R↓]
C C 0 C	[↑] [−] [−] [CLX] [R↓] [R↓]
C C 0 D	[R↓] [CLX] [x≷y] [↑]
C C A 0	[0] [−] [R↓] [CLX] [R↓]
C C A A	[0] [−] [R↓] [CLX] [−]
C C A B	[0] [−] [R↓] [R↓]
C C A C	[x≷y] [↑] [−] [−] [R↓] [R↓]
C C A D	[x≷y] [R↓] [x≷y] [↑]
C C B 0	[↑] [−] [x≷y] [R↓] [R↓]
C C B A	[0] [−] [x≷y] [R↓] [R↓]
C C B B	[↑] [−] [x≷y] [R↓] [−]
C C B C	[↑] [−] [−] [R↓] [R↓]
C C B D	[R↓] [x≷y] [↑]
C C C 0	[↑] [−] [−] [CLX] [R↓]
C C C A	[x≷y] [↑] [−] [−] [R↓]
C C C B	[↑] [−] [−] [R↓]
C C C C	[↑] [−] [−] [CLX] [−]
C C C D	[R↓] [R↓] [↑] [↑]
C C D 0	[CLX] [R↓] [R↓] [↑]
C C D A	[R↓] [R↓] [↑]
C C D B	[x≷y] [R↓] [R↓] [↑]
C C D C	[R↓] [R↓] [↑] [↑] [R↓]
C C D D	[−] [R↓] [↑]

Table 2.5.1 (Continued)

$X\ Y\ Z\ T$	Keystrokes
$C\ D\ 0\ 0$	[CLX] [R↓] [CLX] [R↓]
$C\ D\ 0\ A$	[$x \gtrless y$] [CLX] [R↓] [R↓]
$C\ D\ 0\ B$	[CLX] [R↓] [R↓]
$C\ D\ 0\ C$	[CLX] [R↓] [R↓] [↑] [R↓]
$C\ D\ 0\ D$	[−] [CLX] [R↓] [$x \gtrless y$] [R↓]
$C\ D\ A\ 0$	[R↓] [CLX] [R↓]
$C\ D\ A\ A$	[R↓] [CLX] [−]
$C\ D\ A\ B$	[R↓] [R↓]
$C\ D\ A\ C$	[R↓] [R↓] [↑] [R↓]
$C\ D\ A\ D$	[$x \gtrless y$] [CLX] [−] [R↓] [$x \gtrless y$] [R↓]
$C\ D\ B\ 0$	[CLX] [$x \gtrless y$] [R↓] [R↓]
$C\ D\ B\ A$	[$x \gtrless y$] [R↓] [R↓]
$C\ D\ B\ B$	[R↓] [↑] [R↓] [R↓]
$C\ D\ B\ C$	[$x \gtrless y$] [R↓] [R↓] [↑] [R↓]
$C\ D\ B\ D$	[CLX] [−] [R↓] [$x \gtrless y$] [R↓]
$C\ D\ C\ 0$	[R↓] [CLX] [$x \gtrless y$] [↑] [R↓] [$x \gtrless y$] [R↓]
$C\ D\ C\ A$	[$x \gtrless y$] [R↓] [$x \gtrless y$] [↑] [R↓] [$x \gtrless y$] [R↓]
$C\ D\ C\ B$	[R↓] [$x \gtrless y$] [↑] [R↓] [$x \gtrless y$] [R↓]
$C\ D\ C\ C$	[R↓] [R↓] [↑] [↑] [R↓] [R↓]
$C\ D\ C\ D$	[−] [R↓] [↑] [R↓] [$x \gtrless y$] [R↓]
$C\ D\ D\ 0$	[−] [CLX] [R↓]
$C\ D\ D\ A$	[$x \gtrless y$] [CLX] [−] [R↓]
$C\ D\ D\ B$	[CLX] [−] [R↓]
$C\ D\ D\ C$	[−] [R↓] [↑] [R↓]
$C\ D\ D\ D$	[−] [CLX] [−]

Table 2.5.1 (Continued)

X Y Z T	Keystrokes
D 0 0 0	[CLX] [R↓] [−] [CLX] [R↓]
D 0 0 *A*	[x≷y] [CLX] [R↓] [R↓] [CLX] [x≷y]
D 0 0 *B*	[CLX] [R↓] [R↓] [CLX] [x≷y]
D 0 0 *C*	[CLX] [R↓] [CLX] [R↓] [R↓]
D 0 0 *D*	[−] [CLX] [R↓] [CLX] [R↓] [R↓]
D 0 *A* 0	[R↓] [CLX] [R↓] [CLX] [x≷y]
D 0 *A A*	[R↓] [−] [CLX] [x≷y]
D 0 *A B*	[R↓] [R↓] [CLX] [x≷y]
D 0 *A C*	[x≷y] [CLX] [R↓] [R↓] [R↓]
D 0 *A D*	[R↓] [R↓] [CLX] [x≷y] [↑] [R↓]
D 0 *B* 0	[CLX] [R↓] [R↓] [CLX] [R↓]
D 0 *B A*	[x≷y] [R↓] [R↓] [CLX] [x≷y]
D 0 *B B*	[x≷y] [R↓] [−] [CLX] [x≷y]
D 0 *B C*	[CLX] [R↓] [R↓] [R↓]
D 0 *B D*	[CLX] [R↓] [R↓] [R↓] [↑] [R↓]
D 0 *C* 0	[CLX] [R↓] [CLX] [x≷y] [R↓] [R↓]
D 0 *C A*	[R↓] [R↓] [x≷y] [0] [x≷y]
D 0 *C B*	[CLX] [R↓] [x≷y] [R↓] [R↓]
D 0 *C C*	[CLX] [R↓] [R↓] [↑] [R↓] [R↓]
D 0 *C D*	[−] [CLX] [R↓] [R↓] [R↓]
D 0 *D* 0	[−] [CLX] [R↓] [CLX] [x≷y]
D 0 *D A*	[x≷y] [CLX] [−] [R↓] [CLX] [x≷y]
D 0 *D B*	[CLX] [−] [R↓] [CLX] [x≷y]
D 0 *D C*	[−] [CLX] [x≷y] [R↓] [x≷y]
D 0 *D D*	[−] [−] [CLX] [x≷y]

Table 2.5.1 (Continued)

$X\ Y\ Z\ T$	Keystrokes
$D\ A\ 0\ 0$	[R↓] [CLX] [R↓] [CLX] [R↓]
$D\ A\ 0\ A$	[R↓] [−] [CLX] [R↓] [$x \gtrless y$] [R↓]
$D\ A\ 0\ B$	[R↓] [$x \gtrless y$] [CLX] [R↓] [R↓]
$D\ A\ 0\ C$	[R↓] [CLX] [R↓] [R↓]
$D\ A\ 0\ D$	[R↓] [CLX] [R↓] [R↓] [↑] [R↓]
$D\ A\ A\ 0$	[R↓] [−] [CLX] [R↓]
$D\ A\ A\ A$	[R↓] [−] [CLX] [−]
$D\ A\ A\ B$	[R↓] [$x \gtrless y$] [CLX] [−] [R↓]
$D\ A\ A\ C$	[R↓] [CLX] [−] [R↓]
$D\ A\ A\ D$	[R↓] [−] [R↓] [↑] [R↓]
$D\ A\ B\ 0$	[R↓] [R↓] [CLX] [R↓]
$D\ A\ B\ A$	[R↓] [$x \gtrless y$] [CLX] [−] [R↓] [$x \gtrless y$] [R↓]
$D\ A\ B\ B$	[R↓] [R↓] [CLX] [−]
$D\ A\ B\ C$	[R↓] [R↓] [R↓]
$D\ A\ B\ D$	[R↓] [R↓] [R↓] [↑] [R↓]
$D\ A\ C\ 0$	[R↓] [CLX] [$x \gtrless y$] [R↓] [R↓]
$D\ A\ C\ A$	[R↓] [CLX] [−] [R↓] [$x \gtrless y$] [R↓]
$D\ A\ C\ B$	[R↓] [$x \gtrless y$] [R↓] [R↓]
$D\ A\ C\ C$	[R↓] [R↓] [↑] [R↓] [R↓]
$D\ A\ C\ D$	[$x \gtrless y$] [CLX] [−] [R↓] [R↓] [R↓]
$D\ A\ D\ 0$	[$x \gtrless y$] [R↓] [$x \gtrless y$] [↑] [−] [R↓] [$x \gtrless y$]
$D\ A\ D\ A$	[R↓] [−] [R↓] [↑] [R↓] [$x \gtrless y$] [R↓]
$D\ A\ D\ C$	[$x \gtrless y$] [CLX] [−] [$x \gtrless y$] [R↓] [$x \gtrless y$]
$D\ A\ D\ D$	[R↓] [R↓] [R↓] [↑] [↑] [R↓] [R↓]

Table 2.5.1 (Continued)

X Y Z T	Keystrokes
D B 0 0	[CLX] [x≷y] [R↓] [R↓] [CLX] [R↓]
D B 0 *A*	[x≷y] [R↓] [x≷y] [CLX] [R↓] [R↓]
D B 0 *B*	[R↓] [↑] [R↓] [x≷y] [CLX] [R↓] [R↓]
D B 0 *C*	[CLX] [x≷y] [R↓] [R↓] [R↓]
D B 0 *D*	[CLX] [−] [R↓] [CLX] [R↓] [R↓]
D B A 0	[x≷y] [R↓] [R↓] [CLX] [R↓]
D B A A	[R↓] [x≷y] [CLX] [−] [x≷y]
D B A B	[R↓] [x≷y] [R↓] [↑] [R↓] [x≷y]
D B A C	[x≷y] [R↓] [R↓] [R↓]
D B A D	[x≷y] [R↓] [R↓] [R↓] [↑] [R↓]
D B B 0	[x≷y] [R↓] [−] [CLX] [R↓]
D B B A	[R↓] [R↓] [CLX] [−] [x≷y] [R↓]
D B B B	[x≷y] [R↓] [−] [CLX] [−]
D B B C	[R↓] [↑] [R↓] [R↓] [R↓]
D B B D	[CLX] [−] [R↓] [CLX] [−] [R↓]
D B C 0	[CLX] [x≷y] [R↓] [x≷y] [R↓] [R↓]
D B C A	[x≷y] [R↓] [x≷y] [R↓] [R↓]
D B C B	[R↓] [↑] [R↓] [x≷y] [R↓] [R↓]
D B C C	[CLX] [R↓] [R↓] [R↓] [+]
D B C D	[CLX] [−] [R↓] [R↓] [R↓]
D B D 0	[R↓] [x≷y] [↑] [−] [R↓] [x≷y]
D B D A	[R↓] [R↓] [R↓] [↑] [R↓] [x≷y] [R↓]
D B D B	[CLX] [−] [x≷y] [R↓] [↑] [R↓] [x≷y]
D B D C	[CLX] [−] [x≷y] [R↓] [x≷y]
D B D D	[R↓] [x≷y] [↑] [−] [−] [x≷y]

Table 2.5.1 (Continued)

X Y Z T	Keystrokes
D C 0 0	[CLX] [R↓] [CLX] [R↓] [$x \gtrless y$]
D C 0 *A*	[$x \gtrless y$] [CLX] [R↓] [R↓] [$x \gtrless y$]
D C 0 *B*	[CLX] [R↓] [R↓] [$x \gtrless y$]
D C 0 *C*	[CLX] [R↓] [R↓] [↑] [R↓] [$x \gtrless y$]
D C 0 *D*	[−] [CLX] [$x \gtrless y$] [R↓] [R↓] [R↓]
D C A 0	[R↓] [CLX] [R↓] [$x \gtrless y$]
D C A A	[R↓] [CLX] [−] [$x \gtrless y$]
D C A B	[R↓] [R↓] [$x \gtrless y$]
D C A C	[R↓] [R↓] [↑] [R↓] [$x \gtrless y$]
D C A D	[R↓] [R↓] [$x \gtrless y$] [↑] [R↓]
D C B 0	[CLX] [$x \gtrless y$] [R↓] [R↓] [$x \gtrless y$]
D C B A	[$x \gtrless y$] [R↓] [R↓] [$x \gtrless y$]
D C B B	[R↓] [↑] [R↓] [R↓] [$x \gtrless y$]
D C B C	[$x \gtrless y$] [R↓] [R↓] [↑] [R↓] [$x \gtrless y$]
D C B D	[CLX] [−] [$x \gtrless y$] [R↓] [R↓] [R↓]
D C C 0	[−] [CLX] [−] [R↓] [−] [R↓]
D C C A	[$x \gtrless y$] [R↓] [$x \gtrless y$] [↑] [R↓] [R↓] [R↓]
D C C B	[R↓] [$x \gtrless y$] [↑] [R↓] [R↓] [R↓]
D C C C	[−] [CLX] [−] [R↓] [−] [−]
D C C D	[−] [R↓] [↑] [R↓] [R↓] [R↓]
D C D 0	[−] [CLX] [R↓] [$x \gtrless y$]
D C D A	[$x \gtrless y$] [CLX] [−] [R↓] [$x \gtrless y$]
D C D B	[CLX] [−] [R↓] [$x \gtrless y$]
D C D C	[−] [R↓] [↑] [R↓] [$x \gtrless y$]
D C D D	[−] [CLX] [−] [$x \gtrless y$]

Table 2.5.1 (Continued)

X Y Z T	Keystrokes
D D 0 0	[−] [CLX] [R↓] [CLX] [R↓]
D D 0 *A*	[R↓] [R↓] [CLX] [$x \gtrless y$] [↑]
D D 0 *B*	[CLX] [R↓] [R↓] [R↓] [↑]
D D 0 *C*	[−] [CLX] [R↓] [R↓]
D D 0 *D*	[−] [−] [CLX] [R↓] [R↓]
D D A 0	[R↓] [CLX] [R↓] [R↓] [↑]
D D A A	[R↓] [−] [R↓] [↑]
D D A B	[R↓] [R↓] [R↓] [↑]
D D A C	[$x \gtrless y$] [CLX] [−] [R↓] [R↓]
D D A D	[R↓] [R↓] [R↓] [↑] [↑] [R↓]
D D B 0	[CLX] [−] [R↓] [CLX] [R↓]
D D B A	[$x \gtrless y$] [R↓] [R↓] [R↓] [↑]
D D B B	[CLX] [−] [R↓] [CLX] [−]
D D B C	[CLX] [−] [R↓] [R↓]
D D B D	[CLX] [−] [R↓] [R↓] [↑] [R↓]
D D C 0	[−] [CLX] [$x \gtrless y$] [R↓] [R↓]
D D C A	[R↓] [R↓] [$x \gtrless y$] [↑]
D D C B	[CLX] [−] [$x \gtrless y$] [R↓] [R↓]
D D C C	[−] [R↓] [↑] [R↓] [R↓]
D D C D	[−] [R↓] [$x \gtrless y$] [↑]
D D D 0	[−] [−] [CLX] [R↓]
D D D A	[R↓] [R↓] [R↓] [↑] [↑]
D D D B	[CLX] [−] [R↓] [R↓] [↑]
D D D C	[−] [CLX] [−] [R↓]
D D D D	[−] [−] [CLX] [−]

2.6 ALGEBRAIC MANIPULATIONS IN RPN

Given the problem of translating an algebraic formula into RPN for a calculator, two approaches to optimization (i.e., minimizing the number of keystrokes in the final algorithm) are as follows: (*a*) manipulate the formula algebraically until it is in a form that will yield an optimum algorithm, and translate into RPN; or (*b*) translate into RPN immediately and manipulate the RPN algorithm to optimize it. Of scheme *a*, I would only say that the shortest algebraic expression does not necessarily yield the shortest RPN algorithm. Scheme *b* may seem quite foreign at first because we have all learned well the rules of algebraic manipulation, but the corresponding rules in RPN are strange. I do not wish to promote scheme *b* over *a*, but I do believe that an RPN calculator user should have at least some familiarity with scheme *b* and the rules for manipulations in RPN, and he should be able to use scheme *b* or schemes *a* and *b* together where appropriate.

First consider the dyadic functors operators that take two arguments and give one answer. The common dyadic functors are $+$, $-$, $\times$, and $\div$. Also I define two new symbols ⌡ and ⌊ for the dyadic functors y^x (involution) and $y^{1/x}$ (evolution); that is,

$$a\uparrow b \;⌡\; = a^b, \tag{2.6.1}$$

and

$$a\uparrow b \;⌊\; = a^{1/b} = \sqrt[b]{a}\,. \tag{2.6.2}$$

The symbol ⌡ is intended to suggest a rising power and is pronounced “roop” with rising inflection; ⌊ is pronounced “ramp” with falling inflection. On some RPN calculators ⌡ and ⌊ require more than one keystroke. On an HP-45, for example, ⌡ is [G] [y^x] and ⌊ is [$1/x$] [G] [y^x]. This new notation is more compact and avoids reference to a specific machine.

The following tabulation is the result of translating some of the elementary rules of algebra into RPN.

Cancellation

$$a\uparrow b \times b \div = a;\; b \neq 0, \tag{2.6.3}$$

$$a\uparrow b \div b \times = a;\; b \neq 0, \tag{2.6.4}$$

$$a\uparrow b + b - = a, \tag{2.6.5}$$

$$a\uparrow b - b + = a, \tag{2.6.6}$$

$$a\uparrow b \rfloor b \lfloor = a;\ b \neq 0, \tag{2.6.7}$$

$$a\uparrow b \lfloor b \rfloor = a;\ b \neq 0. \tag{2.6.8}$$

Exchange of order (commutative and associative laws)

$$a\uparrow b + = b\uparrow a +, \tag{2.6.9}$$

$$a\uparrow b \times = b\uparrow a \times, \tag{2.6.10}$$

$$a\uparrow b \times c \times = a\uparrow c \times b \times, \tag{2.6.11}$$

$$a\uparrow b \div c \div = a\uparrow c \div b \div, \tag{2.6.12}$$

$$a\uparrow b \times c \div = a\uparrow c \div b \times, \tag{2.6.13}$$

$$a\uparrow b + c + = a\uparrow c + b +, \tag{2.6.14}$$

$$a\uparrow b - c - = a\uparrow c - b -, \tag{2.6.15}$$

$$a\uparrow b + c - = a\uparrow c - b +, \tag{2.6.16}$$

$$a\uparrow b \rfloor c \rfloor = a\uparrow c \rfloor b \rfloor, \tag{2.6.17}$$

$$a\uparrow b \rfloor c \lfloor = a\uparrow c \lfloor b \rfloor, \tag{2.6.18}$$

$$a\uparrow b \lfloor c \lfloor = a\uparrow c \lfloor b \lfloor, \tag{2.6.19}$$

$$a\uparrow b \times c\uparrow b \times + = a\uparrow c + b \times, \tag{2.6.20}$$

$$a\uparrow b \times c\uparrow b \times - = a\uparrow c - b \times, \tag{2.6.21}$$

$$a\uparrow b \div c\uparrow b \div + = a\uparrow c + b \div, \tag{2.6.22}$$

$$a\uparrow b \div c\uparrow b \div - = a\uparrow c - b \div, \tag{2.6.23}$$

$$a\uparrow b \rfloor c\uparrow b \rfloor \times = a\uparrow c \times b \rfloor, \tag{2.6.24}$$

$$a\uparrow b \rfloor c\uparrow b \rfloor \div = a\uparrow c \div b \rfloor, \tag{2.6.25}$$

$$a\uparrow b \lfloor c\uparrow b \lfloor \times = a\uparrow c \times b \lfloor, \tag{2.6.26}$$

$$a\uparrow b \lfloor c\uparrow b \lfloor \div = a\uparrow c \div b \lfloor. \tag{2.6.27}$$

Replacing extra ↑s

$$a\uparrow b\uparrow c + + = a\uparrow b + c +, \tag{2.6.28}$$

$$a\uparrow b\uparrow c - + = a\uparrow b + c -, \tag{2.6.29}$$

$$a\uparrow b\uparrow c + - = a\uparrow b - c -, \tag{2.6.30}$$

$$a\uparrow b\uparrow c - - = a\uparrow b - c +, \tag{2.6.31}$$

$$a\uparrow b\uparrow c \times \times = a\uparrow b \times c \times, \tag{2.6.32}$$

$$a\uparrow b\uparrow c \div \times = a\uparrow b \times c \div, \tag{2.6.33}$$

$$a\uparrow b\uparrow c \times \div = a\uparrow b \div c \div, \tag{2.6.34}$$

$$a\uparrow b\uparrow c \div \div = a\uparrow b \div c \times, \tag{2.6.35}$$

$$a\uparrow b\uparrow c \times \rfloor = a\uparrow b \rfloor c \rfloor, \tag{2.6.36}$$

$$a\uparrow b\uparrow c \div \rfloor = a\uparrow b \rfloor c \lfloor, \tag{2.6.37}$$

$$a\uparrow b\uparrow c \times \lfloor = a\uparrow b \lfloor c \lfloor, \tag{2.6.38}$$

$$a\uparrow b\uparrow c \div \lfloor = a\uparrow b \lfloor c \rfloor . \tag{2.6.39}$$

The rule to go from the left-hand expression to the right-hand expression is: move the outer (right-most) functor in to replace the second ↑, and if a − thereby crosses a + or a −, reverse it (i.e., $+ \rightleftarrows -$); if a ÷ thereby crosses a × or a ÷, reverse it (i.e., $\times \rightleftarrows \div$); if a $\rfloor$ thereby crosses a × or a ÷, change the × to $\rfloor$ or the ÷ to $\lfloor$; or if a $\lfloor$ thereby crosses a × or a ÷, change the × to $\lfloor$ or the ÷ to $\rfloor$.

Rules of exponents

$$a\uparrow b \rfloor a\uparrow c \rfloor \times = a\uparrow b\uparrow c + \rfloor, \tag{2.6.40}$$

$$a\uparrow b \rfloor a\uparrow c \rfloor \div = a\uparrow b\uparrow c - \rfloor . \tag{2.6.41}$$

Monadic functors

Next consider the monadic functors (one argument, one answer) as defined in Table 2.6.1.

Table 2.6.1 Monadic Functors*

$\mathfrak{s}$	sin	$\bar{\mathfrak{s}}$	$\sin^{-1}$
$\mathfrak{c}$	cos	$\bar{\mathfrak{c}}$	$\cos^{-1}$
$\mathfrak{t}$	tan	$\bar{\mathfrak{t}}$	$\tan^{-1}$
$\mathfrak{l}$	$\ln = \log_e$	^{2}	x^2
$\mathfrak{L}$	$\log = \log_{10}$	$\surd$	$\sqrt{x}$
$\mathfrak{e}$	e^x	$/$	reciprocal ($\boxed{1/x}$)
$\mathfrak{T}$	10^x	$\mathfrak{C}$	negate ($\boxed{\text{CHS}}$)

*Do not confuse the functor $\mathfrak{e}$ with the number $e = 2.71828\cdots$, nor the functors $/$ and $\div$.

Rules of logarithms

$$a\uparrow b \times \mathfrak{l} = a\mathfrak{l}b\mathfrak{l}+;\ a>0,\ b>0, \tag{2.6.42}$$

$$a\uparrow b \div \mathfrak{l} = a\mathfrak{l}b\mathfrak{l}-;\ a>0,\ b>0, \tag{2.6.43}$$

$$a\uparrow b \rfloor \mathfrak{l} = a\mathfrak{l}b\times;\ a>0, \tag{2.6.44}$$

$$a\uparrow b \lfloor \mathfrak{l} = a\mathfrak{l}b\div;\ a>0, \tag{2.6.45}$$

$$a\uparrow b \times \mathfrak{L} = a\mathfrak{L}b\mathfrak{L}+;\ a>0,\ b>0, \tag{2.6.46}$$

$$a\uparrow b \div \mathfrak{L} = a\mathfrak{L}b\mathfrak{L}-;\ a>0,\ b>0, \tag{2.6.47}$$

$$a\uparrow b \rfloor \mathfrak{L} = a\mathfrak{L}b\times;\ a>0, \tag{2.6.48}$$

$$a\uparrow b \lfloor \mathfrak{L} = a\mathfrak{L}b\div;\ a>0. \tag{2.6.49}$$

Change of base (logarithms)

$$a\mathfrak{l} = a\mathfrak{L}e\mathfrak{L}\div, \tag{2.6.50}$$

$$a\mathfrak{L} = a\mathfrak{l}10\mathfrak{l}\div. \tag{2.6.51}$$

Cancellation (of logarithms and exponentials)

$$a\mathfrak{e}\mathfrak{l} = a, \tag{2.6.52}$$

$$a\mathfrak{T}\mathfrak{L} = a, \tag{2.6.53}$$

$$a\mathfrak{l}\mathfrak{e} = a;\ a>0, \tag{2.6.54}$$

$$a\mathfrak{L}\mathfrak{T} = a;\ a>0. \tag{2.6.55}$$

Equivalences among exponentials

$$e\uparrow a \rfloor = a\mathfrak{e}, \tag{2.6.56}$$

$$e\uparrow a \lfloor = a/\mathfrak{e}, \tag{2.6.57}$$

$$10\uparrow a \rfloor = a\mathfrak{T}, \tag{2.6.58}$$

$$10\uparrow a \lfloor = a/\mathfrak{T}. \tag{2.6.59}$$

Rules of exponents

$$a\uparrow b+\mathfrak{e} = a\mathfrak{e}b\mathfrak{e}\times, \tag{2.6.60}$$

$$a\uparrow b-\mathfrak{e} = a\mathfrak{e}b\mathfrak{e}\div, \tag{2.6.61}$$

$$a\uparrow b+\mathfrak{T} = a\mathfrak{T}b\mathfrak{T}\times, \tag{2.6.62}$$

$$a\uparrow b-\mathfrak{T} = a\mathfrak{T}b\mathfrak{T}\div. \tag{2.6.63}$$

Change of base (exponentials)

$$a\mathfrak{T} = 10\mathfrak{l}a\times\mathfrak{e}, \tag{2.6.64}$$

$$a\mathfrak{e} = e\mathfrak{L}a\times\mathfrak{T}. \tag{2.6.65}$$

Equivalences involving squares and square roots

$$a^2 = a\uparrow 2 \rfloor = a\uparrow\times;\ (a>0), \tag{2.6.66}$$

$$a\surd = a\uparrow 2 \lfloor. \tag{2.6.67}$$

Rules of squares and square roots

$$a^2b^2\times = a\uparrow b\times^2, \tag{2.6.68}$$

$$a^2b^2\div = a\uparrow b\div^2, \tag{2.6.69}$$

$$a\surd b\surd\times = a\uparrow b\times\surd;\ a\geqslant 0,\ b\geqslant 0, \tag{2.6.70}$$

$$a\surd b\surd\div = a\uparrow b\div\surd;\ a\geqslant 0,\ b\geqslant 0. \tag{2.6.71}$$

Cancellation (of squares and square roots)

$$a^2\surd = |a|, \tag{2.6.72}$$

$$a\surd^2 = a;\ a \geqslant 0. \tag{2.6.73}$$

Miscellaneous relations involving squares

$$a^2b^2- = a\uparrow b - a\uparrow b + \times, \tag{2.6.74}$$

$$a\uparrow b+^2 = a^2a\uparrow b\times 2\times + b^2+, \tag{2.6.75}$$

$$a\uparrow b-^2 = a^2a\uparrow b\times 2\times - b^2+. \tag{2.6.76}$$

Rules of reciprocals

$$1\uparrow a\div = a/, \tag{2.6.77}$$

$$a// = a;\ a\neq 0, \tag{2.6.78}$$

$$a\uparrow b/\times = a\uparrow b\div, \tag{2.6.79}$$

$$a\uparrow b/\div = a\uparrow b\times;\ b\neq 0, \tag{2.6.80}$$

$$a/b\times = b\uparrow a\div, \tag{2.6.81}$$

$$a/b\div = b/a\div = a\uparrow b\times/. \tag{2.6.82}$$

Rules of sign change

$$a\mathfrak{C}\mathfrak{C} = a, \tag{2.6.83}$$

$$a\uparrow b\mathfrak{C}+ = a\uparrow b-, \tag{2.6.84}$$

$$a\uparrow b\mathfrak{C}- = a\uparrow b+, \tag{2.6.85}$$

$$a\mathfrak{C}\uparrow b+ = b\uparrow a-, \tag{2.6.86}$$

$$a\mathfrak{C}\uparrow b- = b\mathfrak{C}\uparrow a- = a\uparrow b+\mathfrak{C}, \tag{2.6.87}$$

$$a\uparrow b\mathfrak{C}\times = a\uparrow b\times\mathfrak{C} = a\mathfrak{C}\uparrow b\times, \tag{2.6.88}$$

$$a\uparrow b\mathfrak{C}\div = a\uparrow b\div\mathfrak{C} = a\mathfrak{C}\uparrow b\div. \tag{2.6.89}$$

Rules relating reciprocals, exponentials, and sign changes

$$a\uparrow b \rfloor = a\uparrow b / \lfloor, \qquad (2.6.90)$$

$$a\uparrow b \lfloor = a\uparrow b / \rfloor, \qquad (2.6.91)$$

$$a\uparrow b\mathfrak{C} \rfloor = a\uparrow b \rfloor / = a/b \rfloor, \qquad (2.6.92)$$

$$a\uparrow b\mathfrak{C} \lfloor = a\uparrow b \lfloor / = a/b \lfloor, \qquad (2.6.93)$$

$$a\mathfrak{C}\mathfrak{e} = a\mathfrak{e}/, \qquad (2.6.94)$$

$$a\mathfrak{C}\mathfrak{T} = a\mathfrak{T}/, \qquad (2.6.95)$$

$$a/\mathfrak{l} - a\mathfrak{l}\mathfrak{C}, \qquad (2.6.96)$$

$$a/\mathfrak{L} = a\mathfrak{L}\mathfrak{C}. \qquad (2.6.97)$$

Rules relating squares, square roots, logarithms, and exponentials

$$a^2\mathfrak{l} = a\mathfrak{l}2\times;\ a>0, \qquad (2.6.98)$$

$$a^2\mathfrak{L} = a\mathfrak{L}2\times,\ a>0, \qquad (2.6.99)$$

$$a\surd\mathfrak{l} = a\mathfrak{l}2\div, \qquad (2.6.100)$$

$$a\surd\mathfrak{L} = a\mathfrak{L}2\div, \qquad (2.6.101)$$

$$a\mathfrak{e}^2 = a\uparrow 2\times\mathfrak{e} = a\uparrow + \mathfrak{e}, \qquad (2.6.102)$$

$$a\mathfrak{e}\surd = a\uparrow 2\div\mathfrak{e}, \qquad (2.6.103)$$

$$a\mathfrak{T}^2 = a\uparrow 2\times\mathfrak{T} = a\uparrow + \mathfrak{T}, \qquad (2.6.104)$$

$$a\mathfrak{T}\surd = a\uparrow 2\div\mathfrak{T}. \qquad (2.6.105)$$

Trigonometric identities

The tabulation below contains only a select few of the incredible number of trigonometric identities.

$$\csc a = a\mathfrak{s}/ = a\mathfrak{c}a\mathfrak{t}\times/, \qquad (2.6.106)$$

$$\sec a = a\mathfrak{c}/ = a\mathfrak{t}a\mathfrak{s}\div, \qquad (2.6.107)$$

$$\cot a = a\mathfrak{t}/ = a\mathfrak{c}a\mathfrak{s}\div, \qquad (2.6.108)$$

$$\csc^{-1}a = a/\bar{\mathfrak{s}}, \tag{2.6.109}$$

$$\sec^{-1}a = a/\bar{\mathfrak{c}}, \tag{2.6.110}$$

$$\cot^{-1}a = a/\bar{\mathfrak{t}}, \tag{2.6.111}$$

$$\sec^2 a = a\mathfrak{t}^2 1+, \tag{2.6.112}$$

$$\csc^2 a = a\mathfrak{t}^2/1+, \tag{2.6.113}$$

$$a\mathfrak{s}a\mathfrak{c}\div = a\mathfrak{t}, \tag{2.6.114}$$

$$a\mathfrak{s}^2 a\mathfrak{c}^2 + = 1, \tag{2.6.115}$$

$$a\mathfrak{C}\mathfrak{s} = a\mathfrak{s}\mathfrak{C}, \tag{2.6.116}$$

$$a\mathfrak{C}\mathfrak{c} = a\mathfrak{c}, \tag{2.6.117}$$

$$a\mathfrak{C}\mathfrak{t} = a\mathfrak{t}\mathfrak{C}. \tag{2.6.118}$$

Examples of reductions

$$\frac{rs+tu}{su} = r\uparrow s\times t\uparrow u\times + s\uparrow u\times \div \qquad \text{(as formed)},$$

$$= r\uparrow s\times t\uparrow u\times + s\div u\div \qquad \text{(rule 2.6.34)},$$

$$= r\uparrow s\times s\div u\div t\uparrow u\times s\div u\div + \qquad \text{(rule 2.6.22 backward)},$$

$$= r\uparrow u\div t\uparrow u\times u\div s\div + \qquad \text{(rules 2.6.12 and 2.6.3)},$$

$$= r\uparrow u\div t\uparrow s\div + \qquad \text{(rule 2.6.3)}.$$

$$\tfrac{1}{2}\ln(ab^2) - \ln(b) = a\uparrow b^2\times\mathfrak{l}2\div b\mathfrak{l}- \qquad \text{(as formed)},$$

$$= a\mathfrak{l}b^2\mathfrak{l}+2\div b\mathfrak{l}- \qquad \text{(rule 2.6.42)},$$

$$= a\mathfrak{l}b\mathfrak{l}2\times +2\div b\mathfrak{l}- \qquad \text{(rule 2.6.98)},$$

$$= a\mathfrak{l}2\div b\mathfrak{l}2\times 2\div + b\mathfrak{l}- \qquad \text{(rule 2.6.22 backward)},$$

$$= a\mathfrak{l}2\div b\mathfrak{l}+b\mathfrak{l}- \qquad \text{(rule 2.6.3)},$$

$$= a\mathfrak{l}2\div \qquad \text{(rule 2.6.5)},$$

$$= a\sqrt{\mathfrak{l}} \qquad \text{(rule 2.6.100 backward)}.$$

2.7 EXERCISES

2.7.1 (*a*) Write an RPN algorithm to sum two columns of numbers simultaneously. The loop should contain no more than four keystrokes plus data. Do not use [Σ+], [M+], or [STO]. *Hint*: For a less elegant solution to this problem, see the *HP-35 Math Pac*, p. 141. (*b*) Sum three columns simultaneously with a 12-keystroke loop and the same restrictions.

2.7.2 Write an RPN algorithm to use in the supermarket to (*a*) calculate unit prices ($/oz, etc.) given package prices and contents, and (*b*) keep a running total of the price of merchandise purchased, to check against the cash register receipt. Remember that one does not always buy everything for which one calculates a unit price, but keying the package price twice for an item that one does buy should not be necessary. Do not use [Σ+], [M+], or [STO].

2.7.3 Write an algorithm for an HP-45 to do complex multiply with 14 (rather than 21) keystrokes,

$$(a_1+ib_1)(a_2+ib_2)=u+iv.$$

Given a_1, b_1, a_2, and b_2; find u and v; $i^2=-1$.

2.7.4 Write an algorithm for an HP-45 to do complex divide with 15 (rather than 18) keystrokes,

$$\frac{a_1+ib_1}{a_2+ib_2}=u+iv.$$

2.7.5 Write an algorithm for an HP-45 to do complex reciprocal with 7 (rather than 17) keystrokes,

$$\frac{1}{a+ib}=u+iv.$$

2.7.6 Write an algorithm for an HP-45 to do complex square with 10 (rather than 14) keystrokes,

$$(a+ib)^2=u+iv.$$

Hint for problems 2.7.3 *to* 2.7.6: Use [→P] and [→R] and work with the polar-coordinate representation. See the *HP-45 Applications Book*, pp. 48–50.

2.7.7 Fill in the blank lines in Table 2.5.1, namely: *ABCD, ADDB, BADA, BADB, DADB.*

2.7.8 Let us begin with a single fertile pair of rabbits and suppose that at the end of a month, each fertile pair produces another pair. Assume that rabbits become fertile at the age of one month, that foxes do not exist, and that rabbits never die. Thus after one month there will be 2 pairs of rabbits, after 2 months, 3 pairs (because one of the previous pairs is not yet fertile), after 3 months, 5 pairs, and so on. Write a cyclic RPN algorithm to calculate the number of rabbits at the end of each month, and use the algorithm to calculate the number of rabbits at the end of a year. *Hint*: This is the famous Leonardo Fibonacci problem (c. 1202), and the answer is a sequence of Fibonacci numbers. The loop should have no more than five keystrokes.

2.7.9 Convert the following octal numbers to decimal,

43172.,
4.3172,
431.72,

the following hexadecimal numbers to decimal,

AE19C2,
FDA7.CB5,

and the following binary numbers to decimal,

1101011011101,
1011001.1110101101.

Hints: Hexadecimal numbers are base 16 and A=10, B=11, C=12, and so on. For the binary numbers, either convert to octal first, or use base 2 directly.

2.7.10 Without air resistance, the speed of an object falling in a constant gravitational field is

$$v = at,$$

and the distance fallen is

$$s = \tfrac{1}{2}at^2,$$

where a is the acceleration of gravity and t is the time since release. (*a*) Write an RPN algorithm to calculate v and s given a and t (six keystrokes plus data). (*b*) If a stone dropped from a bridge strikes the water 4.2 seconds later, how high is the bridge over the water, and how fast was the stone moving when it hit the water? The acceleration of gravity near the earth's surface is $a \cong 32$ ft/s^2.

2.7.11 The harmonic numbers are defined as

$$H_n = \sum_{k=1}^{n} \frac{1}{k},$$

and are approximately

$$H_n \cong \ln n + \gamma + \frac{1}{2n} - \frac{1}{12n^2} + \frac{1}{120n^4},$$

where $\gamma = 0.5772156649\cdots$ is Euler's constant. Write an algorithm for each of these formulas and compare the answers for n up to 10.

2.7.12 Fresnel's formula for reflection of unpolarized light at the interface of a transparent medium is (FFP, II, p. 367)

$$R = \frac{1}{2}\left[\frac{\sin^2(i-r)}{\sin^2(i+r)} + \frac{\tan^2(i-r)}{\tan^2(i+r)}\right],$$

where i is the angle of incidence, r is the angle of refraction in the medium, and R is ratio of reflected light to incident light. Write an algorithm for an HP-35 or HP-21 to calculate R given i and r. *Hints*: Do not key i and r each more than once. Remember that trigonometric functors lose the contents of T on the HP-35, and that neither the HP-35 nor the HP-21 has an $\boxed{x^2}$ key. The algorithm can be done in 24 keystrokes plus data. *Test case*: $i = 40°$, $r = 24°.5$, get $R = 0.05258 = 5.258\%$.

2.7.13 Write an HP-45 algorithm to sum the first six terms of the series

$$\sum_{k=1}^{\infty} \frac{T_n(x)}{k^n} \qquad (-1 \leqslant x \leqslant 1),$$

where $T_n(x)$ is a Chebyshev polynomial. Evaluate the series for $x = 0.5$ and $n = 1$, 2, and 4. *Hint*: $T_n(\cos\theta) = \cos(n\theta)$.

2.7.14 The length of the arc of a parabola is (CRC SMT, p. 13)

$$s=\sqrt{4x^2+y^2}+\frac{y^2}{2x}\ln\left[\frac{2x+\sqrt{4x^2+y^2}}{y}\right],$$

where x is the depth and y is half the mouth opening. Write an HP-35 algorithm to calculate s given x and y. *Test case*: $x=2$, $y=3$; get $s=7.472$.

2.7.15 The area of the segment of a circle is (CRC SMT, p. 12)

$$A=R^2\cos^{-1}\left(\frac{d}{R}\right)-d\sqrt{R^2-d^2},$$

where R is the radius of the circle and d is the perpendicular distance from the center to the cord. Write an HP-21 algorithm to calculate A given R and d. *Test case*: $d=3$, $R=5$; get $A=11.1824$.

2.7.16 Prof. H. E. Schaffer (1976) challenges anyone to write an HP-25 program for the factorial function (for integers) shorter than

[STO] 1 1 [−] [STO] [×] 1 1 [f] [$x \neq y$] [GTO] 03 [RCL] 1 [GTO] 00,

where $n!=n(n-1)(n-2)\cdots 3\cdot 2\cdot 1$. Try it.

Rewrite and shorten the following five RPN expressions:

2.7.17 $A\uparrow 2\times [2\div B[2\div -$.

2.7.18 $A\uparrow B\times A\uparrow B+\div$.

2.7.19 $A\uparrow 2\div B\uparrow 2\div +C\uparrow 12\div -D\uparrow 12\div +$.

2.7.20 $A\mathrm{t}\uparrow/+A\mathrm{c}\times/$.

2.7.21 $A[B\times \mathrm{e}$.

2.7.22

$$\frac{2x^4+45x^3+381x^2+1353x+1511}{x^3+21x^2+157x+409}=2x+3+\cfrac{4}{x+5+\cfrac{6}{x+7+\cfrac{8}{x+9}}}$$

Write an RPN algorithm to evaluate each side of this equation, and compare the answers for $x=2$.

2.7.23 Convert 3 weeks, 2 days, 9 hours, 22 minutes and 18 seconds, into (*a*) seconds and (*b*) decimal weeks.

2.7.24 Find the "regular continued fraction" expansion of $1/(e-1)$, where $e=2.718281828$.

2.7.25 The normalized poles in the complex s plane of a Chebyshev low-pass filter are at

$$\alpha_k = -\sinh\left(\frac{1}{n}\sinh^{-1}\frac{1}{\varepsilon}\right)\sin\left(\frac{2k-1}{n}90^\circ\right),$$

$$\omega_k = \pm\cosh\left(\frac{1}{n}\sinh^{-1}\frac{1}{\varepsilon}\right)\cos\left(\frac{2k-1}{n}90^\circ\right),$$

where n is the order of the filter, that is, the number of poles; k is the stage number, $k=1$ to $n/2$ if n is even, or $k=1$ to $(n+1)/2$ if n is odd; ε is the peak allowable bandpass ripple; and α_k and ω_k are the real and imaginary parts, respectively, of each pole. So $s_k = \alpha_k + i\omega_k$ where $i=\sqrt{-1}$. If r is the peak-to-peak ripple in decibels (dB), then $\varepsilon^2 = 10^{r/10}-1$. Write an HP-21 algorithm to calculate s_k given n, k, and r. *Hint*: see Section A.5.1. *Test case*: $r=0.5$ dB, $n=3$; get $\varepsilon=0.3493$, $\alpha_1=-0.3132$, $\omega_1=\pm 1.0219$, $\alpha_2=-0.6265$, $\omega_2=0$. *Reference*: Lubkin (1970), Section 4.2.

3
Iterative Solutions of Elementary Transcendental Equations

...damnable iteration...able to corrupt a saint.

—*William Shakespeare (1564–1616)*

3.1 INTRODUCTION

Transcendental equations are equations whose solutions cannot be written explicitly in terms of a finite number of elementary functions. And elementary functions are easy to define—they are functions that appear on the keys of the calculator. The definitions in mathematics textbooks are somewhat different.

Certain functions that are solutions to transcendental equations were considered previously. Bessel functions, for example, are solutions to Bessel's differential equation and are well-known functions in that they have been thoroughly studied, tabulated, and graphed. However they are not elementary functions by the definition just given. Functions such as Bessel's are probably best dealt with in terms of their power series or polynomial approximations, as in Section 2.4.

Consider, instead, elementary algebraic transcendental equations that can be put into the form

$$x=f(x), \tag{3.1.1}$$

where the function f consists of some combination of elementary functions. This is not quite a general form for algebraic transcendental equations, but equations in this form occur very frequently in applications. Methods for dealing with such equations have a long history, going back at least to Newton, and an extensive literature. Almost any book on numerical analysis or numerical methods has a chapter on such equations, and an exhausting treatment can be found in Traub (1964). This section discusses only the iterative techniques that I have found to be most useful.

We can think of f as an operator and apply it repeatedly to an initial guess x_0, to give, we hope, an improved approximation, at each iteration, for the solution nearest x_0. Such a procedure sometimes converges, sometimes diverges, sometimes oscillates between two or more solutions, and sometimes oscillates between solutions of the derived equations

$$\begin{aligned} x_1&=f(x_2),\\ x_2&=f(x_1). \end{aligned} \tag{3.1.2}$$

Even more complex oscillatory behavior occurs in principle, but rarely in practice.

Suppose $x=a$ is a solution to the foregoing transcendental equation, that is,

$$a=f(a). \tag{3.1.3}$$

Then consider a Taylor series expansion of $f(x)$ around the point $x=a$,

$$f(x)=f(a)+(x-a)f'(a)+\tfrac{1}{2}(x-a)^2f''(a)+\cdots. \tag{3.1.4}$$

Suppose x_n is an approximate solution to the equation and let ξ_n be the error in x_n, that is,

$$\xi_n=x_n-a. \tag{3.1.5}$$

If we obtain the next approximation x_{n+1} by iteration, we have

$$x_{n+1} = f(x_n) = f(a) + \xi_n f'(a) + \tfrac{1}{2}\xi_n^2 f''(a) + \cdots, \tag{3.1.6}$$

or

$$\xi_{n+1} = \xi_n f'(a) + \tfrac{1}{2}\xi_n^2 f''(a) + \cdots. \tag{3.1.7}$$

Clearly this process cannot converge unless $|f'(a)| < 1$. The HMF, p. 18, gives the precise criteria of convergence as

$$|f'(x)| \leqslant q < 1 \qquad \text{for} \quad A \leqslant x \leqslant B, \tag{3.1.8}$$

where

$$A \leqslant x_n \pm \frac{|f(x_n) - x_n|}{1-q} \leqslant B. \tag{3.1.9}$$

If $0 < |f'(a)| \leqslant q < 1$, convergence is called "first order"; if $f'(a) = 0$, convergence is called "second order"; if $f''(a) = 0$ also, convergence is called "third order," and so on (see Hartree, 1958, p. 212). With first-order convergence, the magnitude of the error $|\xi_n|$ decreases exponentially with increasing n. With second-order convergence, the number of significant figures in x_n approximately doubles with each iteration. Second-order convergence is thus greatly preferable to first order, and, in any case, $|f'(a)|$ should be small if not zero.

3.2 THE g METHOD

What is wanted is a method for transforming an original function $f(x)$ into a new function $F(x)$ that still satisfies

$$a = F(a) \tag{3.2.1}$$

but also converges and satisfies

$$F'(a) = 0, \tag{3.2.2}$$

at least approximately. Then iteration on $F(x)$ will converge rapidly toward $x = a$, the desired solution of the original equation. Many methods for obtaining an $F(a)$ satisfy these criteria; three particular methods are discussed here.

Consider first

$$F(x)=\frac{f(x)+xg(x)}{g(x)+1}. \tag{3.2.3}$$

If $f(a)=a$, then $F(a)=a$ also, provided only that $g(a)\neq -1$. The function $g(x)$ is otherwise arbitrary and can be chosen so that $F'(a)=0$. The derivative of equation 3.2.3 evaluated at $x=a$ is

$$F'(a)=\frac{f'(a)+g(a)}{g(a)+1}. \tag{3.2.4}$$

Thus we can make $F'(a)=0$ by taking

$$g(a)=-f'(a), \tag{3.2.5}$$

provided $f'(a)\neq 1$. Since we are using an iterative procedure, $g(a)$ needs to have this value only *approximately*. The speed of convergence, but not the ultimate answer, depends on the exact $g(x)$. The simplest iteration scheme, using just $f(x)$, in effect takes $g(x)$ to be zero and works well if $|f'(a)|$ is small compared to unity.

Since we do not know a in advance, we usually cannot set $g(x)=-f'(a)$ $=$constant (and we may not even want to). Therefore try instead

$$g(x)=-f'(x), \tag{3.2.6}$$

since this is obviously one possibility that gives $F'(a)=0$. Other possibilities come from adding to $g(x)$ any function that is zero at $x=a$. With $g(x)$ from equation 3.2.6 however, and with

$$h(x)=x-f(x), \tag{3.2.7}$$

one can show that

$$F(x)=x-\frac{h(x)}{h'(x)}, \tag{3.2.8}$$

and this is the famous Newton-Raphson iterative equation for solving $h(x)=0$. Thus for the particular choice of $g(x)$ in equation 3.2.6, the g method reduces exactly to the Newton-Raphson method.

If we have an $f(x)$ that is difficult to differentiate, or a derivative $f'(x)$

that is difficult to evaluate on a calculator, we might approximate

$$f'(x) \cong \frac{\Delta f}{\Delta x} = \frac{f(x_n) - f(x_{n-1})}{x_n - x_{n-1}}. \tag{3.2.9}$$

With this expression for $-g(x)$,

$$x_{n+1} = F(x_n) = \frac{x_n f(x_{n-1}) - x_{n-1} f(x_n)}{x_n - x_{n-1} - f(x_n) + f(x_{n-1})}$$

$$= x_n - \frac{x_{n-1} - x_n}{\dfrac{x_{n-1} - f(x_{n-1})}{x_n - f(x_n)} - 1}. \tag{3.2.10}$$

This is the Wegstein or modified secant iteration scheme (see the IBM SSP, p. 215), also known as *regula falsi* (rule of false position) as applied to $h(x) = 0$, and is a double-averaging scheme in that x_{n+1} is a function of both x_n and x_{n-1}. Although we need not calculate $f'(x)$ with this scheme, we need to save both x_{n-1} and $f(x_{n-1})$ while calculating $f(x_n)$. Depending on the form of $f(x)$, we probably need at least two storage registers plus the stack. Note also that this expression degenerates into $0/0$ as x_n approaches a. We can rewrite the expression, of course, but such difficulties persist.

A similar but not identical scheme approximates $f'(x)$ as in equation 3.2.9 but takes another x near x_n in place of x_{n-1}. The two xs are not allowed to come together, thus avoiding $0/0$. Such a scheme appears in the *HP-25 Applications Programs*, p. 76.

One way to implement equation 3.2.10 on an HP-45 is as follows:

SETUP: x_0 [STO] 1 [↑] [↑] [↑] [···f···] [−] [STO] 2 [G] [LAST X]

LOOP: [↑] [↑] [↑] [···f···] [−] [x⪸y] [RCL] 1 [x⪸y] [STO] 1 [−] [x⪸y] [RCL] 2 [x⪸y] [STO] 2 [÷] 1 [−] [÷] [−] (see x_n) :|. (3.2.11)

This algorithm keeps x_{n-1} in R_1 and $x_{n-1} - f(x_{n-1})$ in R_2. The symbol [···f···] is a subroutine and represents any sequence of keystrokes that accepts x in X, Y, Z, and T, puts $f(x)$ into X, and leaves x in Y, Z, and T. The symbol :| means loop back to the last preceding colon (:), in this case after the word "loop." The second approximation x_1 is taken to be $f(x_0)$; if a better x_1 is known, substitute it for [G] [LAST X] at the end of the setup.

My experience with this scheme has not been very favorable. If the derivative $f'(x)$ is too much to calculate, try setting $g(x)$ to an approximate

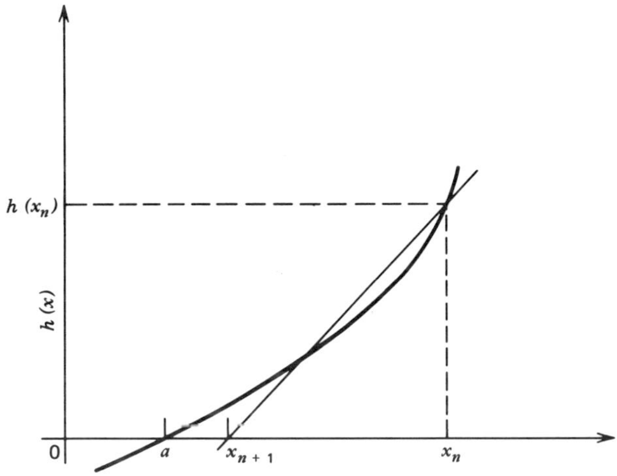

Figure 3.2.1 A graphical interpretation of the g method. The straight line has a slope of $g(x_n)+1$ and goes through the points $(x_n, h(x_n))$, $(x_{n+1}, 0)$.

algebraic expression for $-f'(x)$ or perhaps to a carefully chosen constant; $g(x) = -f'(x_0) = \text{constant}$ is sometimes called the modified Newton method.

A geometric interpretation of the g method appears in Figure 3.2.1. The slope of the straight line is $g(x)+1$, and for second-order convergence this straight line becomes tangent to the curve when $x=a$. If $g(x) = -f'(x)$, as in the Newton-Raphson method, the straight line is tangent to the curve at any x. The g method is substantially identical with Milne's m method (Milne, 1949, Chapter II).

3.3 THE α METHOD

For another possible method, consider

$$F(x) = [f(x)]^{\alpha} x^{1-\alpha} = x\left[\frac{f(x)}{x}\right]^{\alpha}, \tag{3.3.1}$$

where α can be either constant or an arbitrary function of x. This F also satisfies $F(a)=a$, provided $f(a)=a$, and we need only avoid $\alpha=0$. Since for this $F(x)$ we can show that

$$F'(a) = 1 - \alpha(a)[1 - f'(a)], \tag{3.3.2}$$

we can make $F'(a)=0$ by taking

$$\alpha(a)=\frac{1}{1-f'(a)}, \tag{3.3.3}$$

provided $f'(a)\neq 1$.

As before, since we do not know a in advance, we usually cannot set $\alpha(x)=\alpha(a)=\text{constant}$; thus a reasonable choice would be

$$\alpha(x)=\frac{1}{1-f'(x)}. \tag{3.3.4}$$

Other possibilities can be obtained, for example, by adding to $\alpha(x)$ any function that is zero at $x=a$, or by multiplying $f'(x)$ in this expression by any function that is unity at $x=a$. Another expression that sometimes turns out to be simpler is

$$\alpha(x)=\frac{1}{1-xf'(x)/f(x)}. \tag{3.3.5}$$

If we redefine

$$h(x)=\ln[x/f(x)], \tag{3.3.6}$$

and plot $\ln x$ rather than x on the abscissa, we get a geometrical interpretation of the α method, as shown in Figure 3.3.1. The slope of the straight line is $1/\alpha(x_n)$. For second-order convergence, α satisfies equation 3.3.3, and the straight line becomes tangent to the curve when $x=a$. If equation 3.3.5 is also satisfied, the straight line is tangent to the curve at any x.

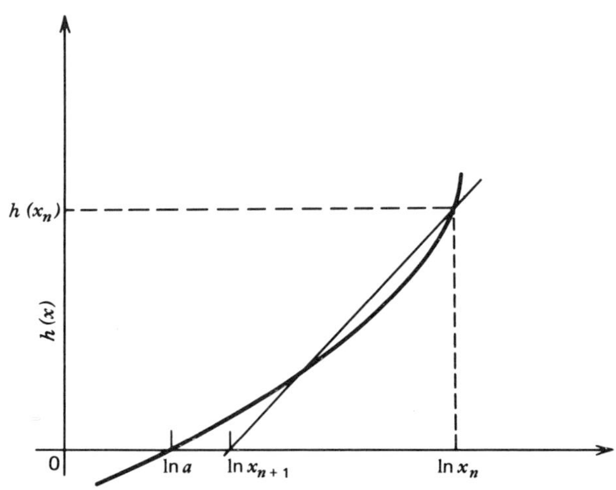

Figure 3.3.1 A graphical interpretation of the α method. The straight line has a slope of $1/\alpha(x_n)$ and goes through the points $(\ln x_n, h(x_n))$, $(\ln x_{n+1}, 0)$.

3.4 THE β METHOD

As yet another possible method, consider

$$\frac{1}{F(x)}=\frac{\beta}{f(x)}+\frac{1-\beta}{x}=\frac{f(x)+\beta\left[x-f(x)\right]}{xf(x)}. \tag{3.4.1}$$

This $F(x)$ satisfies the same conditions as before, and since

$$F'(a)=1-\beta(a)\left[1-f'(a)\right], \tag{3.4.2}$$

we can make $F'(a)=0$ by choosing

$$\beta(a)=\frac{1}{1-f'(a)}. \tag{3.4.3}$$

This is identical to $\alpha(a)$ above, and the same comments apply.

As an exercise, show that the β method with

$$\beta(x)=\frac{1}{1+xg(x)/f(x)}, \tag{3.4.4}$$

is equivalent to the g method. Another expression that sometimes turns out to be simpler is

$$\beta(x)=\frac{1}{1-x^2f'(x)/f^2(x)}. \tag{3.4.5}$$

If we redefine

$$h(x)=\frac{1}{x}-\frac{1}{f(x)}, \tag{3.4.6}$$

and plot $1/x$ rather than x on the abscissa, we get a geometrical interpretation of the β method (Figure 3.4.1). The slope of the straight line is $1/\beta(x_n)$. For second-order convergence, β satisfies equation 3.4.3 and the straight line becomes tangent to the curve when $x=a$. If equation 3.4.5 is also satisfied, the straight line is tangent to the curve at any x.

Mathematicians will recognize that the g method is based on the weighted arithmetic mean of x_n and $f(x_n)$, which would ordinarily be x_{n+1}, the α method on the weighted geometric mean, and the β method on the weighted harmonic mean.

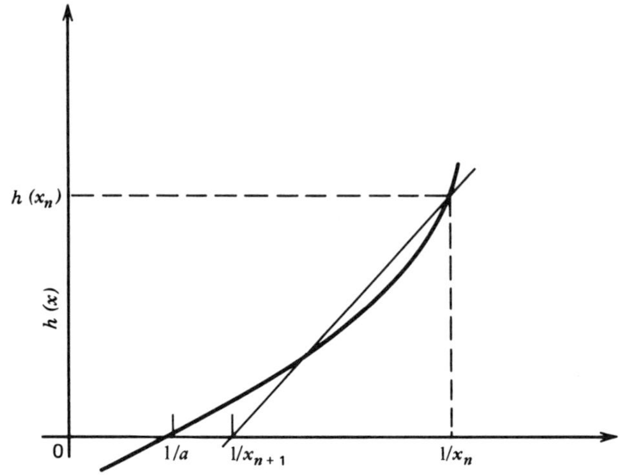

Figure 3.4.1 A graphical interpretation of the β method. The straight line has a slope of $1/\beta(x_n)$ and goes through the points $(1/x_n, h(x_n))$, $(1/x_{n+1}, 0)$.

We should choose among these methods, and select values for $g(x)$, $\alpha(x)$, or $\beta(x)$ with the criterion of minimizing the total number of keystrokes in the procedure—the number of keystrokes per iteration times the number of iterations. Try to find simple, even if approximate, expressions for $g(x)$, $\alpha(x)$, or $\beta(x)$. With programmable calculators, the criterion may sometimes be different; an iteration algorithm that fits the calculator's program memory and converges, however slowly, is preferable to an algorithm that is too large to fit, no matter how fast it might converge.

3.5 EXAMPLE: $x = Ax^{-x}$

Consider next a relatively simple example of a transcendental algebraic equation

$$x = Ax^{-x}. \tag{3.5.1}$$

Given A, the problem is to solve for x. If we think of the equation the other way round—given x, solve for A—the solution is trivial,

$$A = x^{x+1}, \tag{3.5.2}$$

or

$$x \boxed{\uparrow} \boxed{\uparrow} 1 \boxed{+} \boxed{y^x} \qquad \text{(see } A\text{)}. \tag{3.5.3}$$

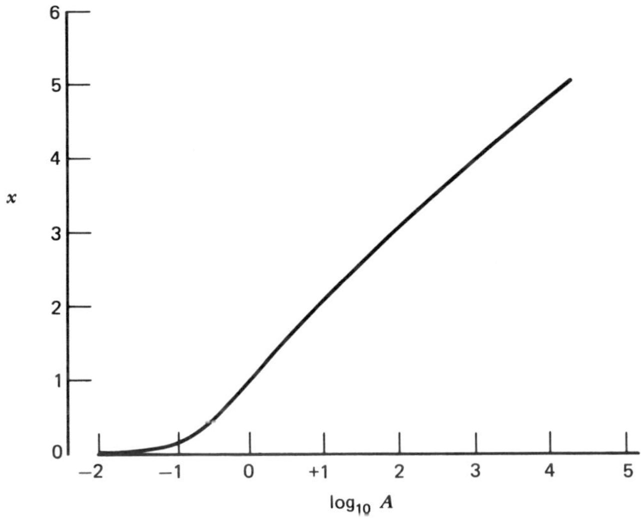

Figure 3.5.1 Example: $x = Ax^{-x}$ with $\log_{10} A$ on the abscissa to compress the scale.

Thus we can plot A against x and use the graph to solve the original problem, or at least to obtain first guesses for iterations. Figure 3.5.1 is such a graph.

This is not a physical problem in that I do not know that the equation occurs in a physical or real-life context; I chose it for its pedagogical value. Section 3.6 contains examples of transcendental equations from the financial world, and Section A.10.7 contains an extensive example from electrical engineering.

First try each of the three methods—g, α, and β—on the exponential form of the equation

$$f(x) = Ax^{-x}. \tag{3.5.4}$$

The derivative is

$$f'(x) = -Ax^{-x}(1 + \ln x). \tag{3.5.5}$$

Note that since $|f'(x)| < 1$ only for $A < 1$ (see Figure 3.5.1), direct iteration on this $f(x)$ could converge only for $A < 1$.

With $g(x) = -f'(x)$, get

$$F(x) = \frac{Ax^{-x} + xAx^{-x}(1 + \ln x)}{1 + Ax^{-x}(1 + \ln x)}, \tag{3.5.6}$$

or

$$F(x)=\frac{1+x(1+\ln x)}{x^x/A+1+\ln x}. \tag{3.5.7}$$

One possible algorithm for an HP-35 would be

SETUP: A [STO] x_0

LOOP: [↑] [↑] [ln] 1 [+] [×] 1 [+] [R↓] [x^y] [RCL] [÷] [$x \gtrless y$] [ln] [+] 1 [+] [÷] (see x_i) :|. (3.5.8)

Since x is in both X and Y on [x^y] in this algorithm, the [x^y] could as well be [y^x]. As a test case, take $A=5$, $x_0=2$, and get

$$x_i=2,\ 1.759340322,\ 1.782767549,\ 1.783016997,\ 1.783017025. \tag{3.5.9}$$

This final answer has an absolute error less than $\pm 2\times 10^{-9}$, as one can verify by substituting back into the original equation, and was achieved by four cycles through a loop of 18 keystrokes, starting from an x_0 with only about one significant figure.

Next try the α method on the same problem, using

$$\alpha(x)=\frac{1}{1+x(1+\ln x)}, \tag{3.5.10}$$

from equation 3.3.5. In this case one should work with $\ln F(x)$, rather than $F(x)$ itself, but this is only a computational convenience and should not be confused with the logarithmic form of the equation considered below. After some manipulation, get

$$\ln F(x)=\frac{\ln A+x(\ln x)^2}{1+x(1+\ln x)}. \tag{3.5.11}$$

Thus an algorithm for an HP-35 would be

SETUP: A [ln] [STO] x_0

LOOP: [↑] [↑] [ln] [↑] [×] [×] [RCL] [+] [$x \gtrless y$] [↑] [ln] 1 [+] [×] 1 [+] [÷] [e^x] (see x_i) :|. (3.5.12)

As a test case, again take $A=5$, $x_0=2$, and get

$$x_i=2,\ 1.796776815,\ 1.783080391,\ 1.783017024. \tag{3.5.13}$$

This algorithm on this test case achieves an absolute error less than $\pm 3 \times 10^{-9}$ in three cycles through a loop of 18 keystrokes—clearly an improvement.

Next try the β method on the same problem, with β from equation 3.4.3, namely,

$$\beta(x) = \frac{1}{1 + Ax^{-x}(1 + \ln x)}. \tag{3.5.14}$$

We avoid using β equal to the form for α above because, as noted previously, the β method would then reduce to the g-method case already done. After some manipulation with this β, get

$$F(x) = \frac{x^x/A + 1 + \ln x}{(x^x/A)^2 + (1 + \ln x)/x}. \tag{3.5.15}$$

This form leads to a rather tortured algorithm for an HP-35, namely,

SETUP: A [STO] x_0

LOOP: [↑] [↑] [ln] 1 [+] [↑] [R↓] [R↓] [R↓] [÷] [$x \gtrless y$] [R↓] [R↓] [x^y] [RCL] [÷] [↑] [R↓] [+] [$x \gtrless y$] [R↓] [R↓] [×] [+] [÷] (see x_i) : |. (3.5.16)

With the same test case, get

$$x_i = 2, 1.67710983, 1.761960934, 1.782130099, 1.783015429, 1.783017024. \tag{3.5.17}$$

Therefore this algorithm on the same test case achieves the same absolute error in five cycles through a loop of 25 keystrokes—worse than either of the two previous algorithms.

Next try Wegstein's iteration scheme with algorithm 3.2.11 and

$$[\cdots f \cdots] = \text{[CHS] [G] } [y^x] \; A \; [\times]. \tag{3.5.18}$$

This is for an HP-45; this iteration is difficult on an HP-35. Normally one would store A in the setup and recall A inside the loop. With the same test case, get

$$x_i = 2, 1.828661057, 1.792159254, 1.782825859, 1.783017815, 1.783017025. \tag{3.5.19}$$

This algorithm on the same test case thus achieves the same absolute error in five cycles through a loop of (about) 27 keystrokes—not very impressive.

Using logarithms, we can manipulate the original transcendental equation into an alternative form. This occurs very frequently with transcendental equations in practice; sometimes many different forms of a given equation exist, all suitable for iteration. Taking logarithms and rewriting gives an alternative

$$f(x)=\frac{\ln(A/x)}{\ln x}=\frac{\ln A}{\ln x}-1. \tag{3.5.20}$$

And we can try the g, α, and β methods in turn on this expression. For a possible g take

$$g(x)=-f'(x)=\frac{\ln A}{x(\ln x)^2}, \tag{3.5.21}$$

and after some manipulation obtain

$$F(x)=\frac{x(1+\ln x)\ln A-x(\ln x)^2}{\ln A+x(\ln x)^2}. \tag{3.5.22}$$

Thus an algorithm for the HP-35 would be

SETUP: A [ln] [STO] x_0

LOOP: [↑] [↑] [ln] [↑] [×] [×] [R↓] [R↓] [ln] 1 [+] [×] [RCL] [×] [$x\gtrless y$] [−] [$x\gtrless y$] [RCL] [+] [÷] (see x_i) : |. (3.5.23)

With the same test case, get

$$x_i=2,\ 1.746507317,\ 1.781769069,\ 1.783015603,\ 1.783017024. \tag{3.5.24}$$

Therefore this algorithm on the same test case achieves the same absolute error in four cycles through a loop of 20 keystrokes—not the best performance.

Next try equation 3.5.20 and the α method with α from equation 3.3.5, namely,

$$\alpha(x)=\frac{1}{1+\ln A/[\ln x\ln(A/x)]}. \tag{3.5.25}$$

This leads to

$$\ln F(x) = \frac{\ln x}{\ln A + \ln x \ln(A/x)} \left\{ \ln(A/x) \ln\left[\frac{\ln(A/x)}{\ln x} \right] + \ln A \right\}, \tag{3.5.26}$$

and this rather unlikely equation yields, as a possible algorithm for the HP-35,

SETUP: A [ln] [↑] [↑] [↑] x_0

LOOP: [ln] [STO] [−] [RCL] [÷] [ln] [$x \gtrless y$] [RCL] [−] [×] [+] [RCL] [×] [$x \gtrless y$] [RCL] [−] [RCL] [×] [R↓] [R↓] [R↓] [+] [÷] [e^x] (see x_i): |. (3.5.27)

With the same test case, get

$$x_i - 2,\ 1.778882554,\ 1.783014341,\ 1.783017024. \tag{3.5.28}$$

Thus this algorithm on this test case achieves the same absolute error in three cycles through a loop of 24 keystrokes. This is not the best overall performance (the α method on the original equation is still the winner because of fewer keystrokes per cycle), but note the very impressive second iteration, which achieved an absolute error of about 3×10^{-6}; the previous best case after two iterations was more than 20 times worse.

Then try the β method on the second form of the transcendental equation with β from equation 3.4.3, namely,

$$\beta(x) = \frac{x(\ln x)^2}{\ln A + x(\ln x)^2}. \tag{3.5.29}$$

After some manipulation, get

$$F(x) = \frac{x\left[\ln A + x(\ln x)^2\right]\ln(A/x)/\ln x}{x^2(\ln x)^2 + \left[\ln A/\ln x - 1\right]\ln A}. \tag{3.5.30}$$

Thus a possible algorithm for the HP-35 is

SETUP: A [ln] [STO] x_0

LOOP: [↑] [↑] [ln] [↑] [×] [×] [RCL] [+] [×] [$x \gtrless y$] [ln] [÷] [$x \gtrless y$] [ln] [RCL] [$x \gtrless y$] [−] [×] [R↓] [ln] [×] [↑] [×] [$x \gtrless y$] [ln] [RCL] [$x \gtrless y$] [÷] 1 [−] [RCL] [×] [+] [÷] (see x_i): |.

(3.5.31)

With the same test case, get

$$x_i = 2,\ 1.678190505,\ 1.756526429,\ 1.78131194,\ 1.783009972,\ 1.783017024. \tag{3.5.32}$$

This algorithm on this test case therefore achieves the same absolute error in five cycles through a loop of 34 keystrokes—very poor performance.

And finally try Wegstein's iteration scheme with algorithm 3.2.11 for the HP-45. Add A [ln] [STO] 3 to the start of the setup, and

$$\boxed{\cdots f \cdots} = \boxed{\text{RCL}}\,3\,\boxed{x \gtrless y}\,\boxed{\ln}\,\boxed{\div}\,1\,\boxed{-}. \tag{3.5.33}$$

With the same test case, get

$$x_i = 2,\ 1.888479701,\ 1.835277232,\ 1.778269426,$$
$$1.783239593,\ 1.783017990,\ 1.783017024. \tag{3.5.34}$$

Thus this algorithm on this test case achieves the same absolute error in six cycles through a loop of (about) 29 keystrokes—very unimpressive.

Figure 3.5.2 gives an error plot for these eight algorithms on this test case. For this particular problem, for this particular test case, and for the particular expressions chosen for g, α, and β, either the original form of the equation with the α method or the logarithmic form of the equation with the α method is preferable. I do not know of any general rule to use to predict which form of the equation, which method, or which expression to choose for g, α, or β. In some cases the shortest algorithm depends on the range of the parameters (see Section A.10.7). A more cheerful way of considering this example is to note that each of the eight algorithms did eventually converge to the right answer. But this is also not a general rule.

3.6 EXAMPLE: INTEREST RATE FOR AN ANNUITY

Consider another example of an elementary transcendental equation—the interest rate for an annuity. An annuity is a series of equal money payments, each of value R (alias PMT for payment) made at the end of each of n periods (e.g., each month or each year). The total value of an annuity can be expressed either by P, its initial value (alias principal or PV for present value) or by S, its terminal value (alias amount of FV for future value). The interest rate per period is i (expressed as a decimal fraction rather than as a percentage), and the interest is normally compounded at the end of each period.

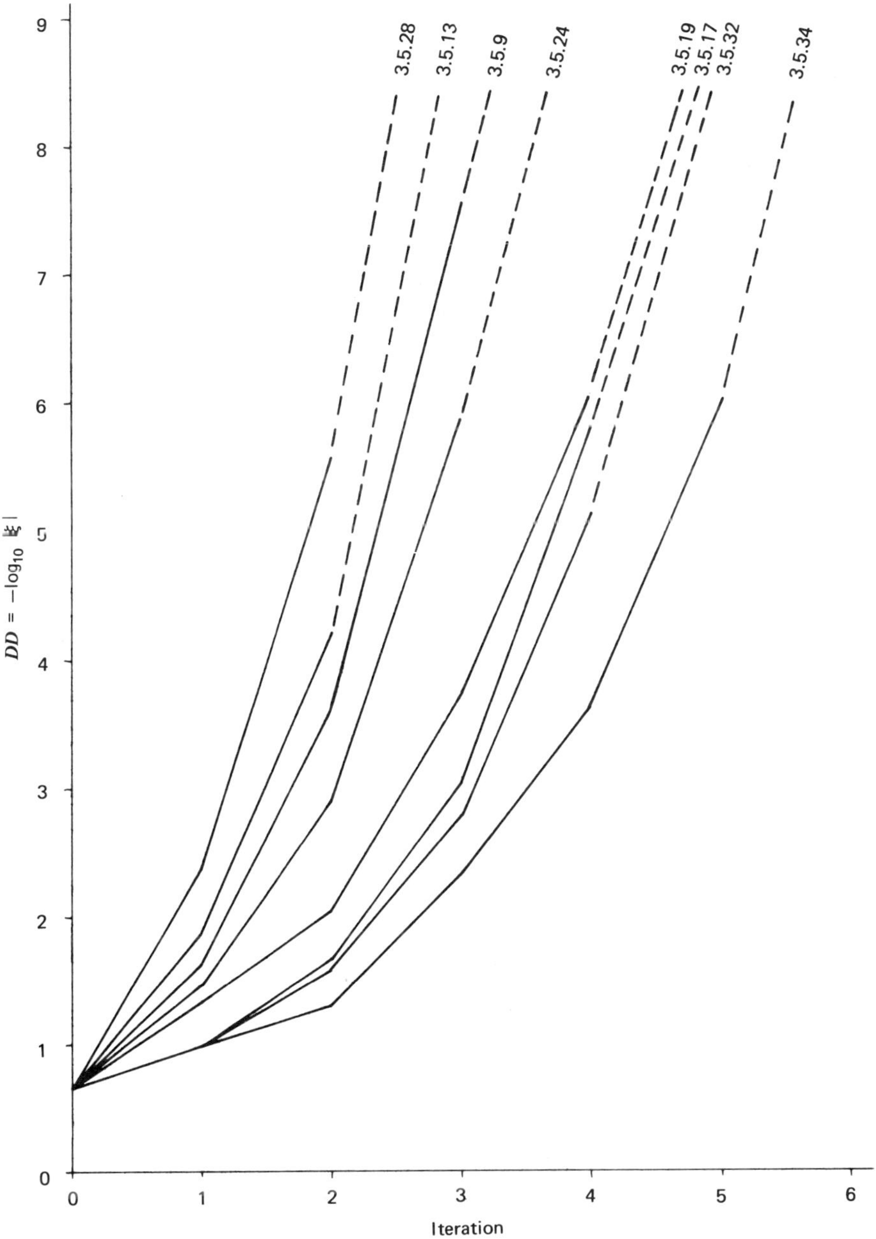

Figure 3.5.2 An error plot for eight different algorithms used to solve $x=5x^{-x}$ starting from $x_0=2$. The abscissa is the iteration number, and the ordinate is the number of correct decimal digits to the right of the decimal point. The labels on the curves correspond to numbered equations in the text.

The equations connecting these four quantities can be written as

$$\frac{R}{P}=i+\frac{i}{(1+i)^n-1},$$

$$\frac{S}{R}=\frac{(1+i)^n-1}{i}. \tag{3.6.1}$$

From these equations, we can easily calculate S, P, R, or n, given the other three parameters in each equation (see Section A.12). But the equations for the interest rate i are transcendental and require an iterative procedure for solution. For example, the interest rate in the *PRin* case can be written as

$$f(i)=\frac{R}{P}\left[1-(1+i)^{-n}\right], \tag{3.6.2}$$

and this expression is used in an iteration scheme in the *HP-45 Applications Book*, p. 142. However $f'(i)$ is not zero, and this $f(i)$ does not converge very rapidly; but then the loop does not have a great many keystrokes because $f(i)$ is rather simple. An alternative $f(i)$ for the same case, obtained from equation 3.6.1, is

$$f(i)=\frac{R}{P}+\frac{i}{1-(1+i)^n}. \tag{3.6.3}$$

For most problems this $f(i)$ has a smaller $|f'(i)|$, as one can verify by calculation. A corresponding algorithm for the HP-45 would be

SETUP: n [STO] 4 R [↑] P [÷] [↑] [↑] [↑] .1

LOOP: [↑] [↑] 1 [+] [RCL] 4 [G] [y^x] 1 [−] [÷] [−] (see i_i) : |. (3.6.4)

The .1 at the end of the setup in this algorithm is a first guess for i. Algorithm 3.6.4 converges slightly faster than HP's for most values of the parameters, has the same number of keystrokes (12) in the loop, and thus is somewhat preferable. But we can do better by getting F' to be zero.

This example does illustrate, however, that sometimes the equation can be manipulated algebraically to obtain an alternative $f(x)$ that has a smaller $|f'(x)|$ and converges faster, even though one does not want to do the work of deriving the algorithms for the g, α, or β methods.

Using equation 3.6.2 for $f(i)$, we can derive

$$f'(i)=n\frac{R}{P}(1+i)^{-n-1}, \tag{3.6.5}$$

and with this for $-g(i)$, after some rearranging, get

$$F(i)=\frac{(1+i)^{1+n}-i(1+n)-1}{(R/P)(1+i)^{1+n}-n}. \tag{3.6.6}$$

So a possible algorithm for the HP-45 would be

SETUP: P [↑] R [÷] [STO] 1 n [STO] 2 1 [+] [STO] 3 .1

LOOP: [↑] [RCL] [×] 3 1 [+] [CHS] [$x \gtrless y$] 1 [+] [RCL] 3 [G] [y^x] [+] [G] [LAST X] [RCL] 1 [×] [RCL] 2 [−] [÷] (see i_i) :|. (3.6.7)

The [RCL] [×] 3 near the beginning of the loop in this algorithm saves a keystroke over [RCL] 3 [×] because another preceding [↑] would otherwise be needed. Note that the loop in this algorithm has twice as many keystrokes (24) as algorithm 3.6.4.

Let $P=\$2000$, $R=\$203.61$, and $n=25$, and try this test case with these three algorithms. First with HP's algorithm, get

$$i=0.101805, 0.092786161, 0.090728834, 0.090194545, 0.090051452, 0.090012818, \text{etc.} \tag{3.6.8}$$

Then with algorithm 3.6.4, get

$$i=0.091636928, 0.090282299, 0.090048094, 0.090007169, 0.090000005, 0.089998750, \text{etc.} \tag{3.6.9}$$

And finally with the g method, algorithm 3.6.7, get

$$i=0.09034752, 0.089999019, 0.089998484. \tag{3.6.10}$$

The last number is the correct i with an absolute error less than about $\pm 2\times 10^{-9}$.

Despite its much more rapid convergence, the g-method algorithm is just marginally preferable in terms of keystrokes, if one only needs, say, four or five significant figures. And, in any case, one might ask whether enough is saved in keystrokes to compensate for the labor of deriving the g-method formulas.

An algorithm using Newton's method for the "direct reduction loan interest rate," which turns out to be the same problem, is given in the *HP-55 Mathematics Programs*, p. 49.

Next consider the problem of finding the interest rate in the *SRin* case. From equation 3.6.1, a possible $f(i)$ is

$$f(i)=\left(1+\frac{iS}{R}\right)^{1/n}-1, \tag{3.6.11}$$

and an algorithm for direct iteration on an HP-45 would be

SETUP: n [1/x] [STO] 4 S [↑] R [÷] [↑] [↑] [↑] .1

LOOP: [×] 1 [+] [RCL] 4 [G] [y^x] 1 [−] (see i_i) :|. (3.6.12)

As a test case, let $S=\$1000$, $R=\$11.81$, $n=25$, and get

$$i_i=0.09408038, 0.091702398, 0.090711015,$$
$$0.090291233, 0.090112309, 0.090035831, \text{etc.} \tag{3.6.13}$$

This is a very short algorithm (only nine keystrokes in the loop), but it converges slowly. As an exercise, show that the alternative

$$f(i)=\frac{R}{S}\left[(1+i)^n-1\right], \tag{3.6.14}$$

does not converge.

Try the g method on equation 3.6.11 with $g(i)=-f'(i)$, and get

$$F(i)=\frac{1+(1-1/n)iS/R-(1+iS/R)^{1-1/n}}{(1+iS/R)^{1-1/n}-S/(nR)}. \tag{3.6.15}$$

An HP-45 algorithm for this iteration is

SETUP: S [↑] R [÷] n [↑] [1/x] 1 [$x \gtrless y$] [−] [STO] 1 [CLX] [+] [÷]
[STO] 2 [CLX] .1

LOOP: [×] [↑] [RCL] [×] 1 1 [+] [$x \gtrless y$] 1 [+] [RCL] 1 [G] [y^x] [−] [G]
[LASTX] [RCL] 2 [−] [÷] (see i_i) :|. (3.6.16)

With the same test case, get

$$i_i=0.09027327, 0.089978898, 0.089978593. \tag{3.6.17}$$

This algorithm on this test case converges to an absolute error less than

$\pm 2\times 10^{-9}$ in three cycles through a loop of 21 keystrokes and is preferable to algorithm 3.6.12, even though the latter has half as many keystrokes per cycle.

An algorithm using Newton's method for the "sinking fund interest rate," which is the same problem, appears in the *HP-55 Mathematics Programs*, p. 61. Perhaps in finance, as in astronomy, much of the mystique involves knowing all 13 names for each entity or process. Further algorithms for money calculations appear in Section A.12.

Some of the material in this chapter was taken from Ball (1976a) with permission of the American Association of Physics Teachers. See also McKelvey (1975).

3.7 EXERCISES

3.7.1 From Figure 3.2.1, show (*a*) that the slope of the straight line is $g(x_n)+1$, and (*b*) that the straight line is tangent to the curve if $g(x)=-f'(x)$.

3.7.2 Show that a way to achieve third-order convergence is to have both $g(a)=-f'(a)$ and also $g'(a)=-\frac{1}{2}f''(a)$.

3.7.3 From Figure 3.3.1, show (*a*) that the slope of the straight line is $1/\alpha(x_n)$, and (*b*) that the straight line is tangent to the curve if equation 3.3.5 is satisfied.

3.7.4 Show that the α method with

$$\alpha(x)=\frac{\ln[f(x)+xg(x)]-\ln[x(1+g(x))]}{\ln f(x)-\ln x},$$

is equivalent to the g method.

3.7.5 From Figure 3.4.1, show (*a*) that the slope of the straight line is $1/\beta(x_n)$, and (*b*) that the straight line is tangent to the curve if equation 3.4.5 is satisfied.

3.7.6 Show that the β method with β from equation 3.4.4 is equivalent to the g method.

3.7.7 Write *approximate* formulas for $f'(i)$ from equations 3.6.2 and 3.6.3, and show that $|f'(i)|$ is usually (but not always) smaller for equation 3.6.3 than for equation 3.6.2.

3.7.8 (*a*) Write an algorithm to iterate on equation 3.6.14. Try a couple of test cases to show that it does *not* converge. (*b*) Show that this equation does *not* meet the convergence criteria of equations 3.1.8

and 3.1.9. (*c*) *Approximate* $f'(i)$ for typical values of the parameters, write an α-method algorithm, and show, with test cases, that this algorithm does converge.

3.7.9 Kepler's equation in celestial mechanics is

$$M = E - e \sin E,$$

where M is the mean anomaly, E is the eccentric anomaly, and e is *not* $2.71828\cdots$ but instead the ellipticity of the orbit (Smart, 1962, p. 113); M and E are angles (in radians) and $0 \leqslant e \leqslant 1$. Write an RPN algorithm to solve Kepler's equation for E given M and e. *Test case*: $M = 69°$, $e = 0.2$; get $E = 80.°2951668$.

3.7.10 Consider a one-dimensional asymmetric potential well in quantum mechanics (Figure 3.7.1). The wave function is $u = 0$ for $x < 0$; $u = \sin(kx)$ for $0 < x < A$, where $k = [2mE/\hbar^2]^{1/2}$; and $u = \exp(-k'x)$ for $x > A$, where $k' = [2m(V-E)/\hbar^2]^{1/2}$, and $V > E$ gives k' real. The derivatives of the logarithms of u must match across $x = A$, and this gives $-k' = k/\tan(kA)$. The problem is to find E (usually a finite set of discrete values for E—energy levels), given V and A.

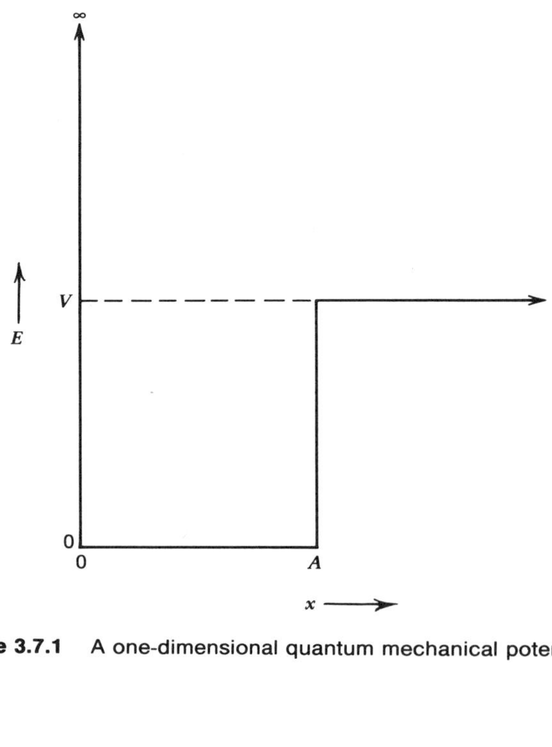

Figure 3.7.1 A one-dimensional quantum mechanical potential well.

Hint: Define $l=[2mV/\hbar^2]^{1/2}$ and $y=Ak$. The problem now is to find y given A and l. *Test case*: $Al=10$; get $y=2.85234190$, 7.06817436, and 8.42320393. *Reference*: Dicke and Wittke (1960), Chapter 3.

Two "Arabian problems" attributed to Beha Eddin Mohammed ben al Hosian al Aamouli (1547–1622):

3.7.11 Express 10 as the sum of two numbers such that if each of them is divided by the other and the resulting quotients added, the sum is equal to one of the original numbers.

3.7.12 Express 10 as the sum of two numbers such that if we add to each number its square root, and form the product of these sums, we obtain a given number A $(0 \leqslant A \lesssim 52.36)$. *Test case*: $A=24$; get 1 and 9.

3.7.13 In Aitken's δ^2 method, one iterates on

$$F(x)=\frac{xf(f(x))-f^2(x)}{f(f(x))-2f(x)+x}.$$

Derive an expression for $g(x)$ that makes the g method equivalent to Aitken's δ^2 method.

3.7.14 Consider the spring pendulum as represented in Figure 3.7.2. The downward force due to the weight is Mg (the rest of the apparatus is massless), and this is balanced by a torque $K\theta$ due to the spring. (If $M=0$, the system balances at $\theta=0$.) Calculate the equilibrium value of θ for a given M, g, K, and R. *Test case*: $MgR/K=3$ rad; get $\theta=1.17012095$ rad.

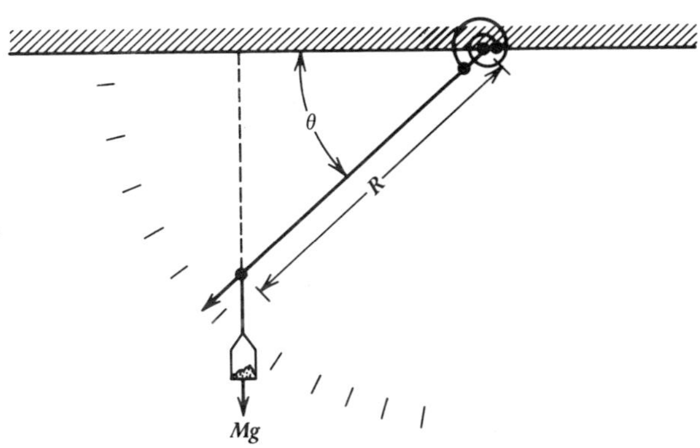

Figure 3.7.2 "No weights; honest spring."

4
Curve Fitting

Dost thou laugh to see how fools are vexed
To add to golden numbers, golden numbers?

—*Thomas Dekker (1570?–1641?)*

The purpose of computing is insight, not numbers.*

—*Richard W. Hamming (1915–   )*

4.1 INTRODUCTION: LEAST SQUARES

The problem of fitting a curve to a set of numbers $Y_i, x_i, \cdots$ is one of the most common problems in applied mathematics. Usually the numbers are data from an experiment or observation. However a mathematical function represented by a table of values to be approximated by a simpler function presents the same problem. The functional form of the curve is specified, but with one or more undetermined parameters to be calculated in the fitting process. The problems considered in this section differ from the problems of polynomial approximation in Chapter 5 in that the fitted

*From Hamming (1973), frontispiece.

curve does not necessarily go through the Y_i, but only near them. And in this section we seek the curve itself rather than some derived property such as the integral or derivative.

To specify the problem precisely, we need a criterion of best fit between curve and data. The most common criterion is least squares: the sum of the squares of the differences between the data and the curve (the sum of the squares of the absolute errors) is minimized. That is, we minimize

$$s^2 = \sum_{i=1}^{n} \left[Y_i - f(x_i, \cdots) \right]^2 \tag{4.1.1}$$

with respect to variations in each of the parameters a_j that characterize the function f. The $\sqrt{s^2/n}$ is called the RMS, for "root mean square," and $\Delta Y_i = Y_i - f(x_i, \cdots)$ is called a residual. But minimizing s^2 is not the only alternative and is usually an arbitrary choice. We might, instead, minimize the sum of the squares of the relative errors, or the sum of the absolute values of the errors, or the maximum absolute error, or the maximum relative error. These last two alternatives, referred to as "minimax" or "best" criteria, are very popular for approximations used for computing transcendental functions in computers (see, e.g., Fike, 1968). Minimax (or near-minimax) polynomial or rational function approximations to certain transcendental functions are presented in Sections A.5.7 and A.5.10. The popularity of the least-squares (absolute-error) criterion probably results in part from the relative simplicity of the mathematics involved. But this criterion can be justified if certain assumptions are made about the statistics of the errors in a measurement or observation (see Whittaker and Robinson, 1944, Chapter IX, or Scarborough, 1962, Chapter 16). We assume that the x_i are known without dispersion.

4.2 LINEAR PROBLEMS

Consider first a linear problem, that is, one in which the function f can be written in the form

$$f(x, \cdots) = \sum_{j=0}^{m} a_j g_j(x, \cdots). \tag{4.2.1}$$

Thus f is a linear function of the a_j but not necessarily of the $x, \cdots$. In this equation each g_j is an arbitrary, but specified, function of $x, \cdots$, which represents all the independent variables. The a_j, for $j=0$ to m, are to be determined from n data $Y_i, x_i, \cdots$, $i=1$ to n, where $n > m$. The case $n = m+1$ is an exact fit (except in pathological cases), and $s=0$.

Combining equations 4.1.1 and 4.2.1 gives

$$s^2 = \sum_i \left[Y_i - \sum_j a_j g_j(x_i, \cdots) \right]^2. \tag{4.2.2}$$

Since we want to minimize s^2 with respect to variations in each of the a_j, we set the partial derivative of s^2 with respect to each a_j to zero. This gives the so-called *normal equations*,

$$\frac{\partial s^2}{\partial a_j} = \sum_i -2\left[Y_i - \sum_k a_k g_k(x_i, \cdots) \right] g_j(x_i, \cdots) \overset{!}{=} 0, \tag{4.2.3}$$

or

$$\sum_i Y_i g_j(x_i, \cdots) = \sum_k a_k \sum_i g_j(x_i, \cdots) g_k(x_i, \cdots), \qquad j = 0, m. \tag{4.2.4}$$

These are $m+1$ linear equations in the $m+1$ unknowns a_k, thus this is, at least in principle, a solvable problem, provided only that the determinant of the coefficients matrix in this equation is not zero.

As a special case of the linear problem, consider a *polynomial* of one independent variable, that is, $g_j(x, \cdots) = x^j$, and

$$f(x) = \sum_{j=0}^{m} a_j x^j. \tag{4.2.5}$$

Then the normal equations become

$$\sum_i Y_i x_i^j = \sum_k a_k \sum_i x_i^{j+k}, \qquad j = 0, m. \tag{4.2.6}$$

The simplest possible problem in this form is for $m = 0$, that is,

$$\sum_i Y_i = a_0 \sum_i 1 = a_0 n, \tag{4.2.7}$$

thus $f = a_0$ is the arithmetic average of the Y_i. With the $\boxed{\Sigma+}$ key on some RPN calculators, this operation is easy to perform.

The next case is a least-squares fit of a straight line or *linear regression*, $m = 1$, and

$$j=0: \qquad \sum Y_i = a_0 \sum 1 + a_1 \sum x_i,$$

$$j=1: \qquad \sum Y_i x_i = a_0 \sum x_i + a_1 \sum x_i^2. \tag{4.2.8}$$

These are two equations in two unknowns and can be solved to give

$$a_1 = \frac{n\Sigma Y_i x_i - (\Sigma x_i)(\Sigma Y_i)}{n\Sigma x_i^2 - (\Sigma x_i)^2},$$

$$a_0 = \frac{\Sigma Y_i - a_1 \Sigma x_i}{n}. \tag{4.2.9}$$

The loop in which the data are keyed must sum and save five quantities—the five sums in these equations. An algorithm for straight-line fitting based on these equations can be found in Section A.6.1. Other straight-line fitting algorithms are given in the *HP-21 Applications Book*, p. 10, the *HP-35 Math Pac*, p. 144, the *HP-45 Applications Book*, p. 77, and the *HP-25 Applications Programs*, p. 87. The HP-55, HP-91, HP-22, and HP-27 calculators have linear regression as an internal function.

The *parabolic* case (least-squares fit of a parabola) is $m=2$, and

$$\begin{aligned} j=0: &\quad \Sigma Y_i = a_0 \Sigma 1 + a_1 \Sigma x_i + a_2 \Sigma x_i^2, \\ j=1: &\quad \Sigma Y_i x_i = a_0 \Sigma x_i + a_1 \Sigma x_i^2 + a_2 \Sigma x_i^3, \\ j=2: &\quad \Sigma Y_i x_i^2 = a_0 \Sigma x_i^2 + a_1 \Sigma x_i^3 + a_2 \Sigma x_i^4. \end{aligned} \tag{4.2.10}$$

These are three equations in three unknowns and can be solved by standard matrix-inversion techniques. Since the loop in which the data are keyed must sum and save eight quantities—the eight different sums in these equations—we need a calculator with about eight [STO] registers. These equations can be solved to give

$$a_2 = \frac{S(r_5 r_8 - r_6 r_3) - (r_5 r_4 - r_7 r_3)(r_5 r_1 - r_7 r_6)}{S(r_5 r_2 - r_6^2) - (r_5 r_1 - r_7 r_6)^2}, \tag{4.2.11}$$

where $S = r_5 r_6 - r_7^2$ and where $r_1 = \Sigma x_i^3$, $r_2 = \Sigma x_i^4$, $r_3 = \Sigma Y_i$, $r_4 = \Sigma Y_i x_i$, $r_5 = n = \Sigma 1$, $r_6 = \Sigma x_i^2$, $r_7 = \Sigma x_i$, and $r_8 = \Sigma Y_i x_i^2$. These symbols correspond to the [STO] register numbers in the algorithm for this problem in Section A.6.3. Then

$$a_1 = \frac{r_5 r_4 - r_7 r_3 - a_2(r_5 r_1 - r_7 r_6)}{S},$$

$$a_0 = \frac{r_3 - a_2 r_6 - a_1 r_7}{r_5}. \tag{4.2.12}$$

In addition to the algorithm for this problem in Section A.6.3, a similar least-squares–parabola algorithm is given in the *HP-45 Applications Book*, p. 82.

The next case, $m=3$ for a *cubic* fit, requires 11 quantities to be summed and saved; thus it can be done in an elegant way only on a calculator that has this many [STO] registers, such as an HP-55. In case the x_i are equally spaced, an alternative fitting procedure for cubics and higher-order polynomials employs orthogonal polynomials. The Chebyshev-Gram polynomials, for example, are orthogonal over a finite sum of equally spaced x_i. This fitting procedure is discussed in HMF 22.19, and in Hildebrand (1956), Sections 7.10 and 7.11. See also Lanczos (1956), Chapter 5; Milne (1949), Section 72; or Davidon (1977).

A comment on *weighting factors* is appropriate here. Suppose that among the data Y_i, x_i, a given pair of values occurs w times over (w different i with the same Y_i and x_i). Then we could combine or group these w data, renumber the indices i, and perhaps save some keystrokes in the sums. Specifically, we multiply each Y_i and each x_i by its w_i. Then Σx_i would become $\Sigma w_i x_i$, Σx_i^2 would become $\Sigma w_i x_i^2$ (*not* $\Sigma w_i^2 x_i^2$), ΣY_i would become $\Sigma w_i Y_i$, and so on. Note particularly that $\Sigma 1$ becomes Σw_i, rather than just n. These w_i are called weighting factors; they occur frequently, and for various reasons, in least-squares fitting problems. The point is that adding weighting factors to the problem complicates it only slightly, and we can leave them out of the equations until needed.

Another fitting problem that occurs frequently and can be put into a linear form is the so-called *polar problem* (because it has something vaguely to do with polar coordinates)

$$f = A\sin(\theta + B) + C. \tag{4.2.13}$$

This problem occurs, for example, in radio astronomy in determining the linear polarization parameters from data taken at a series of arbitrary polarization position angles $\theta_i/2$ (see Ball, 1975c, p. 213). Using a trigonometric identity, this equation can be rewritten as

$$f = C + A_x y + A_y x, \tag{4.2.14}$$

where

$$\begin{aligned} A_x &= A\cos B, \\ A_y &= A\sin B, \\ x &= \cos\theta, \\ y &= \sin\theta. \end{aligned} \tag{4.2.15}$$

Then the problem is to fit equation 4.2.14 in a least-squares sense to a set of Y_i, θ_i, to find A_x, A_y, and C, from which A and B can be calculated if needed. In equation 4.2.1, $m=2$, $g_0(x,y)=1$, $g_1(x,y)=y$, and $g_2(x,y)=x$. Then equation 4.2.5 gives

$$\begin{aligned} j=0: &\quad \Sigma Y_i = C\Sigma 1 + A_x \Sigma y_i + A_y \Sigma x_i, \\ j=1: &\quad \Sigma Y_i y_i = C\Sigma y_i + A_x \Sigma y_i^2 + A_y \Sigma x_i y_i, \qquad (4.2.16) \\ j=2: &\quad \Sigma Y_i x_i = C\Sigma x_i + A_x \Sigma x_i y_i + A_y \Sigma x_i^2. \end{aligned}$$

This time the loop in which the data are keyed should sum and save nine quantities—the nine different sums in these equations—which calls for a calculator with about nine [STO] registers. Furthermore, if we start with θ_i (rather than x_i and y_i), we need either a [→R] or some trigonometric functors in this loop, thus destroying the contents of [STO] register 9 on an HP-45 or HP-65. We can squeeze out of this difficulty by noting that since $\sin^2\theta_i = 1 - \cos^2\theta_i$ and

$$\Sigma y_i^2 = \Sigma 1 - \Sigma x_i^2, \qquad (4.2.17)$$

we avoid calculating Σy_i^2 explicitly.

After some manipulation, equation 4.2.16 can be solved to give

$$A_x = \frac{(r_5 r_2 - r_3 r_8)S - (r_5 r_4 - r_3 r_7)(r_5 r_1 - r_7 r_8)}{(r_5^2 - r_5 r_6 - r_8^2)S - (r_5 r_1 - r_7 r_8)^2}, \qquad (4.2.18)$$

where $S = r_2 r_6 - r_7^2$, and where $r_1 = \Sigma x_i y_i$, $r_2 = \Sigma Y_i y_i$, $r_3 = \Sigma Y_i$, $r_4 = \Sigma Y_i x_i$, $r_5 = n = \Sigma 1$, $r_6 = \Sigma x_i^2$, $r_7 = \Sigma x_i$, and $r_8 = \Sigma y_i$. These symbols again correspond to the [STO] register numbers in the algorithm for this problem. Then

$$A_y = \frac{r_5 r_4 - r_3 r_7 - A_x(r_1 r_5 - r_8 r_7)}{S},$$

$$C = \frac{r_3 - A_x r_8 - A_y r_7}{r_5}. \qquad (4.2.19)$$

An algorithm based on these equations for the polar problem is given in Section A.6.4. In case the θ_i are equally spaced over a whole cycle, this problem is somewhat easier to do using Fourier series techniques (see Section A.5.5).

Other examples of linear least-squares fitting problems that can be done fairly simply on an RPN calculator are $m=2$, $g_0(x,y)=1$, $g_1(x,y)=x$, $g_2(x,y)=y$, that is,

$$f=a_0+a_1x+a_2y. \tag{4.2.20}$$

This is called multiple linear regression, and it is obviously very similar to the polar problem. An explicit algorithm can be found in the *HP-45 Applications Book*, p. 79. If $m=1$, $g_0(x)=x^\alpha$, $g_1(x)=x^\beta$, where α and β are *given* constants, then

$$f=a_0x^\alpha+a_1x^\beta. \tag{4.2.21}$$

An algorithm for this fitting problem is in the *HP-45 Applications Book*, p. 84. If $m=1$, $g_0(x)=1$, $g_1(x)=\ln x$, then

$$f=a_0+a_1\ln x. \tag{4.2.22}$$

This is called a logarithmic curve fit and is equivalent to a linear fit but with x replaced by $\ln x$. Thus the linear algorithm in Section A.6.1 can be used (add [ln] after x_i), or an explicit algorithm can be taken from the *HP-21 Applications Book*, p. 16, the *HP-55 Statistics Programs*, p. 79, or the *HP-25 Applications Programs*, p. 95. If $m=1$, $g_0(x)=1$, $g_1(x)=e^x$, then

$$f=a_0+a_1e^x. \tag{4.2.23}$$

This would be called an exponential curve, but that name applies to a different formula discussed below. Nameless equation 4.2.23 is equivalent to a linear fit but with x replaced by e^x. Thus the linear algorithm in Section A.6.1 can be used; add [e^x] after x_i.

There are almost endless variations on this theme; the essence of this class of problems is that one specifies the functions $g_j(x,\cdots)$, thus making it a linear problem. Most such problems can be done fairly easily if $m\leqslant 2$. For larger m, we run out of [STO] registers on most calculators, and the calculations are longer, of course, but there are no essential additional difficulties.

4.3 AN APPROACH TO CERTAIN NONLINEAR PROBLEMS

The nonlinear least-squares fitting problem, represented by equation 4.1.1 when it cannot be reduced to equation 4.2.2, is much more difficult. This nonlinear problem is a multidimensional analogue of a transcendental algebraic equation discussed in Chapter 3, and a popular method of solution is a multidimensional generalization of Newton's method. The problem can be linearized using derivatives, but then it must be iterated because the derivatives change as the solution is approached. Another method of solution involves "relaxing" the unknown a_j, a process that is not relaxing at all for the calculator (or calculator operator). Several such techniques are discussed in Demidovich and Maron (1973), Chapter 13, and in the IBM SSP (see especially pp. 221–226). See also Ball (1975c), p. 210.

But consider a nonlinear problem that can be linearized by some operation F performed on both the Y_i and f. Suppose $Ff(x,\cdots)$, rather than $f(x,\cdots)$, is a linear function of the a_j,

$$Ff(x,\cdots)=\sum_{j=0}^{m} a_j g_j(x,\cdots), \tag{4.3.1}$$

as in equation 4.2.1. Then we would be tempted to apply F to the Y_i and proceed as before with a linear problem. Unfortunately, such a fit is no longer least squares in ΔY as advertised. Specifically, the fit would then be done in FY space, and we would minimize

$$s^2=\sum_i \left[FY_i - Ff(x_i,\cdots)\right]^2, \tag{4.3.2}$$

rather than the s^2 in equation 4.1.1. This can be dramatically different, especially with "noisy" Y_i.

Consider a specific example. Suppose

$$f=\exp\left(\sum_{j=0}^{m} a_j x^j\right) \tag{4.3.3}$$

Then with $F=\ln$,

$$\ln f=\sum_{j=0}^{m} a_j x^j \tag{4.3.4}$$

is a polynomial as in equation 4.2.5. For $m=0$ this is similar to the previous $m=0$ case (equation 4.2.7); for $m=1$, equation 4.3.3 becomes the exponential or geometric curve,

$$f=\exp(a_0+a_1x), \tag{4.3.5}$$

and for $m=2$, the Gaussian or normal curve of error

$$f=\exp(a_0+a_1x+a_2x^2). \tag{4.3.6}$$

The exponential curve occurs, for example, in electronics in the model of a semiconductor diode (see Section A.10.7), and the Gaussian, disguised by various aliases, is an extremely important function in a half-dozen different fields. We want to be able to fit these functions in a least-squares sense.

The $m=0$ case is not very interesting in itself but serves to illustrate the difference between fitting in Y space versus fitting in $\ln Y$ space. We noted previously (equation 4.2.7), that a_0 for $m=0$ is just the arithmetic average

$$f=a_0=\frac{\Sigma Y_i}{n}. \tag{4.3.7}$$

If we fit equation 4.3.4 with $m=0$ to the $\ln Y_i$, we get instead

$$a_0=\frac{\Sigma \ln Y_i}{n}, \tag{4.3.8}$$

or

$$f=e^{a_0}=\exp\left(\frac{1}{n}\Sigma \ln Y_i\right). \tag{4.3.9}$$

Depending on the Y_i, a_0 from equation 4.3.7 can be very different indeed from e^{a_0} from equation 4.3.9. A further difficulty occurs with negative values for Y_i; they are no longer permitted.

We can use *weighting factors* w_i to ameliorate this problem somewhat. We can choose w_i such that the fit to $\ln Y$ is least squares to first order in ΔY. That is, we choose w_i such that

$$w_i[FY_i-Ff(x_i,\cdots)]\cong\Delta Y_i=Y_i-f(x_i,\cdots), \tag{4.3.10}$$

as closely as possible. We cannot make this equation exact because we do not know $f(x_i,\cdots)$ until the a_j are calculated; but to the extent that the fit is a good one, Y_i is an approximation to $f(x_i,\cdots)$. With $F=\ln$, after some manipulation, get

$$w_i\cong Y_i, \tag{4.3.11}$$

independent of the form of $f(x_i,\cdots)$, but not, of course, independent of F. Fortunately these are easy w_i to work with.

An algorithm to fit a power-law curve

$$f = a_0 x^{a_1} \tag{4.3.12}$$

to $\ln Y_i$ using these w_i appears in Section A.6.5. Algorithms for the same problem but without w_i are given in the *HP-21 Applications Book*, p. 14, the *HP-45 Applications Book*, p. 86, the *HP-25 Applications Programs*, p. 98, and the *HP-55 Statistics Programs*, p. 82. For an algorithm to fit an exponential curve, equivalent to equation 4.3.5, using w_i, see Section A.6.5. Algorithms for the same problem but without w_i appear in the *HP-21 Applications Book*, p. 12, the *HP-45 Applications Book*, p. 88, the *HP-25 Applications Programs*, p. 92, and the *HP-55 Statistics Programs*, p. 76. An algorithm to fit a Gaussian curve, modified from equation 4.3.6, using w_i, is given in Section A.6.7.

Fitting to $\ln Y_i$ without w_i is an approximation to fitting by minimizing the sum of the squares of the *relative* errors $\Delta Y_i / Y_i$.

A test case is shown in Table 4.3.1. Pseudo-random noise with an RMS of about 0.1 was added to the values of the function $3x^3$ for $x = 0.2(0.2)1$. The noise was obtained by averaging 10 numbers from a table of random numbers evenly distributed between 0 and 1, and subtracting $\frac{1}{2}$ from the average. These noisy data were then fit with a power-law curve using two algorithms: (*a*) from Section A.6.5, which has w_i, and (*b*) from the *HP-45 Applications Book*, p. 86, which has no w_i. The function values from these fitted curves, and the RMS of the residuals in the fits, are also listed in Table 4.3.1. The RMS for the fit using w_i is less than half the RMS for the fit without w_i.

Least-squares fitting is discussed in more detail by Hamming (1973), Chapter 17 and 18, and Hildebrand (1956), Chapter 7.

Table 4.3.1. Power-Law Curve Fitting—Test Case

x_i	$3x_i^3$	Noise	Y_i	Fit with w_i		Fit Without w_i	
			$(3x_i^3 + \text{noise})$	$f(x_i)$	ΔY_i	$f(x_i)$	ΔY_i
0	0	—	—	0	—	0	—
0.2	0.024	0.102756	0.126756	0.0658	−0.0610	0.0961	−0.0306
0.4	0.192	0.063622	0.255622	0.3256	0.0699	0.3733	0.1177
0.6	0.648	−0.001678	0.646322	0.8297	0.1834	0.8253	0.1789
0.8	1.536	0.099074	1.635074	1.6115	−0.0236	1.4490	−0.1860
1.0	3.	−0.189405	2.810595	2.6966	−0.1140	2.2424	−0.5682
RMS	—	0.1098	—	—	0.1056	—	0.2843

4.4 EXERCISES

4.4.1 Fit a straight line to the following data:

x_i	0	0.5	1.0	1.5	2.0	2.5
Y_i	0.1	0.5	1.5	2.8	4.2	5.9

4.4.2 Fit a parabola to the data of the preceding problem.

4.4.3 Fit a polar curve to the following data:

θ_i	0	60°	120°	180°	240°	300°
Y_i	6	2	0	2	0	2

4.4.4 Write an algorithm for a least-squares fit of

$$f = A\cos^2(\theta) + B.$$

4.4.5 Use the algorithm from problem 4.4.4 to fit the data of problem 4.4.3.

4.4.6 Write an algorithm for a least-squares fit of

$$f = A\cos(2\theta) + B\cos(\theta) + C.$$

Hint: Use a trigonometric identity on $\cos(2\theta)$.

4.4.7 Use the algorithm from problem 4.4.6 to fit the data of problem 4.4.3.

4.4.8 Let $Y_i = i$ for $i = 1(1)10$; calculate f from equation 4.3.7 and compare with f from equation 4.3.9.

4.4.9 Show that equation 4.3.11 leads to approximate least squares in fitting to $\ln Y_i$.

4.4.10 Fit a power law to most of the data of problem 4.4.1.

4.4.11 Fit a Gaussian to the following data:

x_i	1	2	3	4	5
Y_i	0.87	2.18	3.12	2.55	1.19

4.4.12 If $F = \exp$ (instead of ln), what w_i give approximate least squares?

4.4.13 Using the answer from problem 4.4.12, write an algorithm to fit

$$f=\ln(a_0+a_1x),$$

in an approximate least-squares sense.

4.4.14 Use the algorithm from problem 4.4.13 to fit the following data:

x_i	0	1	2	3	4
Y_i	0.7	1.4	1.8	2.1	2.3

4.4.15 Show that fitting to $\ln Y_i$ without w_i is an approximation to fitting by minimizing the sum of the squares of the relative errors $\Delta Y_i / Y_i$.

4.4.16 The following data were taken on a particular semiconductor diode:

I (amperes)	10^{-9}	10^{-8}	10^{-7}	10^{-6}	10^{-5}	10^{-4}	10^{-3}	10^{-2}
V (volts)	0.153	0.212	0.270	0.329	0.388	0.447	0.505	0.564

The engineer suspects that the m factor of the diode may not be constant. Split the data into two parts, fit each part separately, and determine two different m factors. Compare also the two values for I_0. *Hint*: Use the equations and discussion in Section A.10.7.

4.4.17 The flux of many continuum sources in radio astronomy varies as

$$S_\nu \propto \nu^{-\alpha}$$

where ν is the frequency, S_ν is the flux, and α is the spectral index. The following fluxes were measured on the radio source Taurus S:

ν(MHz)	200	1000	2000	10000	20000
S_ν (janskys)	1010	340	220	75	47

Using a least-squares procedure, determine α.

4.4.18 The human population on the earth increases more rapidly than exponentially. In fact the growth rate (%/year) is approximately proportional to the population, which suggests a curve of the form

$$P(t)=\frac{K}{t_d-t},$$

where $P(t)$ is the population at time t, K is a constant, and t_d is a specific time called "doomsday." Fit the following data to this expression to determine K and t_d.

t_i (A.D.)	1650	1750	1850	1900	1930	1940	1950	1960	1964	1967	1970	1972
P_i (millions)	490	690	1080	1550	2050	2280	2520	3005	3225	3420	3615	3730

Hint: Use $Fx = 1/x$ (F is the reciprocal operator) and choose w_i so that the fit is approximately least squares in ΔP_i. This corresponds to the higher fractional error in our knowledge of the population in the past.

4.4.19 Write an HP-45 algorithm for an approximate least-squares fit of

$$Y = \sin(\omega t + \theta),$$

given a set of Y_i, t_i. Use the algorithm to estimate ω and θ from

Y_i	0.07	0.36	0.62	0.82	0.95
t_i	1	2	3	4	5

Hint: Assume that all the data lie in the first quadrant, that is, $0 \leqslant \omega t + \theta \leqslant 90°$. *Answers*: $\omega = 16.995$ (not 16.957), $\theta = -12.°892$ (not $-12.°806$). (The units of ω are degrees per unit of t.)

5
Numerical Integration, Differentiation, and Interpolation

In order to solve this differential equation you look at it till a solution occurs to you.*

—*Anonymous mathematics professor quoted by George Pólya (1887–)*

It has been usual in discussing properties of matter to regard the medium as continuous, to set up differential equations, look at them for awhile, give up, and replace them by difference equations. The difference equations are then solved, and no attention is paid to their physical significance, if any. An alternative procedure is to handle the problem discretely from the beginning, lumping the "molecules" together in groups as small as the computing equipment can handle.†

—*John Todd (1911–)*

The better part of valour is discretion.

—*William Shakespeare (1564–1616)*

*From Pólya (1973), p. 208.
†From Todd (1962), p. 17.

5.1 INTRODUCTION

The techniques of numerical integration, differentiation, and interpolation are a last resort, to be used in desperation. Suppose, for example, one has a differential equation to solve. The ideal would be to find a solution in terms of elementary functions, using the analytic techniques of the theory of differential equations. Failing this, one can try for a solution in the form of a power series, either directly from the differential equation, or from a power series expansion of well-known functions. Failing this as well, one can try for a solution in terms of well-known (i.e., tabulated) functions and use numerical interpolation to calculate intermediate values from the table. Only if all these methods fail should the use of numerical integration on the differential equation be considered. An exception to this desperation statement occurs in work with data from an experiment or observation. In such cases either curve fitting (Chapter 4) or the numerical techniques of this chapter may be unavoidable.

Numerical *interpolation* is the process of estimating intermediate values from a table of numbers. A typical table shows $f(x)$ for equal increments of x; but one may need to know $f(x)$ for values of x between those in the table. Numerical *differentiation* is the process of estimating the derivative of $f(x)$ from the same kind of table, for values of x not necessarily coincident with the tabulated values.

Numerical *integration* of a specified function of x is called "quadrature." This name is at least as old as Archimedes, and it originally referred to the process of squaring (constructing a square with the same area as) a given area bounded by a curve. There are two categories, and some subcategories, of quadrature: one may need to know a *definite integral* of a function $f(x)$ that cannot be integrated analytically. The function may be available as a table with equally spaced arguments, in which case one can choose from several possible formulas for tabular quadrature; or if an algorithm is available for calculating $f(x)$ at any x, one of the so-called Gaussian quadrature formulas is appropriate.

The result of doing a definite integral is a number; the result of doing an *indefinite* or *primitive integral* is a function in principle, or a table of values in practice. Indefinite numerical quadrature differs only slightly from another category of problems—*numerically integrating differential equations*. For example, a typical quadrature problem would be

$$y=\int_{x_L}^{x_u} f(x)\,dx. \tag{5.1.1}$$

If y is needed for a single set of values of x_L and x_u, this is a problem of definite integration or quadrature. If y is needed for a range of x_L or x_u, this is a problem of indefinite or primitive integration or quadrature.

For a typical first-order ordinary differential equation (initial-value problem)

$$y'(x)=f(x,y), \qquad y(a)=b, \tag{5.1.2}$$

the formal solution is

$$y(x)=b+\int_a^x f(x',y)\,dx'. \tag{5.1.3}$$

This integration differs from quadrature in that the integrand is also a function of y. Analytically this is a very important distinction, but numerically it is not particularly significant.

A friend of mine once expressed the view that numerical integration of differential equations is rather much for pocket calculators and seemed surprised that I was able to do it at all. If a FORTRANable computer is available, perhaps it should be used for this purpose. However differential equations were being integrated numerically before the invention of electronic calculators or computers, and in some cases without using even a mechanical calculator.

The basic idea of any of these numerical techniques is that a polynomial (or very rarely some other function) is fitted through $f(x_i)$ at a finite number of selected x_i, then the value, the derivative, or the integral of the polynomial is used to approximate the corresponding operations on $f(x)$. The approximating polynomial usually does not appear explicitly. Instead, the formula for finding the value, derivative, or integral of the polynomial is combined with the formula for finding the polynomial itself, to give a composite formula for an approximation for the value, derivative, or integral of $f(x)$, in terms of selected $f(x_i)$.

This technique succeeds to the extent that the polynomial is a good approximation to $f(x)$ over the range of interest. A continuous function can be approximated well by a polynomial over a limited range of x, but, generally speaking, the wider the range of x, the poorer the fit obtainable with a polynomial of given order. For interpolation and differentiation, the polynomial is required to fit only over a narrow range of x, but for definite quadrature the polynomial must fit over the range of integration. And in indefinite quadrature, errors usually accumulate as the integration proceeds. Therefore quadrature typically requires a higher-order formula for the same relative error compared with interpolation or differentiation. By contrast, tabular quadrature is less sensitive to "noise" or round-off errors in the values of the function to be integrated, because such errors tend to average out.

5.2 TABULAR INTERPOLATION

In classical numerical analysis, higher-order interpolation of tabular values was usually done by calculating and manipulating differences of various orders. Some help is provided by an RPN calculator, but this process is strongly oriented toward the use of pencil and paper. As an alternative, one can use a so-called Lagrange interpolation formula. These formulas give interpolated values directly in terms of sums of weighted tabular values; in effect, the differences are taken implicitly. The Lagrange interpolation formulas are much better suited to RPN calculators, especially since the weighting factors or coefficients need not be from a table but can be calculated as required.

For example, the Lagrange three-point interpolation formula can be written as (HMF 25.2.11)

$$f(x_0+ph) \cong \tfrac{1}{2}p(p-1)f_{-1}+(1-p^2)f_0+\tfrac{1}{2}p(p+1)f_1, \tag{5.2.1}$$

where f_{-1} is $f(x_{-1})$, f_0 is $f(x_0)$, and so on; h is the spacing between x_i, that is

$$h=x_1-x_0=x_0-x_{-1}=\text{etc.}; \tag{5.2.2}$$

and $p=(x-x_0)/h$. The x_i should be chosen so that $|p| \leqslant \frac{1}{2}$, that is, x_0 should be the closest x_i to x. An algorithm for this formula for the HP-35 is

$$\begin{gathered}
x_0 \boxed{\uparrow}\boxed{\uparrow} x \boxed{-}\boxed{x \gtrless y} x_1 \boxed{-} \quad (\text{see } -h) \quad \boxed{\div} \\
\boxed{\uparrow}\boxed{\uparrow}\boxed{\uparrow} \quad (\text{see } p) \quad 1 \boxed{-} f_{-1} \boxed{\times}\boxed{x \gtrless y} \\
1 \boxed{+} f_1 \boxed{\times}\boxed{+}\boxed{\times} 2 \boxed{\div}\boxed{x \gtrless y}\boxed{\uparrow} \\
\boxed{\times} 1 \boxed{-} f_0 \boxed{\times}\boxed{-} \quad (\text{see } \cong f(x)).
\end{gathered} \tag{5.2.3}$$

Algorithms for several other interpolation formulas are given in Section A.2.

Lagrange interpolation formulas were criticized, for example, by Hartree (1958), p. 75. The problem is that one cannot easily detect an error or determine whether the formula being used is of too low (or too high) an order. But in my view the advantages of using Lagrange formulas on an RPN calculator outweigh the disadvantages. Doing Lagrange interpolation to two different orders and comparing the results is usually easier than doing one interpolation using a traditional method. In some cases tables have footnotes indicating what order of interpolation is necessary for a specified precision, but a higher-order formula is necessary if x is midway between tabulated points ($|p|=\frac{1}{2}$) than if x is near a tabulated point ($|p|\ll\frac{1}{2}$).

Table 5.2.1 Abstract from HMF Table 5.1

x	$xe^xE_1(x)$	x	$xe^xE_1(x)$
7.5	0.892687854	8.0	0.898237113
7.6	0.893846312	8.1	0.899277888
7.7	0.894979666	8.2	0.900297306
7.8	0.896088737	8.3	0.901296033
7.9	0.897174302	8.4	0.902274695

$$\begin{bmatrix} (-6)3 \\ 5 \end{bmatrix}*$$

*Somewhat different values are given in HMF, p. 243.

Let us follow an example given in HMF, p. XI. Suppose we need to know $xe^xE_1(x)$ for $x=7.9527$. An abstract from HMF Table 5.1 is reproduced here as Table 5.2.1. The footnote in square brackets means that the maximum absolute error in linear interpolation is 3×10^{-6} and that a five-point interpolation formula is needed to achieve full tabular accuracy —nine decimal digits. Table 5.2.2 shows the results of applying several Lagrange interpolation formulas from Section A.2 to this problem. The last column in this table is $-\log_{10}$ of the magnitude of the absolute error and is the number of correct decimal digits in the corresponding value. In HMF, p. XI, the same example is worked using traditional methods employing high-order differences. Another method for calculating $E_1(x)$ appears in Section A.5.11.

Table 5.2.2 Example of Numerical Interpolation

Lagrange Formula	$\simeq xe^xE_1(x)$ for $x=7.9527$	$DD=-\log_{10}\lvert\xi\rvert$
Linear	0.8977344034	5.6
Three-point	0.8977371499	7.4
Four-point	0.8977371942	9?
Five-point	0.8977371929	9?
Six-point	0.8977371933	9?
From HMF, p. XI	0.897737193	9?
From nine-term continued fraction (cf. Section A.5.11)	0.8977371938	10?

5.3 TABULAR DIFFERENTIATION

Almost exactly the same comments made about tabular numerical interpolation in Section 5.2 apply to tabular numerical differentiation. The Lagrange three-point differentiation formula can be written as (HMF 25.3.4)

$$f'(x) \cong \frac{\left(p-\frac{1}{2}\right)f_{-1} - 2pf_0 + \left(p+\frac{1}{2}\right)f_1}{h}, \tag{5.3.1}$$

where the symbols have the same meaning as in equation 5.2.1. An algorithm for this formula for the HP-35 is

$$\begin{array}{ll}
x_0 \boxed{\uparrow}\boxed{\uparrow} x \boxed{-} \boxed{x \gtrless y} x_1 \boxed{-} & (\text{see } -h) \\
\boxed{\text{STO}}\boxed{\div} \quad (\text{see } p) & \boxed{\uparrow}\boxed{\uparrow}\boxed{\uparrow}\boxed{+} \\
f_0 \boxed{\times}\boxed{x \gtrless y} .5 \boxed{-} f_{-1} \boxed{\times}\boxed{-}\boxed{x \gtrless y} & \\
.5 \boxed{+} f_1 \boxed{\times}\boxed{-}\boxed{\text{RCL}}\boxed{\div} & (\text{see } \cong f'(x)).
\end{array} \tag{5.3.2}$$

Algorithms for several other Lagrange differentiation formulas are given in Section A.2.

Table 5.3.1 Abstract of HMF Table 6.1

x	$\Gamma(x)$
1.455	0.88562 20800
1.460	0.88560 43364
1.465	0.88560 80495
1.470	0.88563 31217

As a numerical example, suppose we need to know the derivative of the gamma function $\Gamma'(x)$ for $x = 1.461632145$. An abstract from HMF Table 6.1 is reproduced here as Table 5.3.1; and Table 5.3.2 displays the results of

Table 5.3.2 Example of Numerical Differentiation

Lagrange Formula	$\cong \Gamma'(x)$ for $x = 1.461632145$	$DD = -\log_{10}\lvert\xi\rvert$
Linear	7.4262×10^{-4}	3.1
Three-point	-2.24×10^{-6}	5.6
Four-point	0	(10)

applying linear, three-point, and four-point Lagrange differentiation for this x. Numerical differentiation is discussed further by Kopal (1955), Chapter III.

5.4 TABULAR QUADRATURE

The best-known formulas for numerically integrating a function specified by a table of values with equally spaced arguments are the so-called Newton-Cotes formulas (HMF 25.4). The lower-order forms of these formulas are sometimes referred to by other names: the two-point form is called the trapezoidal rule, the three-point form is called Simpson's rule, the four-point form is called the $\frac{3}{8}$ rule, and the five-point form is called Bode's rule. The coefficients of f_i in these formulas are all rational numbers, and RPN algorithms are easy to write; algorithms for up to eight points are contained in Section A.3.1.

Simpleton's formula (not to be confused with Simpson's rule) gives an approximation to the integral by just adding up all the f_i, or if both end points of the interval are included, weighting these end points by $\frac{1}{2}$. A simpler formula is hard to imagine. For most cases, Simpleton's formula is much less precise than the Newton-Cotes formula of the same order. For certain special cases, however, Simpleton's formula is exact and Newton-Cotes formulas are inexact! Such cases occur whenever the function to be integrated has a Fourier transform that is band limited and the function is tabulated at the Nyquist interval or closer (see Bracewell, 1965, Chapter 10, and Hamming, 1973, Section 21.5). Alternatively we can derive or justify Simpleton's formula by assuming that $f(x)$ is to be approximated by a truncated Fourier series rather than by a polynomial. This procedure often works well for a cyclic function that is to be integrated over a complete cycle.

As an example of the precision obtainable with these formulas, consider two functions whose integral is known analytically: the notoriously intractable $\sqrt{x}$, and the much easier e^x. The trouble with $\sqrt{x}$ is that polynomial approximations do not fit very well even over a limited range of x. Try doing a MacLaurin series expansion of $\sqrt{x}$! If a polynomial is used as a test function, when the order of the Newton-Cotes formula equals or exceeds the order of the polynomial, quadrature becomes exact, except possibly for round-off errors.

The results of integrating $\int_0^1 \sqrt{x}\, dx$ and $\int_0^1 e^x dx$ with a variety of quadrature formulas are presented in Figure 5.4.1. Fortunately, most integrands behave more like e^x than $\sqrt{x}$. Nevertheless, numerical integration over a significant range of x is a tricky business. Tabular quadrature is

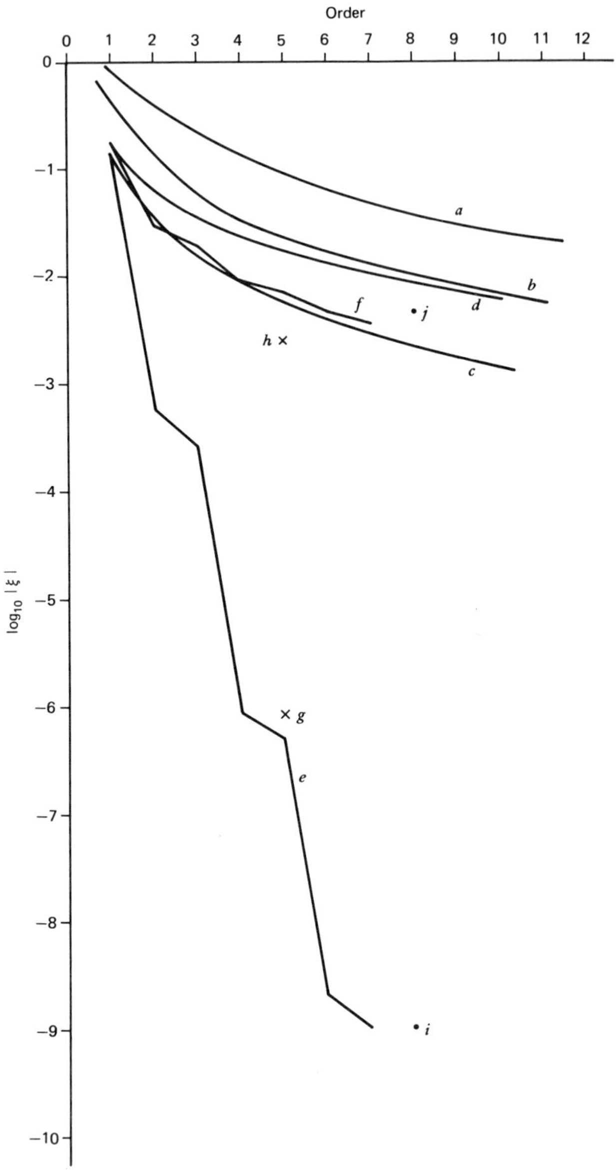

Figure 5.4.1. Error resulting from applying a variety of quadrature formulas to two test cases: I, $\int_0^1 e^x\,dx$; and II, $\int_0^1 \sqrt{x}\,dx$. The ordinate is $\log_{10}$ of the absolute error, and is the negative of the number of correct decimal digits to the right of the decimal point. The abscissa is the order of the formula. Then (*a*) shows the results of applying the Gauss-Chebyshev algorithm (Section A.3.2) to test case I; (*b*) shows the same algorithm applied to II; (*c*) shows Simpleton's formula (Section A.3.1) applied to I; (*d*) shows the same algorithm applied to II; (*e*) shows the Newton-Cotes formulas (Section A.3.1) applied to I; (*f*) shows the same algorithms applied to II; (*g*) shows the three-point Gauss-Legendre formula (Section A.3.2) applied to I; (*h*) shows the same algorithm applied to II; (*i*) shows the five-point Lobatto-Radau formula (Section A.3.2) applied to I; and (*j*) shows the same algorithm applied to II.

discussed further by Hartree (1958), Section 6.5, and by Hamming (1973), Chapter 12.

5.5 GAUSSIAN QUADRATURE

The term "Gaussian quadrature" applies to a variety of methods in which the x_i are not uniformly spaced, but are assigned by the method. For an n-point formula—that is, for n different x_i—one has $2n-1$ degrees of freedom in assigning the x_i and the weights w_i. (It is $2n-1$ rather than $2n$ because the sum of the w_i is fixed.) For this reason an n-point Gaussian quadrature formula is exact for polynomials up to degree $2n-1$; thus in some sense it is a $(2n-1)$-order formula. This is sometimes a great advantage.

The disadvantage is that one must be able to calculate $f(x_i)$ for the x_i given in the formula. If $f(x)$ is specified by a table of values, rather than by an algorithm, the additional effort needed to interpolate in the table to obtain the $f(x_i)$ usually wipes out the advantage of Gaussian quadrature formulas.

Such formulas exist in several forms, depending on whether the range of integration is finite (Gauss-Legendre or Gauss-Chebyshev), 0 to ∞ (Gauss-Laguerre), or $-\infty$ to ∞ (Gauss-Hermite). The second half of each hyphenated name designates the type of the associated orthogonal polynomial used in the derivation of the formulas. The x_i are zeros of the associated polynomials, and the w_i can be calculated from them.

As a simple example, consider the three-point Gauss-Legendre formula for integrating

$$\int_{-1}^{1} f(x)\,dx \cong \sum_{i=0}^{3} w_i f(x_i), \tag{5.5.1}$$

as given in HMF 25.4.29. The x_i are the zeros of Legendre polynomial $P_3(x)$, namely, 0 and $\pm\sqrt{3/5}$. The w_i are given by

$$w_i = \frac{2}{(1-x_i^2)[P_n'(x_i)]^2}, \tag{5.5.2}$$

and for $n=3$ these are $\frac{8}{9}$ for $x_i=0$, and $\frac{5}{9}$ for $x_i=\pm\sqrt{3/5}$. (The w_i add to 2 because of the range of integration.) Thus the formula can be written

$$\int_{-1}^{1} f(x)\,dx \cong \frac{5f(-\sqrt{3/5}\,)+8f(0)+5f(\sqrt{3/5}\,)}{9}. \tag{5.5.3}$$

An algorithm for an HP-35 would be

$$.6\boxed{\sqrt{x}}\boxed{\uparrow}\boxed{\uparrow}\boxed{\uparrow}\boxed{\cdots f\cdots}\boxed{\text{STO}}\boxed{\text{R}\downarrow}$$

$$\boxed{\text{CHS}}\boxed{\uparrow}\boxed{\uparrow}\boxed{\uparrow}\boxed{\cdots f\cdots}\boxed{\text{RCL}}\boxed{+}$$

$$5\boxed{\times}\boxed{\text{STO}}\boxed{\text{CL}X}\boxed{\uparrow}\boxed{\uparrow}\boxed{\uparrow}\boxed{\cdots f\cdots} \tag{5.5.4}$$

$$8\boxed{\times}\boxed{\text{RCL}}\boxed{+}9\boxed{\div} \qquad \left(\text{see} \cong \int \cdots\right).$$

In this algorithm $\boxed{\cdots f\cdots}$ is a subroutine and is any sequence of keystrokes that accepts x in X, Y, Z, and T, calculates $f(x)$ and leaves it in X, and retains x in Y (Z and T are irrelevant on return from $\boxed{\cdots f\cdots}$).

As a test case try $f(x)=e^x$, $\boxed{\cdots f\cdots}=\boxed{e^x}$, and get 2.350336929. The precise answer is $e-1/e=2.350402387$.

Section A.3.2 contains a three-point Gauss-Legendre algorithm for an arbitrary finite range of x, a five-point Lobatto-Radau algorithm (in which the ends of the range of integration are taken for x_i but the other x_i are set by the formula), an n-point Gauss-Chebyshev algorithm useful for a finite range when the integrand has $1/\sqrt{x}$ singularities at the ends, a two-point Gauss-Laguerre algorithm for integrating from 0 to ∞, and a three-point and five-point Gauss-Hermite algorithm for integrating from $-\infty$ to ∞. These particular formulas were selected from the myriad available to give relatively simple algorithms. The *HP-45 Applications Book* has algorithms for three-point Gauss-Laguerre quadrature (p. 119) and six-point Gauss-Legendre quadrature (p. 117).

The results of using the applicable algorithms on the same two test cases used with tabular quadrature are also plotted in Figure 5.4.1. The poor showing of the Gauss-Chebyshev formula is not an indication of poor quality; rather, it points up the inappropriateness of these test cases to this formula.

Gaussian quadrature is discussed in more detail in Kopal (1955), Chapter VII; Hildebrand (1956), Chapter 8; and Hartree (1958) Section 6.6; for collections of formulas, see HMF 25.4.28 to 25.4.46 and CRC HTM, pp. 839–847.

5.6 INDEFINITE QUADRATURE AND DIFFERENTIAL EQUATIONS

This section deals with both indefinite quadrature and numerically integrating certain kinds of ordinary differential equation. Like the traditional approach to numerical interpolation and differentiation, traditional methods in numerical solutions of differential equations usually employed tables of differences and minimized the number of points at which $f(x,y)$ was evaluated. Among traditional methods, the *predictor-corrector* methods associated with Milne, Adams, and Hamming are probably most popular. These methods are oriented toward the use of either pencil and paper or a fair amount of numerical storage, and they are not very appropriate for RPN calculators.

As an alternative, *Runge-Kutta* integration formulas use no previous information, but start anew at each point and calculate a polynomial to fit only over the interval from one point to the next. These formulas are an advantage with fast-changing functions because the approximating polynomial needs to fit only over this narrow interval. With Runge-Kutta methods, one does not need to remember previous values, and the technique requires no previous information to get started. The disadvantage is that $f(x,y)$ is calculated n times for each step in x, where n is the order of the formula, rather than only once or twice per step, regardless of n, in most traditional methods.

Consider a first-order* ordinary differential equation of the form in equations 5.1.2 and 5.1.3. The simplest possible formula that might be used to carry the solution away from $x=a$ is the Euler or point-slope formula (HMF 25.5.1), which is the same as the first-order Runge-Kutta formula,

$$y_{n+1} \cong y_n + hf(x_n, y_n), \tag{5.6.1}$$

where h, as usual, is the step size in x. This is a first-order formula, but other first-order formulas are possible; for example, one might use $f(x_n+h/2, y_n)$ or even $f(x_n+h/2, (3y_n-y_{n-1})/2)$ in the formula. The integration begins from $x_0=a$, $y_0=b$; h can be either positive or negative and can change in size from one step to the next.

An HP-35 algorithm for the point-slope formula would be

SETUP: a [↑] [↑] [↑] b [STO]

LOOP: [$\cdots f \cdots$] h [×] [RCL] [+] (see y_{n+1}) [STO] [$x \gtrless y$] h [+]
(see x_{n+1}) [↑] [↑] [R↓] [R↓] [R↓] :| . (5.6.2)

In this algorithm [$\cdots f \cdots$] is any sequence of keystrokes that accepts y in X, x in Y, Z, and T, produces $f(x,y)$ in X and retains x in Y (Z and T are irrelevant). Note that auto-enter will be disabled on the first call to subroutine [$\cdots f \cdots$], but not on subsequent calls. The requirement to key h

*Distinguish between the order of the differential equation itself, which refers to the order of the highest derivative that appears, and the order of the method used to numerically integrate the equation, which refers to the order of the polynomial used in the approximation, or equivalently, to the number of different evaluations of $f(x,y)$ appearing in the formula.

twice in each loop is very undesirable, but the disadvantage can be lessened by choosing h to be a number such as 0.1 or 0.01 that is easy to key. On some calculators one could $\boxed{\text{STO}}$ and $\boxed{\text{RCL}}$ h.

There are several possible formulas for second-order Runge-Kutta for this problem; one is (HMF 25.5.6),

$$y_{n+1} \cong y_n + \frac{k_1 + k_2}{2}, \tag{5.6.3}$$

where

$$k_1 = hf(x_n, y_n),$$

$$k_2 = hf(x_n + h, y_n + k_1). \tag{5.6.4}$$

One can easily see why this formula is a good approximation; k_1 is $hy'(x_n)$, and k_2 is an estimate of $hy'(x_{n+1})$ based on the previous point-slope formula. Then the average of k_1 and k_2, which is used to step from y_n to y_{n+1}, should be a good approximation for the average $hy'(x)$ over the interval x_n to x_{n+1}. This form of second-order Runge-Kutta is identical with the simplest possible predictor-corrector formula and is also known as the Euler-Cauchy method or as Euler's second improved method.

Another second-order Runge-Kutta formula, also known as Euler's first improved method, is (HMF 25.5.7)

$$y_{n+1} \cong y_n + k_2, \tag{5.6.5}$$

where

$$k_1 = hf(x_n, y_n),$$

$$k_2 = hf\left(x_n + \frac{h}{2}, y_n + \frac{k_1}{2}\right). \tag{5.6.6}$$

This formula is clearly a good approximation: since k_2 is an estimate of $hy'(x_{n+1/2})$, (at the midpoint of the interval), it should also be a good approximation for the average $hy'(x)$ over the interval x_n to x_{n+1}. Higher order Runge-Kutta formulas are less obvious.

An HP-35 algorithm for formulas 5.6.5 and 5.6.6 would be as follows:

SETUP: a $\boxed{\uparrow}$ $\boxed{\uparrow}$ $\boxed{\uparrow}$ b $\boxed{\text{STO}}$

LOOP: $\boxed{\cdots f \cdots}$ $(h/2)$ $\boxed{\times}$ $\boxed{\uparrow}$ $\boxed{\uparrow}$ $\boxed{\text{RCL}}$ $\boxed{+}$ $\boxed{\text{STO}}$ $\boxed{+}$ $\boxed{x \gtrless y}$ h $\boxed{+}$
(see x_{n+1}) $\boxed{\uparrow}$ $\boxed{\uparrow}$ $\boxed{\text{R}\downarrow}$ $\boxed{\text{R}\downarrow}$ $\boxed{\text{R}\downarrow}$ $\boxed{\cdots f \cdots}$ $(h/2)$ $\boxed{\times}$ $\boxed{\text{RCL}}$ $\boxed{+}$
(see y_{n+1}) $\boxed{\text{STO}}$:| . (5.6.7)

In this algorithm both h and $h/2$ need to be keyed in each loop, thus the comments under algorithm 5.6.2 apply. Subroutine $\boxed{\cdots f\cdots}$ is the same as before, but note that auto-enter will be sometimes enabled, sometimes disabled on the call to $\boxed{\cdots f\cdots}$.

Section A.4.1 contains other algorithms for second-, third-, and fourth-order Runge-Kutta applied to this problem. To test and compare these algorithms, we need a differential equation whose solution is known analytically, but it must not be so simple that any of the methods become exact. One possibility is

$$f(x,y)=2xy, \quad y(0)=1. \tag{5.6.8}$$

With $h=0.1$ except as otherwise noted, the results of numerically integrating this differential equation from $x=0$ to 1 or more are shown in Figure 5.6.1. The "hole" in two of the curves is a common occurrence and corresponds to a change of sign of ξ. Note that second-order Runge-Kutta (with $h=0.1$) is much better than the point-slope method even with $h=0.05$, and requires the same number of evaluations of $f(x,y)$. Other generalizations, such as choosing between formulas of the same order, are not possible from this figure because this comparison depends on the particular differential equation.

Since the error term in each method is of order h^{n+1}, where n is the order of the method, choosing a smaller h will improve the higher-order methods by a larger amount and increase the differences between methods of different order.

Consider next a second-order ordinary differential equation (initial value problem) of the form

$$y''=f(x,y), \; y(a)=b, \; y'(a)=c. \tag{5.6.9}$$

A possible third-order Runge-Kutta formula for this problem is (HMF 25.5.22)

$$\begin{aligned} y_{n+1} &\cong y_n + h\left(y'_n + \frac{k_1+2k_2}{6}\right), \\ y'_{n+1} &\cong y'_n + \frac{k_1+4k_2+k_3}{6}, \end{aligned} \tag{5.6.10}$$

where

$$\begin{aligned} k_1 &= hf(x_n, y_n), \\ k_2 &= hf\left(x_n + \frac{h}{2}, \; y_n + \frac{h(y'_n + k_1/4)}{2}\right), \\ k_3 &= hf\left(x_n + h, \; y_n + h\left(y'_n + \frac{k_2}{2}\right)\right). \end{aligned} \tag{5.6.11}$$

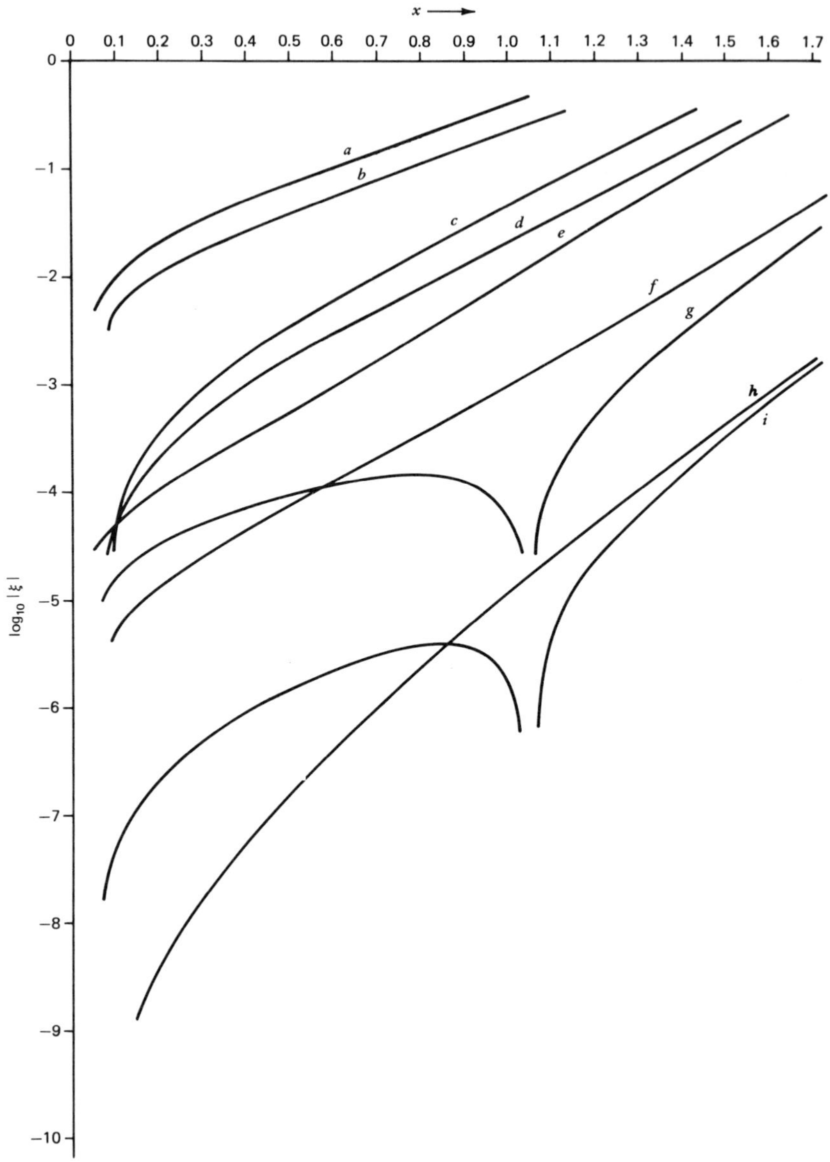

Figure 5.6.1 Error resulting from applying a variety of numerical integration formulas (Section A.4.1) to the test case $y' = 2xy$ through $y(0) = 1$. The ordinate is $\log_{10}$ of the absolute error and is the negative of the number of correct decimal digits to the right of the decimal point. The abscissa is x. Then (a) shows the results of using the point-slope formula with $h = 0.1$; and (b) with $h = 0.05$. For all the following cases $h = 0.1$. Then (c) shows the first-order bastard formula; (d) second-order Runge-Kutta from HMF 25.5.7; (e) second-order Runge-Kutta from HMF 25.5.6; (f) third-order Runge-Kutta from HMF 25.5.9; (g) third-order Runge-Kutta from HMF 25.5.8; (h) fourth-order Runge-Kutta from HMF 25.5.10 or Gill's modified fourth-order Runge-Kutta from HMF 25.5.12; and (i) fourth-order Runge-Kutta from HMF 25.5.11.

An algorithm for this formula for the HP-45 is given in Section A.4.2. This is a third-order formula and the three ks are calculated at the starting point (k_1), half-way along to the next point (k_2), and approximately at the next point (k_3). The weighted average of the three ks is an estimate of the average hy'' over this interval. Note two interesting characteristics of this formula: the y argument of f in k_3 is very nearly y_{n+1}, and we could use y_{n+1} to calculate k_3 because y_{n+1} does not depend on k_3. If this substitution is made, k_3 for one point is identical with k_1 for the following point—a considerable simplification. Section A.4.2 provides an algorithm for this simplified formula.

As a test case, we again need a differential equation whose solution is known analytically. One possibility is

$$f(x,y)=(2+4x^2)y,\ y(0)=1, \qquad y'(0)=0. \tag{5.6.12}$$

With $h=0.1$, we can numerically integrate this differential equation with these two algorithms, to obtain the results in Figure 5.6.2. The simplified formula is not as good (on this test case), but the difference is not very dramatic.

Suppose, instead, that we need to solve a second-order ordinary differential equation of the form

$$y''=f(x,y,y'), \qquad y(a)=b,\ y'(a)=c. \tag{5.6.13}$$

This is a more difficult problem because of the appearance of y' in f. A possible fourth-order Runge-Kutta formula for this problem is given in HMF 25.5.20, and a corresponding RPN algorithm can be found in Section A.4.2. The results of applying this algorithm to the test case

$$y''=2(y+xy'), \qquad y(0)=1,\ y'(0)=0 \tag{5.6.14}$$

are given in Figure 5.6.3.

Consider finally a set of two first-order ordinary differential equations

$$\begin{aligned} y'&=f(x,y,z),\\ z'&=g(x,y,z), \end{aligned} \tag{5.6.15}$$

through

$$y(a)=b, \qquad z(a)=c. \tag{5.6.16}$$

A second-order Runge-Kutta formula for this case is contained in HMF 25.5.17, and Section A.4.3 presents a corresponding RPN algorithm as well

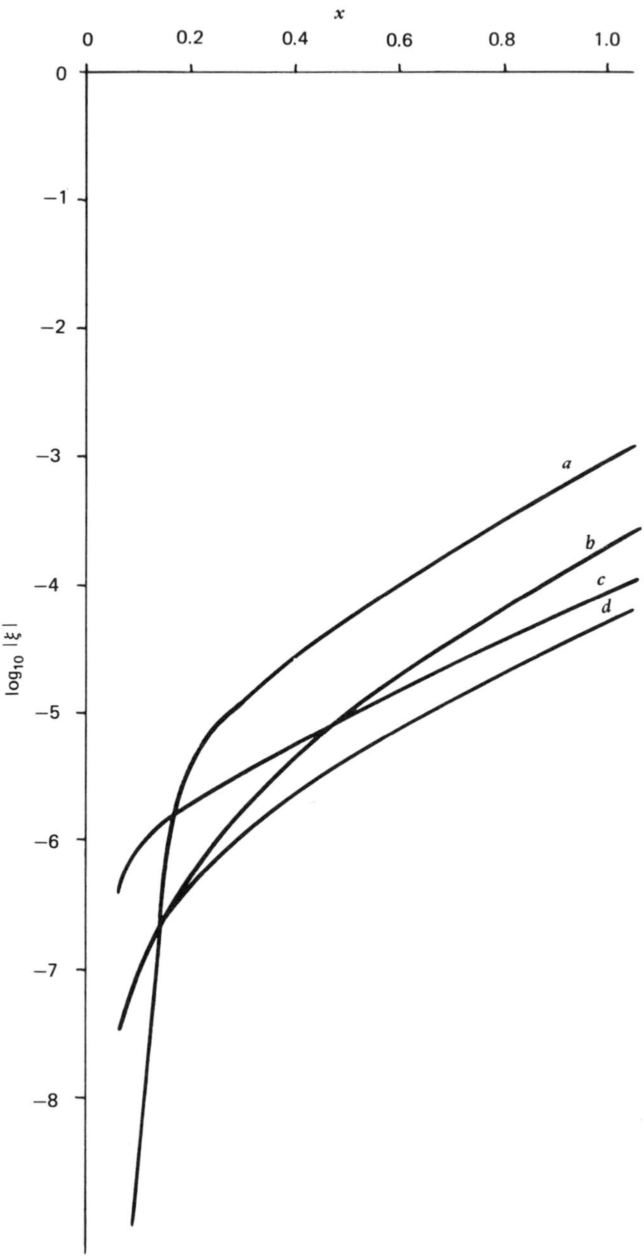

Figure 5.6.2 Results of applying two numerical integration formulas (Section A.4.2) to the test case $y'' = (2+4x^2)y$, through $y(0)=1$ and $y'(0)=0$ with $h=0.1$. (*a*) y', (*b*) y from the simplified third-order Runge-Kutta, (*c*) y', (*d*) y from the unmodified third-order Runge-Kutta.

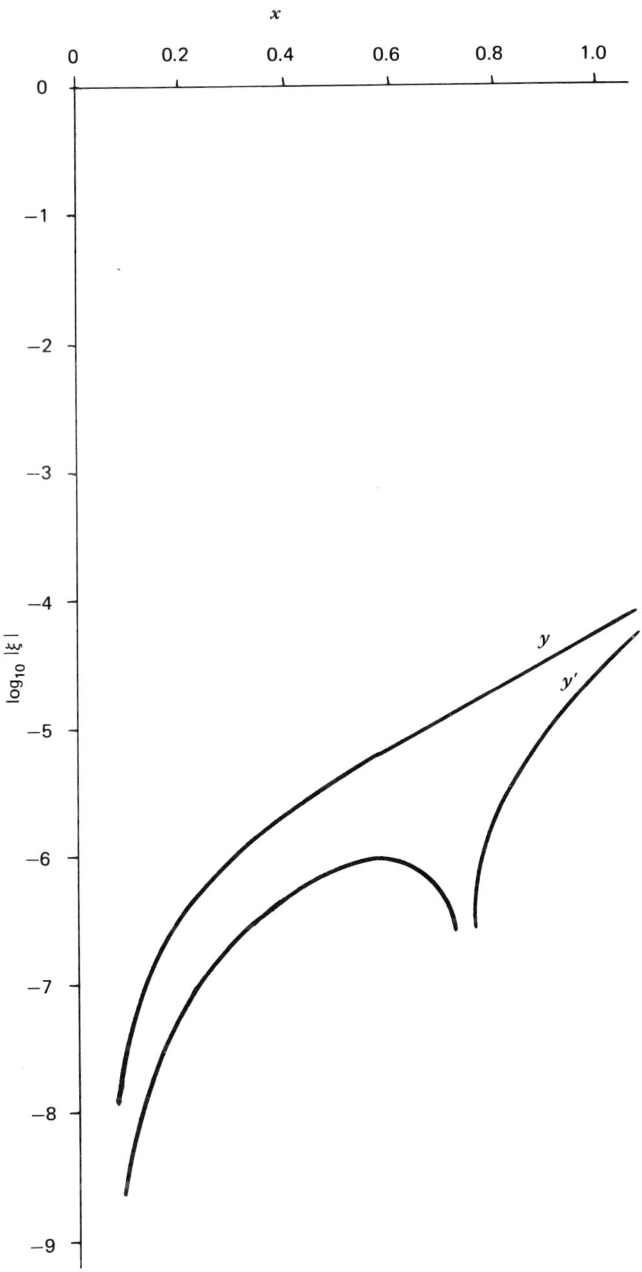

Figure 5.6.3 Results of applying fourth-order Runge-Kutta from HMF 25.5.20 (Section A.4.2) to the test case $y'' = 2(y + xy')$ through $y(0) = 1$ and $y'(0) = 0$ with $h = 0.1$.

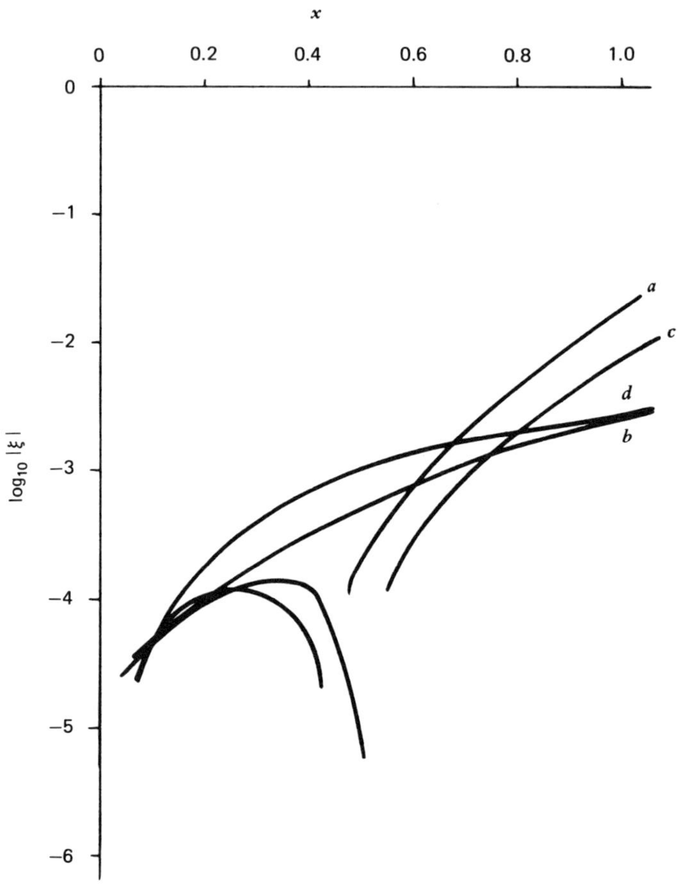

Figure 5.6.4 Results of applying two numerical integration formulas (Section A.4.3) to the test case $y'=2x/z$, $z'=-2x/y$ through $y(0)=1$ and $z(0)=1$ with $h=0.1$. (a) and (b) are y and z, respectively, from second-order Runge-Kutta. (c) and (d) are y and z, respectively, from the modified second-order Runge-Kutta.

as an algorithm for a slightly modified version of this formula. Figure 5.6.4 shows the results of applying these two algorithms to the test case

$$y' = \frac{2x}{z}, \qquad z' = \frac{-2x}{y}, \tag{5.6.17}$$

through

$$y(0) = 1, \qquad z(0) = 1. \tag{5.6.18}$$

As a special case of equations 5.6.15, we could have

$$y' = f(x,y,z) = z, \tag{5.6.19}$$

thus

$$z' = y'' = g(x,y,y'). \tag{5.6.20}$$

Therefore the previous second-order problems can be dealt with as special cases of the present problem, equations 5.6.15 and 5.6.16. But this is not usually a simplification.

The Runge-Kutta formulas for numerically integrating differential equations are discussed in more detail in Hildebrand (1956), Sections 6.15 and 6.16, and a collection of formulas appears in CRC HTM, p. 619, and, of course, in HMF 25.5.

5.7 EXERCISES

5.7.1 $\int_0^{\pi} \sin^2 x\, dx.$

5.7.2 $F(x) = \int_0^{\infty} t^{x-1} e^{-t}\, dt$ for test case $x=4$.

5.7.3 Given the following table of data from an experiment, estimate $\int_1^{1.7} y(x)\, dx$ using the eight-point Newton-Cotes formula, estimate $y(1.32)$ using three-point and four-point Lagrange interpolation, and estimate $y'(1.32)$ using three-point Lagrange differentiation.

x	y
1.0	0.
1.1	0.0007
1.2	0.0064
1.3	0.0228
1.4	0.0562
1.5	0.1133
1.6	0.2009
1.7	0.3261

5.7.4 $\displaystyle\int_{-1}^{1} \frac{x^2 \cos^{-1}(x)}{\sqrt{1-x^2}}\, dx.$

5.7.5 $F(x) = \dfrac{1}{\pi} \displaystyle\int_0^{\pi} \cos(x \sin\theta)\, d\theta$ for test case $x=2$.

5.7.6 $F(x)=\frac{2}{\sqrt{\pi}}\int_0^x e^{-t^2}\,dt$ for test case $x=1$.

5.7.7 $F(x)=e^{-x^2}\int_0^x e^{t^2}\,dt$ for test case $x=1$.

5.7.8 $F(x)=\int_0^x \sin(\pi t^2/2)\,dt$ for test case $x=1$.

5.7.9 Integrate the differential equation $y''=-xy$ from $y(0)=0.3550280539$, $y'(0)=0.2588194038$, in the positive x direction up to about the first zero. *Hint:* The first zero is around $x=2.3$. Try $h=0.1$.

5.7.10 Integrate $y'=\sqrt{\pi}-2xy-1/x$ from $y(1)=0.6051336526$ to $x=2$.

5.7.11 Calculate

$$y(x)=\int_0^\infty \frac{e^{-t^2}}{x+t}\,dt,$$

for $x=2$ and compare with problem 5.7.10.

5.7.12 Consider a simple pendulum (Figure 5.7.1). A mass m is connected

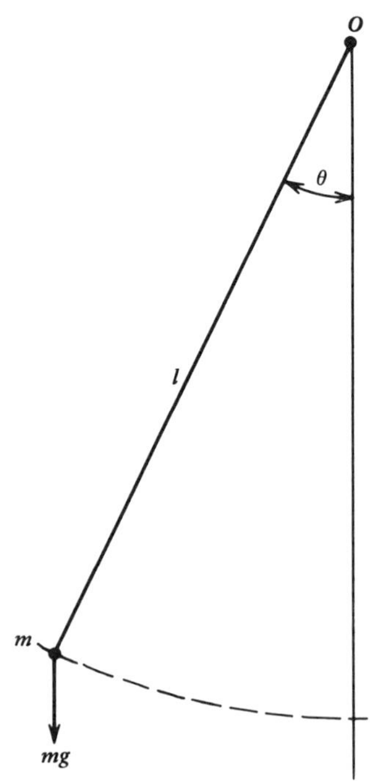

Figure 5.7.1 A simple pendulum.

by a massless rod of length l to a pivot O. The force of gravity mg acts directly downward and causes the mass to swing back and forth. The equation of motion can be written as follows:

$$\theta'' = -\frac{g}{l}\sin\theta,$$

where $\theta = \theta(t)$ and t is the time. We multiply this equation by $2\theta'$ and integrate to give

$$(\theta')^2 = \frac{2g}{l}(\cos\theta - \cos\theta_0),$$

where θ_0 is a constant and is the maximum value attained by θ. The well-known first approximation for the period P comes from taking $\sin\theta \cong \tan\theta \cong \theta$ and is $P \cong 2\pi\sqrt{l/g}$. Beginning from $\theta(0) = 0$, integrate the θ' equation until θ reaches θ_0, that is, for a quarter period, and compare the period so derived with the approximation above. Take $\theta_0 = 44°$ and $g/l = \frac{1}{2}$.

5.7.13 One can show that

$$-\int_0^\theta \ln\left[2\sin\left(\frac{t}{2}\right)\right]dt = \sum_{k=1}^{\infty}\frac{\sin k\theta}{k^2}.$$

Write an algorithm to sum the first six terms of the series, approximate the integral numerically, and compare the two answers for $\theta = 50°$.

5.7.14 Chebyshev's formula for the approximate number of primes less than or equal to n is

$$P(n) = \int_2^n \frac{dx}{\ln x}.$$

Evaluate this integral for $n = 1000$ and compare with the known number of primes less than 1000.

6
Suggestions for Future Developments

Every thesis should have a chapter of suggestions for future developments.

—*Prof. Alan H. Barrett (1927–   )*

Conservatism discards Prescription, shrinks from Principle, disavows Progress; having rejected all respect for antiquity, it offers no redress for the present, and makes no preparation for the future.

—*Benjamin Disraeli (1804–1881)*

Old men and comets have been reverenced for the same reason; their long beards, and pretences to foretell events.

—*Jonathan Swift (1667–1745)*

This chapter contains my ideas and recommendations (not predictions) for the design of future RPN calculators. I am more concerned with the design of inexpensive and portable models than with elaborate programmable calculators. The challenge is to do a lot with a little; to do a lot with a lot is easier.

6.1 GENERAL SUGGESTIONS

To begin, I have a few suggestions for almost all future RPN calculators. The first concerns the [CHS] key. In Section 1.3.1 I pointed out that this key can be confusing because it performs two functions. But the confusion can be reduced and both functions retained by a simple change—[CHS] *should enable auto-enter* (but *not* terminate a number). Then the sequence

A [↑] [CHS] B

would leave three numbers on the stack: B, $-A$, and A; [CHS] within numbers would work just as it does now.

Second, I suggest that all RPN calculators would benefit from a [R↑] key. Only a few RPN calculators now have such a key. The [R↑] operation is, of course, equivalent to [R↓] [R↓] [R↓]; the usefulness of [R↑] is indicated by the frequency with which these three keystrokes occur in algorithms.

Third, I suggest that all RPN calculators have *recall arithmetic* as well as storage register arithmetic. Only the HP-45 and HP-46 now have this feature. In Section 1.3.1 I pointed out how useful this feature is, and Appendix A contains many examples. Sometimes an HP-45 algorithm is difficult to convert to work on a nominally more powerful calculator such as the HP-27, because the HP-27 lacks recall arithmetic.

With programmable RPN calculators that merge keystrokes (e.g., the HP-25, HP-67, and HP-97), recall arithmetic would have an additional advantage. Sequences such as [RCL] [×] 4 would be merged into a single program step, but [RCL] 4 [×] takes two. To see how important this saving might be, I checked 17 programs from the *HP-65 Standard Pac*—all the programs therein except the two user diagnostics. These 17 programs total 1465 program steps. For each occurrence of storage register arithmetic, two program steps would be saved by merging; and for each place at which recall arithmetic could be used, one or sometimes two program steps would be saved. For these 17 programs, I counted 48 program steps that would be saved by merging storage register arithmetic, and 63 to 66 program steps that would be saved by going to merged recall arithmetic.

An ideal calculator would give round$(A+B,n)$ as the answer for A [↑] B [+], where round(x,n) is the round-off function, that is, x rounded to n digits, and n is the number of digits of precision in the calculator (see Knuth, 1969, Section 4.2.2). Most RPN calculators do *not* do this. Instead, they calculate the sum to a larger number of digits (usually 13 if n is 10), truncating any other digits in the smaller number, and finally round the answer to n digits. This procedure is nearly, but not exactly, the same (see exercise 6.6.10). I recommend the precise round$(A+B,n)$ because it leads to fewer difficulties, as pointed out by Knuth (*op. cit.*).

My fifth suggestion concerns a special symbol ȼ for "clear," distinct from zero. When the calculator is turned on, or when [CLR] or [CLX] is used, the appropriate registers should be set to ȼ rather than 0. The display should distinguish among three cases—true zero, shown as a single, full-sized zero; ȼ, shown as a half-sized zero; and "rounded off to zero," shown with a decimal point and as many (full-sized) zeros as are needed to correspond to the selected display format. The distinction between true zero and rounded off to zero is made in this way, for example, by most FORTRAN systems.

The clear symbol ȼ behaves like a zero if it occurs in arithmetic calculations. Functors accept ȼ as a parameter and treat ȼ as zero. A functor can produce zero but not ȼ for an answer. However ȼ differs from zero on the display and in *not entering Y from Z on a normal pop*. Thus if Z contains ȼ, any dyadic functor leaves Y (and also Z and T) unchanged. This is an extension of the idea that T should not change on pop, a remarkably useful feature, as noted in Section 2.4. Whenever ȼ is in X, and a new number is to be put into X, the new number *overwrites ȼ rather than pushing*; in effect, whenever ȼ is in X, auto-enter is disabled. Thus [CLX], which writes ȼ into X, automatically disables auto-enter. Using the ȼ system increases the value of the otherwise rather useless [CLX] and [CLR] keys.

The idea of the ȼ system came from noting that in computer systems that use RPN, the active register (corresponding to X) is logically separate from the stack or stacks, and the last-in–first-out register in a stack (corresponding to Y) does not clear on pop. In computer systems this enables a stack to be used in place of a [STO] register, because an indefinite number of [RCL]s can follow a single [STO].

With the ȼ system and a [R↑] key, [R↑] can be used as a number separator instead of [↑]. On pressing [R↑], a ȼ appears in X, *provided the stack is not full*. By watching for the ȼ after each [R↑], one can ensure that no numbers are lost off the top of the stack. The ȼ in X after [R↑] is overwritten by the next number key, which means that after the number key is pressed, the effect is the same as if [↑] had been used.

So should [↑] then enable auto-enter and serve only as a "duplicate" key? I think not. The advantages are slight, and this system would lead to errors by persons familiar with more conventional RPN systems. And using [R↑] as a number separator necessitates [CLR] between problems. The option of using [↑] as a number separator should be retained. In certain cases one *wants* to lose numbers from the top of the stack.

The ȼ system has some of the advantages of the HP-35, which allowed [CHS] to precede a number (see Section 1.3.1), without the disadvantages.

Since ¢ cannot be negated, if CHS is pressed with ¢ in X, a new number is begun with a minus sign, overwriting ¢ into X. The display should show just the minus sign until a number key is pressed.

More than four registers in the stack would be desirable for many problems. To "fill up" the stack to save a number with the ¢ system requires only one ↑ regardless of the length of the stack, provided it was initially cleared. The difficulty occurs with R↓ and R↑ or with whatever

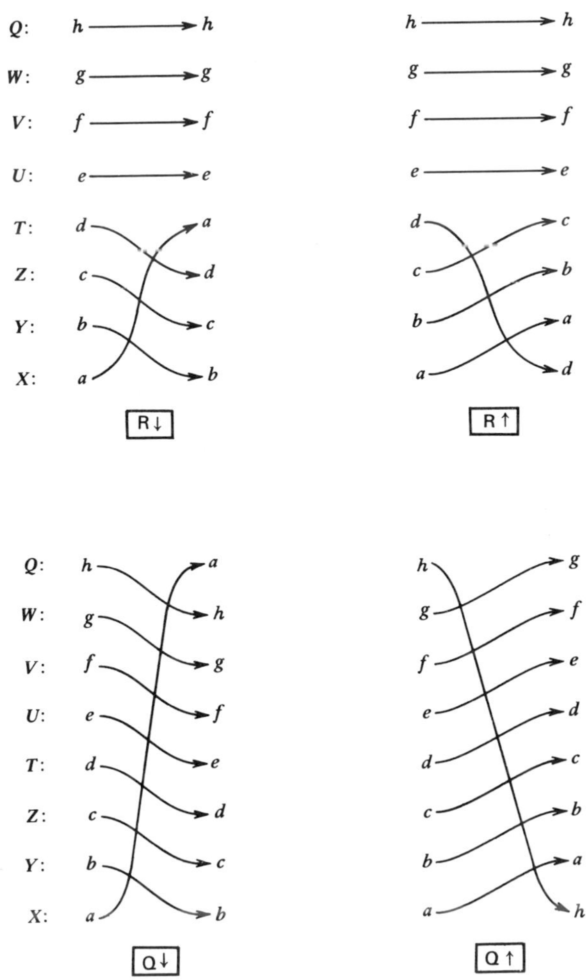

Figure 6.1.1 A possible implementation of an eight-register stack. $x \gtrless y$ would function as usual.

new keys are invented to replace [R↓] and [R↑]. With upward compatibility in mind, I suggest that [R↓] and [R↑] be retained with the same function as on traditional RPN calculators—[R↓] and [R↑] should affect the bottom four registers only. A different key or keys would rotate the whole stack. One such scheme is represented in Figure 6.1.1. This scheme has rotates containing two ([$x \gtrless y$]), four, and eight registers.

With the ȼ system, each [STO] register can be a stack also. Two [STO]s with different numbers in X can be followed by two [RCL]s to recover the two numbers in reverse order. However an indefinite number of [RCL]s could follow a single [STO] as usual.

For new programmable RPN calculators, I recommend [GSR] ("go to subroutine") and [RTN] keys for subroutines. There should be at least one internal subroutine return-address register to be set by [GSR], so that the [RTN] at the end of a subroutine would return control to the instruction following the [GSR]. The HP-65 and HP-67 have similar capabilities.

Finally, I recommend that a variable-length field be assigned to the exponent—both internally and on the display. The characteristics of one example of such a system are given in Table 6.1.1. The table is based on a $12\frac{1}{2}$ digit display (12 digits plus a preceding minus sign) and an internal representation of 12 digits plus a bit for the sign of the number, a bit for the sign of the exponent, and two bits to define the start of the exponent field.

Table 6.1.1 Example of Variable-Length Field

	Number of Figures	
Range	Internally	Displayed
10^{-999} to 10^{-99}	9	8
10^{-99} to 10^{-9}	10	9
10^{-9} to 10^{-2}	11	10
10^{-2} to 10^{-1}	11	11
10^{-1} to 1	11*	(12)†
1 to 10	12	12
10 to 10^{10}	11	(12)
10^{10} to 10^{13}	10	(12)
10^{13} to 10^{100}	10	9
10^{100} to 10^{1000}	9	8

*Or 12, using the convention that −0 for the exponent is interpreted as −1 instead.
†The parentheses indicate that the display could show more figures than are available internally.

6.2 0, U, AND ∞

"By every number canst thou divide, but by 0 shalt thou not divide!" said the Lord when he placed Adam in the Garden of Eden.*

—*Rózsa Péter (1905–)*

6.2.1 ∞ Is a Number

In ordinary arithmetic, zero is a number with special properties and must be treated as a special case in a calculator. For example, we cannot divide by zero or find the logarithm of zero. Calculator designers are forced to deal with zero, however, and if one tries to take the reciprocal of zero, the punishment is a flashing display or "Error."

Now infinity (∞) is also a number with special properties and should be treated as a special case in a calculator. With a representation for ∞ both internally and on the display (perhaps two half-size zeros), some of the difficulties with zero would be eliminated. We would then have, for example, a representation for tan 90° and $1/0$. An overflow from 10^{99} to ∞ is no worse (and no better) than an underflow from 10^{-99} to zero. Either condition results in inaccuracies whose severity depends on the particular problem. For no very good reason, most calculator designs condone underflow but not overflow.

A calculator with a representation for ∞ would still have forbidden operations: those that are indeterminate (e.g., $0/0$ or $0\times\infty$) and those that should be complex numbers (e.g., $\log(-2)$ and $\cos^{-1}(4)$); but such a calculator would be, in my view, a considerable improvement over current models.

6.2.2 And So Is U

In addition to the new symbols (numbers) $\not{0}$ and ∞, I recommend a third new symbol U, for "undefined," to represent the answer to an indeterminate operation. A U on the display as the answer to a calculation is essentially equivalent to flashing zero or "Error" on current models. But U is a number; thus it can be manipulated on the stack, stored and recalled, and *used as an argument in functors*. Most but not all functors operating on U give U for the answer. If one thinks of U as an arbitrary, possibly complex, *finite* number, then $0\times U=0$, $U/\infty=0$, $U/0=\infty$, $U\times\infty=\infty$, $U^0=1$, and so on. However $U/U=U$ (not 1) because the two Us can

*From Péter (1976), p. 138.

represent two different undefined numbers. The rules for manipulating 0, ∞, and U are given in Table 6.2.1.

With the U system, an undefined operation can occur in the middle of a chain calculation and yet, at the end, one has either a U still in the display to show the error, or maybe the right answer, in case the U did not matter.

I recommend that 0, U, and ∞ all be *signed numbers*. I agonized long over this recommendation. The question is not whether to carry and display the sign—that obviously is desirable—but what use to make of the sign. The criterion is to *minimize the number of special cases for which one gets the wrong answer*; but such cases cannot be eliminated altogether even by going to full complex arithmetic. If we carry along the signs according to the usual rules of arithmetic, we get *fewer* wrong answers. The sign convention for $\div$ and $\times$ is the same as in ordinary arithmetic. For $\pm$, the sign convention is indicated in Table 6.2.1; essentially we take the sign of

Table 6.2.1 Rules for Arithmetic with 0, U, **and** ∞

$0 \div 0 = U$	$0 \times 0 = 0$	$0 \pm 0 = 0$	$\ln(0) = -\infty$	
$0 \div \infty = 0$	$0 \times \infty = U$	$0 \pm \infty = \pm\infty$	$\ln(\infty) = \infty$	
$0 \div U = 0$	$0 \times U = 0$	$0 \pm U = \pm U$	$\ln(U) = U$	
$\infty \div 0 = \infty$	$\infty \times 0 = U$	$\infty \pm 0 = \infty$	$\ln(A) = \begin{cases} \text{as usual} \\ U \end{cases}$	for $A > 0$ for $A < 0$
$\infty \div \infty = U$	$\infty \times \infty = \infty$	$\infty \pm \infty = \infty$	$e^0 = 1$	
$\infty \div U = \infty$	$\infty \times U = \infty$	$\infty \pm U = \infty$	$e^{+\infty} = \infty$	
$\infty \div A = \infty$	$\infty \times A = \infty$	$\infty \pm A = \infty$	$e^{-\infty} = 0$	
$U \div 0 = \infty$	$U \times 0 = 0$	$U \pm 0 = U$	$e^U = U$	
$U \div \infty = 0$	$U \times \infty = \infty$	$U \pm \infty = \pm\infty$	$\sin(\infty) = U$	
$U \div U = U$	$U \times U = U$	$U \pm U = U$	$\cos(\infty) = U$	
$U \div A = U$	$U \times A = U$	$U \pm A = U$	$\tan(\infty) = U$	
$A \div 0 = \infty$	$A \times 0 = 0$	$A \pm 0 = A$	$\sin(U) = U$	
$A \div \infty = 0$	$A \times \infty = \infty$	$A \pm \infty = \pm\infty$	$\cos(U) = U$	
$A \div U = U$	$A \times U = U$	$A \pm U = U$	$\tan(U) = U$	
			$\tan(90°) = \infty$	
			$\tan(-90°) = -\infty$	
			$\sin^{-1}(\infty) = U$	
			$\cos^{-1}(\infty) = U$	
			$\tan^{-1}(+\infty) = 90°$	
			$\tan^{-1}(-\infty) = -90°$	
			$\sin^{-1}(U) = U$	
			$\cos^{-1}(U) = U$	
			$\tan^{-1}(U) = U$	
			$\sin^{-1}(A) = U$	if $\lvert A\rvert > 1$
			$\cos^{-1}(A) = U$	if $\lvert A\rvert > 1$

the largest quantity whenever it is known. Sometimes the sign is arbitrary (e.g., $0-0$).

All the rules for powers and roots (y^x and $\sqrt{x}$) follow from the rules for logarithms and exponentials in Table 6.2.1, except that integer exponents (0 and ∞ are even integers, but U is not) should be handled as special cases, permitting one to do integer powers of negative numbers.

Most calculators use binary coded decimal (BCD) arithmetic internally, and each digit is represented by four bits (16 states). The states corresponding to 10 through 15 usually see little service; these unused states can be pressed into action to represent ȼ, U, and ∞.

The numbers ∞ and U were suggested for computers by K. Zuse (as reported in Knuth, 1969, Vol. II, p. 192), and actually used in the IBM 7030 (stretch) and CDC 6600 computers.

6.3 EL CHEAPO, MODEL A

This section contains rough design specifications for an RPN calculator affectionately known as El Cheapo, model A. It is to be relatively inexpensive, an alternative to the "four-function" AES calculators that are now so popular.

One would hope that such a calculator would cost no more than a textbook, so that schools, for example, could afford to buy one for each student. However this leads to some conflicting design goals. For a student, a calculator should be (*a*) as simple as possible, but (*b*) upward compatible, so that he need not relearn any of the simple operations on a more powerful calculator. Thus one might just cut off the top half of the keyboard of a scientific RPN calculator—that is, remove all the keys such as [SIN], [COS], [ln], and [y^x], but retain all the usual arithmetic operations including the four-register stack. Such a topless calculator, (Figure 6.3.1) would still be impressively powerful and would, for example, run rings around a "four-function" AES calculator, but would probably cost more, too.

For the model A, the two design goals above were relaxed to try to make a calculator still less expensive but about equally powerful. The cost of a calculator in this price range is a strong function of the number of mechanical parts (e.g., keys and digits in the display) but not so strong a function of the complexity of the integrated circuit chips. For a calculator that sells in large quantities, the engineering costs (divided by the number of units sold) and the incremental cost of a complex chip over a simpler chip are usually small compared to the costs of the keyboard, display, battery, and case. This suggests that the designer should extend the idea of

Figure 6.3.1. A 20-key topless calculator made by removing all the keys such as [SIN], [COS], [ln] and [y^x] from a scientific RPN calculator.

keys doing multiple duty, as already done in the [↑] and [CHS] keys, and in calculators with shift keys.

Now consider the display on the El Cheapo, model A (Figure 6.3.2). There are eight digits plus a preceding half-digit that can display a minus sign and a decimal point. For numbers too large or too small (more than three leading zeros) for the decimal point to be on the display, the model A automatically switches to scientific notation, but with room for only five significant figures. The HP-35 has a similar display system, but it is inefficient in its use of display digits. The HP-35 actually has a total of 15 digits (two of which display only minus signs) in its display, but only 10 significant figures. The model A always keeps (at least) eight significant figures internally, regardless of the decimal point.

The two unusual keys on the model A, [↔] (pronounced "interchange"

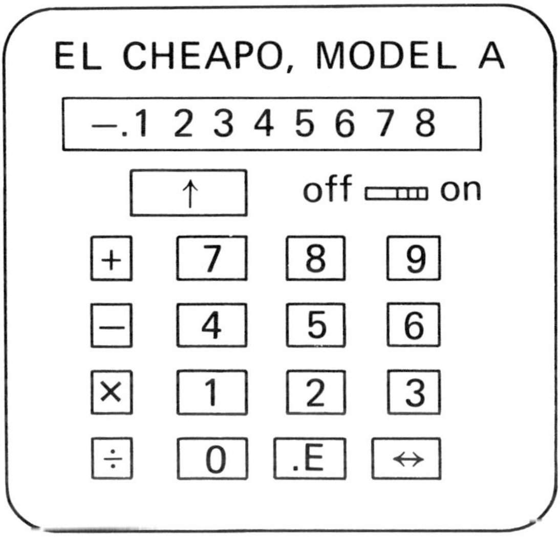

Figure 6.3.2. El Cheapo, model A, has 17 keys and RPN with a stack of four registers. The display shows either eight significant figures in floating decimal format or five significant figures in scientific notation. The [.E] and [↔] keys are explained in the text.

to distinguish it from [$x \gtrless y$], "exchange") and [.E] (pronounced "dotty") take the place (more or less) of [$x \gtrless y$], [R↓], [·], [EEX], and with [−], of [CHS] also. But of course a price must be paid for this reduction in keys—some operations require both more keystrokes and more thought with the model A than with a conventional RPN calculator.

Figure 6.3.3 illustrates the effect on the four-register stack of repeatedly pressing [↔]. By pressing [↔] the correct number of times, the effect of many (but not all) combinations of [$x \gtrless y$] and [R↓] can be duplicated. Missing, for example, is the effect of [$x \gtrless y$] [R↓]. One need not memorize this whole table. One press of [↔] is equivalent to [$x \gtrless y$], two presses of [↔] are equivalent to [R↓] (from the *original* arrangement), and then the process starts over from this new arrangement. Each pair of pushes of [↔] corresponds to a [R↓]; thus after eight [↔]s (equal to four [R↓]s), we get back the original arrangement as shown in Figure 6.3.3. If any other keystroke occurs between [↔]s, the pattern restarts at the beginning.

The [.E] key functions as follows. The first time [.E] is pressed it is equivalent to a [·] (decimal point) and thus can begin a number or appear within a number. The second time [.E] is pressed *within the same number*, the effect is the same as [EEX], which means that the following digits are taken as the exponent. If more than five number keys were pressed before this second pressing of [.E], some of the digits can no longer be seen on the display, but they are kept internally. Therefore one can key, but not

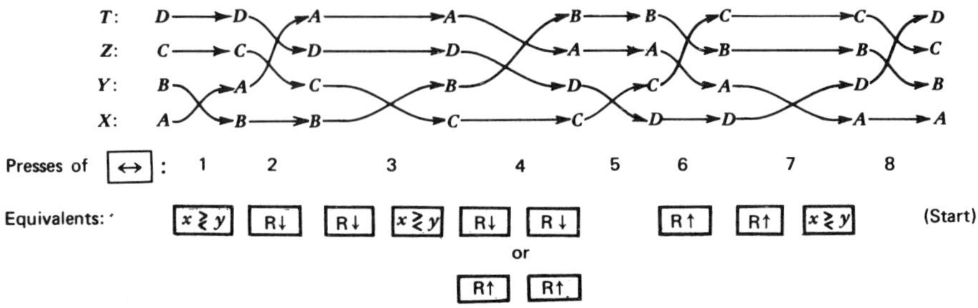

Figure 6.3.3 The effect of repeatedly pressing [↔] on the model A.

see, an eight-digit number with a two-digit exponent in scientific notation.

The [−] key functions with [.E] as follows: if [−] does *not* immediately follow [.E], the [−] means subtract as usual. Thus [−] without [.E] immediately preceding is a number terminator and a dyadic functor. But if [−] immediately follows [.E], the [−] is taken as [CHS] instead, and either the number or the exponent can be negated. One can use [−] in this way because neither [·] nor [EEX] in the conventional RPN system is ever needed at the end of a number.

Finally if [.E] is pressed a third time *within the same number*, [.E] functions as [CLX]; the display shows ⊄. Another way to correct a number-keying error is [↔] [↔] (= [R↓]).

This explanation of the model A system is specifically intended for persons who have read the preceding chapters of this book; a very different explanation would be needed in the instruction booklet supplied with the calculator.

The following tabulation supplies examples of keying numbers in the model A system.

Number	Key
3.	[3]
3.3	[3] [.E] [3]
−3.	[3] [.E] [−]
−3.3	[3] [.E] [−] [3]
1000. or 10^3	[1] [0] [0] [0] or [.E] [.E] 3
−1000. or -10^3	[1] [0] [0] [0] [.E] [−] or [.E] [−] [.E] [3]
−0.001 or -1×10^{-3}	[.E] [−] [0] [0] [1] or [.E] [−] [.E] [−] [3]
1.6×10^4	[1] [.E] [6] [.E] [4]
-1.6×10^4	[1] [.E] [−] [6] [.E] [4]
1.6×10^{-4}	[1] [.E] [6] [.E] [−] [4]
-1.6×10^{-4}	[1] [.E] [−] [6] [.E] [−] [4]

This tabulation illustrates some simple arithmetic in the model A system.

Arithmetic	Key
$3+1.6$	[3] [↑] [1] [.E] [6] [+]
$3-1.6$	[3] [↑] [1] [.E] [6] [−]
$-3-1.6$	[3] [.E] [−] [↑] [1] [.E] [6] [−]
$3+(-1.6)$	[3] [↑] [1] [.E] [−] [6] [+]
	(or [3] [↑] [1] [.E] [6] [−])
$3-(-4)$	[3] [↑] [4] [.E] [−] [−] (or [3] [↑] 4 [+])

This system has, of course, some minor annoyances. Some operations require more keystrokes than are needed on a conventional RPN calculator. To negate a number already on the display from a previous calculation requires three keystrokes, [0] [↔] [−] (unless Y already contains 0 or ⊄), and also loses the contents of T. Reciprocal also takes three keystrokes, [1] [↔] [÷], and loses the contents of T. And for a [CLR], turn the calculator off, then on again. But [CLR] is not good for much anyway.

Most serious of all, of course, are missing functors such as [$\sqrt{x}$], [SIN], and [y^x]. These can be done on the model A using power series expansions or iterative techniques, but if such functions are really necessary perhaps one should consider the model B.

6.4 THE MODEL B

The next major step forward for RPN calculators is probably *complex arithmetic*. I believe that the best way to implement complex arithmetic is to have two stacks, one for the real and the other for the imaginary parts of numbers. The [STO] registers should also be double. In a simple implementation, one needs about two extra keys and two extra bits of information in the display, but not two separate displays.

Two important design goals can be stated. (1) A person unfamiliar with complex arithmetic, but familiar with traditional RPN calculators, should be able to turn on the model B, use it just like an ordinary calculator, and get all the right answers; he would have to avoid only pushing any unfamiliar keys. If one tries a combination of keystrokes that yields a complex number (which on an ordinary calculator would give a flashing display or "Error") the "other-non-zero" light (described below) will indicate a complex answer, which would be interpreted as an error. Thus the model B is upward compatible from an ordinary scientific RPN calculator. (2) The object of many of the following rules is to ensure that the real and imaginary parts of a given number are never separated accidentally.

The model B has two registers labeled X_r and X_i (for real and imaginary), two labeled Y_r and Y_i, and so on. The display shows the contents of X_r when the calculator is initially turned on. In this state (r stack active), numbers keyed go into X_r following the usual rules. [$x \gtrless y$], [R↑], and [R↓] operate as usual but on *both stacks* simultaneously. [CLX] writes ₡ into the active X only, and [↑] duplicates X_r into Y_r and X_i into Y_i (push both stacks) and disables auto-enter as usual. The swap key [S] swaps the *display and the active stack* from r to i or from i to r; [S] terminates a number and enables auto-enter (by analogy with [$x \gtrless y$]); but whenever ₡ is in the display, in effect auto-enter is disabled (see Section 6.1).

Whenever one stack is pushed by a number entry (auto-enter), the other stack pushes also with a ₡ written into the inactive X. This rule is essential to avoid separating the real and imaginary parts of a number. Whenever a number is overwritten in the active stack, either because auto-enter was disabled or because the active X contained ₡, the other stack is unaffected.

The only other new key on the model B is [$\gtrless$] (i conjugate), which exchanges the contents of X_r and X_i and enables auto-enter, but has no effect on which stack is active. As the name suggests, [$\gtrless$] is equivalent to conjugation (changing the sign of the imaginary part) followed by multiplication by i.

[CHS] operates as usual on the active X only. Pushing [CHS] when the i stack is active gives conjugation. [STO], [RCL] and [LASTX] operate as usual, but each of these registers is double, to hold the real and imaginary parts of a number.

The following functors should be implemented in full complex form, but producing, of course, the principal value: [+], [−], [×], [÷], [$1/x$], [x^2], [$\sqrt{x}$], [ln], [e^x], and [y^x] (or [x^y]). As an aid in calculating alternative values for some of these functions, a register, perhaps [STO] register 0, should be set to $\exp(2\pi ix)$ on either [e^x] or [y^x], and to -1 on [$\sqrt{x}$]. An example below shows the usefulness of this feature for nonprincipal values.

As an exception to the rule that a noncomplex user should be able to get all the right answers, I suggest that [→R] and [→P] should operate as usual but on the X_r and X_i registers. Then, for example, [→P] gives the modulus or absolute value and the phase of a complex number, and differs from [ln], which gives the logarithm of the modulus and the phase. All these functors should reset the active stack to r.

I feel less strongly about these recommendations, but I think that the trigonometric functors ([SIN], [COS], [TAN], [SIN^{-1}], [COS^{-1}], and [TAN^{-1}]) should be implemented in real numbers only, but on the *active* X; they would have no effect on the inactive stack.

The display on the model B is conventional except that it must show two additional bits of information: the stack that is active (r or i), and whether the inactive X contains 0 or ₡, or not ("other-non-zero" light). Both these

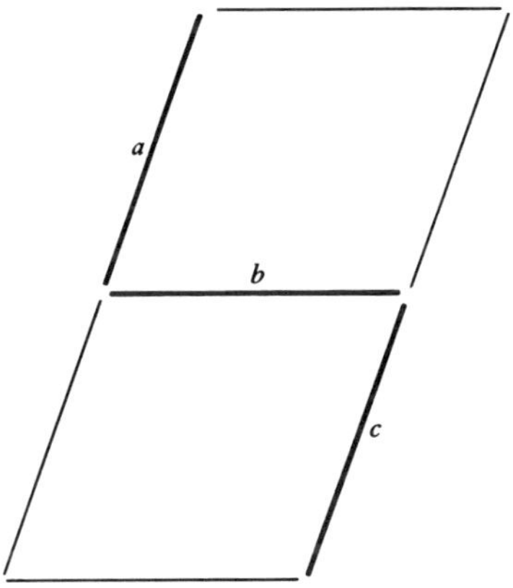

Figure 6.4.1 The left-most digit in the model B display shows three bits of information: a is called the "other-non-zero" light and is lighted whenever the inactive X contains anything other than 0 or ȼ; b is the minus sign as usual; and c represents i for imaginary and is lighted whenever the i stack is active (the contents of X_i are displayed). Whenever the r stack is active (the contents of X_r are displayed), c is dark.

additional bits can be incorporated into the left-most digit of the display (Figure 6.4.1).

Consider some examples on the model B. Each example starts with the r stack active and auto-enter enabled (or else ȼ in X_r and X_i); a, b, c, and so on, are real numbers.

$$(a+ib)(c+id)=u+iv \qquad a\boxed{\text{S}}b\boxed{\text{S}}c\boxed{\text{S}}d\boxed{\times} \qquad \text{(see } u\text{)}$$

$$\boxed{\text{S}} \qquad \text{(see } v\text{).}$$

When a is keyed, both stacks push, ȼ is written into X_i, and this ȼ appears in the display after the $\boxed{\text{S}}$ following a, causing b to overwrite this ȼ. The $\boxed{\text{S}}$ following b puts a back into the display (X_r again), but with auto-enter enabled, c pushes the stack and writes ȼ into X_i again. Then $\boxed{\text{S}}$ after c reveals ȼ in X_i for d to overwrite. The functor $\boxed{\times}$ resets the active register to r so that u rather than v is displayed first.

$$a\boxed{\gtrless}b\boxed{\gtrless}d\boxed{\gtrless}c\boxed{\times} \qquad \text{(see } u\text{)}$$

$$\boxed{\gtrless} \qquad \text{(see } v\text{).}$$

This alternative algorithm for this problem requires the same number of keystrokes, but is, in my view, less obvious. There are also several other alternatives.

$$\frac{a+ib}{a+ic}=u+iv \qquad a\boxed{\text{S}}b\boxed{\uparrow}c\boxed{\div} \qquad \text{(see } u\text{)}$$

$$\boxed{\text{S}} \qquad \text{(see } v\text{).}$$

$$\frac{a+ib}{a-ib}=u+iv \qquad a\boxed{\text{S}}b\boxed{\uparrow}\boxed{\text{CHS}}\boxed{\div} \qquad \text{(see } u\text{)}$$

$$\boxed{\text{S}} \qquad \text{(see } v\text{).}$$

As with real numbers, one can often use [↑] to avoid rekeying a number.

$(a+ia)^3 = u+iv$ a[↑][≷][+]3[y^x] (see u)

[S] (see v).

$\frac{1}{a+ia/2} = u+iv$ a[↑][↑]2[÷][≷][+][$1/x$] (see u)

[S] (see v).

This is another trick to avoid rekeying a number.

$\sqrt[4]{16}$ 16[↑]4[$1/x$][y^x] (see 2).

But the equation $s^4 = 16$ has three other solutions. To get the others, remember that storage register 0 contains $\exp(2\pi ix)$ after [y^x], and $\exp(2\pi i/4) = i$. Thus

[RCL] 0 [×] (see 0) [S] (see 2; i.e., another s is $2i$),
[RCL] 0 [×] (see -2) [S] (see 0; i.e., another s is -2),
[RCL] 0 [×] (see 0) [S] (see -2; i.e., another s is $-2i$),
[RCL] 0 [×] (see 2; we are back to the first answer, $s = 2$).

This scheme works for all *real rational* x on [y^x] or [e^x]. If x is irrational (or nearly so) or complex, the contents of storage register 0 after [y^x] or [e^x] are less obviously useful.

6.5 A PLOTTER!

Lest anyone think that this book is not perfectly serious, I have a final suggestion for the design of future programmable RPN calculators. The calculator should have a *plotter*, not as an attachment but built into the case. The bottom of the calculator would have a retractable pen point and three wheels (Figure 6.5.1). The two wheels in the back are idler wheels and are not driven or steered (though they might want a parking brake when the calculator is not plotting). The front wheel is steerable through 180° and can be driven forward or backward by the program.

The pen can move in any direction by any distance using a combination of steering and driving the front wheel. One needs to assume that steering and driving are independent so that the pen does not move on the paper if the wheel is steered but not driven, and the drive wheel does not slip. This would clearly be an approximation.

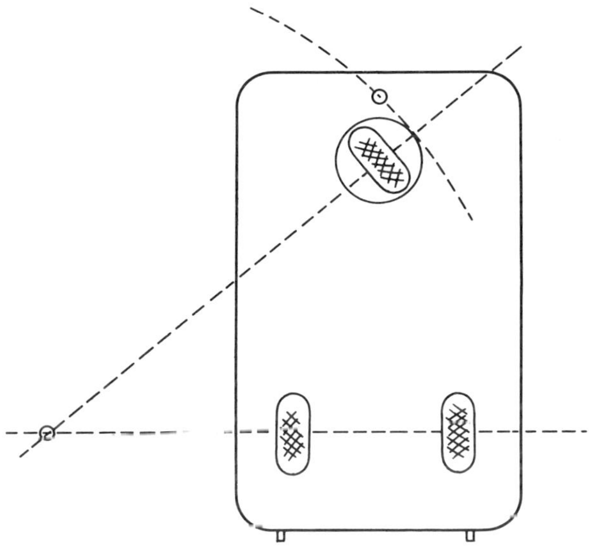

Figure 6.5.1 A plotting calculator (bottom view) showing the retractable pen (top), the steerable drive wheel, and the two idler wheels. The circled point outside the calculator is the pivot point for the indicated steering position, and the dashed line through the pen shows where it would move.

The calculations necessary to convert from *xy*-coordinate commands to steering and drive commands are not trivial. But this conversion would be done by a preprogrammed internal function and would not concern the user. He needs only program the function to be plotted, place the calculator on a sheet of paper on a level surface, initially oriented and positioned to define the coordinate axes, and push [R/S].

I recommend that the calculator not be left alone while plotting, for it might try to spill coffee on itself or jump off the desk and run away.

6.6 EXERCISES

Write model A algorithms for each of the following four problems:

6.6.1 $ax^4 + bx^3 + cx^2 + dx + e.$

6.6.2 $$\frac{1}{1/R_1 + 1/R_2 + 1/R_3}.$$

(This is the formula for three electrical resistances in parallel.)

6.6.3 Lagrange three-point interpolation. *Hint*: Convert the algorithm in Section A.2.

6.6.4 All the tabular-quadrature formulas in Section A.3.1.

Write model B algorithms for problems 6.6.5 to 6.6.8.

6.6.5 $$\frac{\sqrt{a+ib}-\sqrt{c+id}}{\sqrt{c+id}}=u+iv.$$

6.6.6 $$\frac{(1+i2)(1-i2)}{(1+i3)(2-i4)}-\frac{5}{14+i2}=u+iv.$$

6.6.7 Find the three cube roots of -1.

6.6.8 $\left|\frac{e}{a+ib}\right|^{(c+id)}=u+iv$, where $e=2.71828\cdots$.

6.6.9 Will the algorithm in Section A.5.3, "Real Roots of a Cubic by Iteration," work on the model B to find complex roots also?

6.6.10 Design and perform an experiment to determine how a calculator handles round-off in $\boxed{+}$ or $\boxed{-}$. In what way does the answer from A $\boxed{\uparrow}$ B $\boxed{+}$ differ from round$(A+B,\ n)$?

APPENDIX A
Algorithms

Who can count the sands of the sea,
the drops of rain, or the days of eternity?
Who can measure the height of the heavens,
the breadth of the earth, or the depth of the abyss?

—*Ecclesiasticus I, 2.*

CONTENTS

A.1 INTRODUCTION, RULES, AND NOTES

A.1.1 Introduction

No collection of algorithms can ever be all inclusive. The algorithms in this appendix deal with a wide variety of topics and fields (e.g., finance, music); but astronomy, radio astronomy, and astrophysics are more comprehensively represented.

To some extent this appendix overlaps collections of algorithms available from HP Company (see references in Appendix C). HP's algorithms can be criticized in some cases for being plodding and mundane, but they are remarkably error-free and reliable. Nothing succeeds like success—they give the right answer! Since HP's algorithms are now widely available, I edited the algorithms in this appendix to eliminate most duplications, except for cases in which my algorithm is preferable to HP's for some reason (e.g., fewer keystrokes). Although I am critical of some of HP's algorithms in these pages, I strongly recommend the HP books and "pacs."

The algorithms in this appendix are either designated for a specific calculator—most frequently the HP-45 or HP-35—or they will work on either the HP-45 or HP-35. Section A.1.2 contains rules for adapting algorithms to other calculators. I considered and rejected two alternatives to this scheme: (*a*) write all the algorithms in a universal RPN notation such as was used in Section 2.6, or (*b*) write each algorithm in 13 different versions to fit all the currently popular RPN calculators. I rejected the first alternative because algorithms in a universal notation are difficult to *use*; in effect one must adapt such algorithms for every calculator. I feel that the *practical* value of this appendix would have been reduced by choosing alternative *a*. I rejected alternative *b* because of space limitations. Such a scheme is inevitably obsolete by the time it is published, but so also is the scheme I chose. At least I cannot be accused of trying to sell HP-45s and HP-35s: both models have been discontinued.

The user of this appendix should be familiar with the instruction booklet with the calculator, but not necessarily with the preceding parts of this book.

Notes on notation

Keystroke symbols in solid boxes (e.g., [+]) are printed on the top of the key, those in dashed boxes (e.g., [→R]) on the side of the key or on the land area above or below the key. An exception is the original HP-35, which has all the labels on the top half of the keyboard printed on the land area above the keys. Keystroke symbols in dashed boxes are always preceded by a shift keystroke explicitly in the algorithm. Numbers in the algorithm are printed without boxes. Key numbers exactly as printed.

[G] represents an unlabeled gold-colored key, [B] an unlabeled blue key, and [↑] stands for [ENTER↑]. The symbol :| is analogous to a musical repeat symbol and means loop back to the last preceding colon (:) not in parentheses.The subroutines [···*f*···] and [···*g*···] are explained where they occur.

A symbol for a parameter (e.g., *A*) occurs in an algorithm where the numerical value should be keyed. When units or explanations are needed, I enclose the symbol and its unit or explanation in parentheses. Calculated answers are identified by "see..." in parentheses, except in a few short algorithms where this is unnecessary. After "(see...)" just look, do not key.

When a parameter is to be keyed, use number keys, [·], [EEX], and [CHS] as necessary. Use [CHS] for negative parameters, not [−]. After a miskeyed number, use [CLX] or [R↓]; then key the correct number. Do not press [↑] or any other function key unless it is explicitly in the algorithm.

The symbol DD.MMSS means degrees, minutes, and seconds of arc, with two digit locations for each. The decimal point after DD must be keyed. Any digits following SS will be taken for a decimal fraction of a second. In answers I write this as, for example, DD.MMSS‸SS, where the caret (‸) means an implied but unseen decimal point. Similarly, HH.MMSS means hours, minutes, and seconds of time or in time units. For example, −5.420202 would mean −5°42′02.″02 with DD.MMSS or $-5^{h}42^{m}02\overset{s}{.}02$ with HH.MMSS. Details are in the instruction booklet with the calculator.

A period ends an algorithm. This is important whenever one algorithm follows another with no text between. Since this period is not beside a number, it cannot be confused with a decimal point.

The HP-45 is to be in the mode in which it turns on, that is, degrees. After using radians, for example, change back to degrees, or turn the calculator off, then on, before going to another algorithm with trigonometric functors. The display mode can be changed as needed, except in algorithms that specify the display mode. No other clearing or presetting operation is necessary unless it is in the algorithm; do not [CLEAR] between problems unless the algorithm so specifies.

A.1.2 Converting to Other Calculators

This section contains rules for converting the algorithms in Appendix A that are written for the HP-45 or HP-35 to work on other RPN calculators. This section was rather difficult to write. One would like to have the converted algorithms not only work but be elegant and concise. In many cases, however, the peculiarities of a calculator would allow some simplifications, so that the algorithms resulting from converting directly, using the rules in this section, are not optimum. For example, [R↓] [R↓] [R↓] should be replaced by [R↑] on any calculator that has [R↑]. I hope that the reader takes this as a challenge for creative thinking rather than as a criticism of the book. In other cases, the algorithm must be rewritten rather than just converted. This occurs, for example, with most of the algorithms that employ [Σ+] because the effect of this key differs markedly among calculator models, as shown in Table A.1.1.

Table A.1.1 Comparison of $\boxed{\Sigma+}$ Capabilities on Various RPN Calculators

	Register Numbers						
Calculator	$\Sigma 1$	Σx	Σx^2	Σy	Σy^2	Σxy	X
National Semiconductor 4640	3	1	2	—	—	—	x
Corvus 500 / APF Mark 55	7	9	8	†	—	—	x*
HP-25	3	7	6	4	—	5	$\Sigma 1$*
HP-27	4	5	6	7	8	9	$\Sigma 1$*
HP-45	5	7	6	8	—	—	$\Sigma 1$*
HP-46	5	7	6	8	—	—	$\Sigma 1$*
HP-55	.0	.1	.2	.3	.4	.5	$\Sigma 1$*
HP-91	.0	.1	.2	.3	.4	.5	$\Sigma 1$*
HP-67 / HP-97	S9	S4	S5	S6	S7	S8	$\Sigma 1$*

*With auto-enter disabled.†On the Corvus 500 and APF Mark 55, Σy is saved in an arcane register, not accessible by number. Instead, $\boxed{\text{RCL}}$ $\boxed{\Sigma+}$ puts Σx into X and Σy into Y. This sequence gives Σx and Σy on many other calculators also, but on the Corvus 500 and APF Mark 55 this is the only way to get Σy.

The following notation is used throughout this section:

X	No go; will not work,
$\boxed{\bullet}$	Any of $\boxed{+}$, $\boxed{-}$, $\boxed{\times}$, or $\boxed{\div}$,
n	Any number key, 1 through 9.

Novus Scientist

Although the Novus 4520 Scientist (see Figure 1.3.4) has RPN with a stack of four registers and is superficially about as powerful as an HP-35, enough differences exist to prevent the easy adaptation of many HP-35 algorithms to work on the Novus Scientist. Among the significant differences are: T clears on pop; most monadic and all dyadic functors cause T to be cleared. Furthermore, $\boxed{\ln}$, $\boxed{\log}$, and $\boxed{y^x}$ cause Y and Z also to be cleared. Only $\boxed{1/x}$ and $\boxed{\sqrt{x}}$ can be done without losing the contents of T. $\boxed{\text{SIN}}$, $\boxed{\text{COS}}$, and $\boxed{\text{TAN}}$ do not work for angles >90° or <0. And $\boxed{\text{ARC}}$ $\boxed{\text{SIN}}$ and $\boxed{\text{ARC}}$ $\boxed{\text{COS}}$ do not work for negative arguments. Instead of assuming a 1, an $\boxed{\text{EE}}$ without a preceding digit is ignored.

Some of these shortcomings can be understood as design economies, but there is no excuse for clearing T; this can only be described as a design error.

Whenever these features are not needed, HP-35 algorithms will work on the Novus Scientist with the following changes in notation:

Change (HP-35)	to (Novus Scientist)
[x^y]	[$x \leftrightarrow y$] [y^x]
[CLR]	Turn calculator off, then on
[$\sqrt{x}$]	[$\sqrt{\ }$]
[$x \gtrless y$]	[$x \leftrightarrow y$]
[R↓]	[ROLL↓]
[STO]	[MS]
[RCL]	[MR]
[↑]	[ENT↑]
[EEX]	[EE]
[CLX]	[C↓] sometimes, or X

National Semiconductor 4640

In the design of this calculator from the same company, most, but not quite all, of the shortcomings of the Novus Scientist were avoided. The 4640 has capabilities intermediate between the HP-45 and HP-35 including rectangular-to-polar coordinate conversion and three storage registers with [MS] (storage register) arithmetic. The principal shortcoming of the National Semiconductor 4640, in my view, is that T clears on pop. Why?!

Most HP-35 and some HP-45 algorithms will work on the 4640 with the following changes in notation:

Change (HP-35)	to (National Semiconductor 4640)
[x^y]	[$x \leftrightarrow y$] [F] [y^x]
[e^x]	[F] [e^x]
[CLR]	[F] [CA]
[$\sqrt{x}$]	[$\sqrt{\ }$]
[ARC]	[F]
[$x \gtrless y$]	[$x \leftrightarrow y$]
[R↓]	[ROLL]
[STO]	[MS] 1 (or 2 or 3)
[RCL]	[MR] 1 (or 2 or 3)
[↑]	[EN]

Change (HP-35)	to (National Semiconductor 4640)
[CLX]	[C] sometimes, or X
[π]	[F] [π]

Change (HP-45)	to (National Semiconductor 4640)
[G]	[F] (exceptions below)
[e^x]	[F] [e^x]
[FIX]	[F] [DS]
[x^2]	[F] [x^2]
[$x \gtrless y$]	[$x \leftrightarrow y$]
[R↓]	[ROLL]
[→P]	[→P] [$x \leftrightarrow y$]
[G] [→R]	[$x \leftrightarrow y$] [F] [→R]

(The extra [$x \leftrightarrow y$]s are necessary because on the 4640, θ is in X and R in Y.)

[STO] n	[MS] n if $n = 1, 2, 3$, X otherwise

(HP-45 algorithms in this book that use only one [STO] register, use number 4. Such algorithms can be converted for the 4640 by changing n from 4 to 1 (or 2 or 3) wherever it occurs.)

[STO] [*] n	[MS] [*] n if $n = 1$, 2, or 3, X otherwise
[RCL] n	[MR] n if $n = 1$, 2, or 3, X otherwise
[RCL] [*] n	X
[↑]	[EN]
[CLX]	[C] sometimes, or X
[Σ+]	[Σ+] sometimes, or X

(Algorithms employing [Σ+] usually have to be rewritten to work with the 4640; see Table A.1.1.)

[G] [log]	[log]
[SCI]	No exact equivalent; try [F] [DS] [·] and drop the n after [SCI]
[G] [$\sqrt{x}$]	[$\sqrt{\ }$]
[$n!$]	[$x!$]
[$\bar{x}, s$]	[SD] [F] [$\bar{x}$] sometimes or X; see note under [Σ+]

Change (HP-45)	to (National Semiconductor 4640)
G →D.MS	→DMS } Provided mode is degrees
D.MS→	→D } Provided mode is degrees
G Δ%	x↔y F Δ% sometimes, or X
GRD	GRAD
CLEAR	CA
LASTX	X
Σ−	Σ− sometimes, or X; see note under Σ+

Corvus 500 and APF Mark 55

The Corvus 500 (Figure 1.3.7) and the APF Mark 55 are calculators with most of the capabilities of the HP-45 (missing only STO and RCL arithmetic, →D.MS and D.MS→ conversions, and GRD mode), and they have some additional features. Most of the algorithms herein designated for the HP-45 will work on the Corvus 500 or APF Mark 55 with the following changes in notation:

Change (HP-45)	to (Corvus 500)	or to (APF Mark 55)
G	G (exceptions below)	Shift (exceptions below)
RCL * n	X	X
STO * n	X	X
$1/x$	G $1/x$	Shift $1/x$
e^x	INV ln	INV ln
FIX	G INV SCI DSP	Shift INV sci DPS
x^2	INV G $\sqrt{x}$	INV Shift $\sqrt{x}$
→P	G →POL	Shift →pol
R↓	R↓	↓
%	G %	Shift %
↑	ENT	ENT
$x \gtrless y$	$y \gtrless x$	x/y
G SCI	G SCI DSP	Shift sci DPS
CHS	CHS	+/−
EEX	EE	EE
G y^x	y^x	y^x
G 10^x	G INV log	Shift INV log
G →R	INV G →POL	INV Shift →pol
G SIN^{-1}	INV SIN	INV SIN
G COS^{-1}	INV COS	INV COS
G TAN^{-1}	INV TAN	INV TAN

Change (HP-45)	to (Corvus 500)	or to (APF Mark 55)
n!	x!	x!
→D.MS	X	X
	(or convert separately)	
D.MS→	X	X
	(or convert to decimal separately and ENT if followed by a number or parameter)	
G DEG	INV G RAD	INV Shift rad
GRD	X	X
CLEAR	CLR	clr
G LASTX	RCL 0	RCL 0
G Σ−	INV Σ+	INV Σ+

On the Corvus 500 the keys STO and RCL, and on the APF Mark 55 the keys HYP and INV, are marked on the land area above the corresponding keys but in a box, to prevent confusion with shifted functions. LASTX over 0 is a fake; use RCL 0 instead as indicated.

The keys Σ+ and Σ− on these calculators do not use the same register numbers as the HP-45, and Σ1 does not appear in X after Σ+. Thus HP-45 algorithms involving Σ+ usually need to be rewritten for the Corvus 500 or the APF Mark 55. See Table A.1.1.

HP-21

The HP-21 (Figure 1.3.2) is a calculator with capabilities intermediate between the HP-35 and HP-45. All the HP-35 and some of the HP-45 algorithms herein will work on the HP-21 with the following changes in notation:

Change (HP-35)	to (HP-21)
x^y	$x \gtrless y$ B y^x
log	B LOG
ln	B LN
$\sqrt{x}$	B $\sqrt{x}$
ARC	B
π	B π

(Always use DEG mode on the HP-21 unless otherwise noted.)

Change (HP-45)	to (HP-21)
[ln]	[B] [LN]
[FIX]	[DSP] [·]
[G]	[B] (exceptions below)
[x^2]	[↑] [×] sometimes, or X
[→P]	[B] [→P]
[STO] 4	[STO]
[STO] *n*	X if $n \neq 4$
[RCL] 4	[RCL]
[RCL] *n*	X if $n \neq 4$
[RCL] [•] *n*	X
[STO] [•] *n*	X if $n \neq 4$
[STO] [+] 4	[B] [M+]
[STO] [−] 4	[B] [M−]
[STO] [×] 4	[B] [M×]
[STO] [÷] 4	[B] [M÷]
[Σ+]	X
[$n!$]	X
[$\bar{x}, s$]	X
[G] [D.MS→]	X (or convert to decimal separately, and [↑] if followed by a number or parameter)
[G] [→D.MS]	X (or convert separately)
[G] [DEG]	Switch to "DEG" mode
[G] [RAD]	Switch to "RAD" mode
[CLEAR]	[CLR]
[LASTX]	X
[Σ−]	X

HP-25

The HP-25 (Figure 1.3.6) is a programmable RPN calculator with more capabilities than the HP-45, and missing only [RCL] arithmetic and [$n!$]. Almost all algorithms designated for the HP-45 will work on the HP-25 in *nonprogrammed form*. Converting to HP-25 *programs* is possible for algorithms that are short enough to fit into the 49 steps available. The following tabulation shows the necessary changes in notation:

Change (HP-45)	to (HP-25)
[$1/x$]	[g] [$1/x$]
[ln]	[f] [ln]

Change (HP-45)	to (HP-25)
[e^x]	[g] [e^x]
[FIX]	[f] [FIX]
[x^2]	[g] [x^2]
[→P]	[g] [→P]
[SIN]	[f] [sin]
[COS]	[f] [cos]
[TAN]	[f] [tan]
[STO] n [STO] [*] n [RCL] n	No change unless $n=8$, then change to $n=0$; or unless $n=9$, then X
[RCL] [*] n	X
[%]	[g] [%]
[G] [y^x]	[f] [y^x]
[G] [log]	[f] [log]
[G] [10^x]	[g] [10^x]
[G] [SCI]	[f] [SCI]
[G] [$\sqrt{x}$]	[f] [$\sqrt{x}$]
[G] [→R]	[f] [→R]
[G] [SIN^{-1}]	[g] [sin^{-1}]
[G] [COS^{-1}]	[g] [cos^{-1}]
[G] [TAN^{-1}]	[g] [tan^{-1}]
[G] [$n!$]	X (but see exercise 2.7.16)
[G] [$\bar{x}$,s]	[f] [s] [f] [$\bar{x}$] (but see note on [Σ+] below)
[G] [→D.MS] [G] [D.MS→]	[f] [→H.MS] [g] [→H] — Provided mode is DEG
[G] [Δ%]	X
[G] [DEG]	[g] [DEG]
[G] [RAD]	[g] [RAD]
[G] [GRD]	[g] [GRD]
[G] [CLEAR]	[f] [REG] [f] [STK]
[G] [LASTX]	[f] [LASTX]
[G] [π]	[g] [π]
[G] [Σ−]	[f] [Σ−] (but see note on [Σ+] below)

Algorithms employing [Σ+] usually have to be changed to work on the HP-25. In particular, [RCL]s in an algorithm containing [Σ+] must have their register numbers changed. Such algorithms should be rewritten to take advantage of the increased capabilities of [Σ+] on the HP-25. See Table A.1.1.

HP-27

The HP-27 is an RPN calculator with more capabilities than the HP-45 and lacks only [RCL] arithmetic. Essentially all algorithms designated for the HP-45 will work on the HP-27 with the following changes in notation:

Change (HP-45)	to (HP-27)
[1/x]	[g] [1/x]
[ln]	[f] [ln]
[e^x]	[g] [e^x]
[FIX]	[f] [FIX]
[x^2]	[g] [x^2]
[→P]	[g] [→P]
[SIN]	[f] [sin]
[COS]	[f] [cos]
[TAN]	[f] [tan]
[RCL] [•] n	X
[G] [y^x]	[y^x]
[G] [log]	[f] [log]
[G] [10^x]	[g] [10^x]
[G] [SCI]	[f] [SCI]
[G] [$\sqrt{x}$]	[f] [$\sqrt{x}$]
[G] [→R]	[f] [→R]
[G] [SIN^{-1}]	[g] [sin^{-1}]
[G] [COS^{-1}]	[g] [cos^{-1}]
[G] [TAN^{-1}]	[g] [tan^{-1}]
[G] [$n!$]	[f] [$n!$]
[G] [$\bar{x}, s$]	[g] [s] [$\bar{x}$] (but see note on [Σ+] below)
[G] [→D.MS]	[f] [→H.MS] } Provided mode is DEG
[G] [D.MS→]	[g] [→H] } Provided mode is DEG
[G] [Δ%]	[g] [Δ%]
[G] [DEG]	[g] [DEG]
[G] [RAD]	[g] [RAD]
[G] [GRD]	[g] [GRD]
[G] [CLEAR]	[f] [REG] [f] [STK]
[G] [LASTX]	[f] [LASTX]
[G] [π]	[g] [π]
[G] [Σ−]	[f] [Σ−] (but see note on [Σ+] below)

As a rule, algorithms employing [Σ+] have to be changed to work on the HP-27. In particular, [RCL]s in an algorithm containing [Σ+] must have their register numbers changed. Such algorithms should be rewritten to take advantage of the increased capabilities of [Σ+] on the HP-27. See Table A.1.1.

HP-46

Although there are some minor differences, the HP-46 is essentially a desktop printing version of the HP-45. Therefore all algorithms designated for the HP-45 will work on the HP-46 with the following changes in notation:

Change (HP-45)	to (HP-46)
[G]	Rust-orange key, top center
[SIN⁻¹]	[ASN]
[COS⁻¹]	[ACS]
[TAN⁻¹]	[ATN]
[→P]	[TO POL]
[→R]	[REC]
[cm/in]	[C/I]
[kg/lb]	[K/L]
[ltr/gal]	[L/G]
[D.MS→]	[DM→]
[→D.MS]	[DM←]
[G] [LAST*X*]	[LAST*X*] (but also works with [G] preceding)

HP-55

The HP-55 is a programmable RPN calculator with more capabilities than the HP-45, and missing only [RCL] arithmetic. Essentially all algorithms designated for the HP-45 will work on the HP-55 in *nonprogrammed form*. Converting to HP-55 *programs* is possible for algorithms that are short enough to fit into the 49 steps available. The following changes in notation are necessary:

Change (HP-45)	to (HP-55)
[ln]	[g] [ln]
[e^x]	[g] [e^x]

Change (HP-45)	to (HP-55)
[x^2]	[g] [x^2]
[→P]	[g] [→P]
[SIN]	[f] [sin]
[COS]	[f] [cos]
[TAN]	[f] [tan]
[G] [y^x]	[y^x]
[G] [log]	[f] [log]
[G] [10^x]	[g] [10^x]
[G] [$\sqrt{x}$]	[f] [$\sqrt{x}$]
[G] [→R]	[f] [R←]
[G] [SIN^{-1}]	[g] [sin^{-1}]
[G] [COS^{-1}]	[g] [cos^{-1}]
[G] [TAN^{-1}]	[g] [tan^{-1}]
[G] [$n!$]	[f] [$n!$]
[G] [$\bar{x}, s$]	[g] [s] [f] [$\bar{x}$] (but see note on [Σ+] below)
[G] [→D.MS]	[g] [→H.MS] } provided mode is DEG
[G] [D.MS→]	[f] [H←] } provided mode is DEG
[G] [Δ%]	X
[G] [DEG]	[f] [DEG]
[G] [RAD]	[f] [RAD]
[G] [GRD]	[f] [GRD]
[G] [CLEAR]	[f] [CLR]
[G] [LASTX]	[f] [LASTX]
[G] [π]	[f] [π]
[G] [Σ−]	[f] [Σ−] (but see note on [Σ+] below)
[RCL] [*] n	X

Algorithms employing [Σ+] usually have to be changed to work on the HP-55. In particular, the register numbers of [RCL]s in an algorithm containing [Σ+] must be changed. Such algorithms should be rewritten to take advantage of the increased capabilities of [Σ+] on the HP-55. See Table A.1.1.

HP-65

The HP-65 (Figure 1.3.8) is a programmable RPN calculator with more capabilities than the HP-45 and lacks only [Σ+] and [RCL] arithmetic. Most algorithms designated for the HP-45 will work on the HP-65 in *nonprogrammed form*. Converting to HP-65 *programs* is possible for algorithms that are short enough to fit into the 100 steps available. The following changes in notation are necessary:

Change (HP-45)	to (HP-65)
1/x	g 1/x
ln	f LN
e^x	f^{-1} LN
FIX	DSP ·
x^2	f^{-1} $\sqrt{x}$
→P	f R→P
SIN	f SIN
COS	f COS
TAN	f TAN
$x \gtrless y$	g $x \gtrless y$
R↓	g R↓
RCL • *n*	X
%	X
Σ+	X
G y^x	g y^x
G log	f LOG
G 10^x	f^{-1} LOG
G SCI	DSP
G $\sqrt{x}$	f $\sqrt{x}$
G →R	f^{-1} R→P
G SIN^{-1}	f^{-1} SIN
G COS^{-1}	f^{-1} COS
G TAN^{-1}	f^{-1} TAN
G *n*!	g *n*!
G $\bar{x},s$	X
G →D.MS	f →D.MS
G D.MS→	f^{-1} →D.MS
G Δ%	X
G DEG	g DEG
G RAD	g RAD
G GRD	g GRD
G LAST*X*	g LST*X*
G π	g π
G Σ−	X

HP-67

The HP-67 (Figure 1.3.9) is a very impressive programmable RPN calculator with more capabilities than the HP-45 and missing only RCL arithmetic. Most algorithms designated for the HP-45 will work on the HP-67 in *nonprogrammed* form using the tabulation below. Converting to HP-67 programs is also possible.

Change (HP-45)	to (HP-67)
1/x	h 1/x
ln	f LN
e^x	g e^x
FIX	DSP ·
x^2	g x^2
→P	g →P
SIN	f SIN
COS	f COS
TAN	f TAN
$x \gtrless y$	h $x \gtrless y$
R↓	h R↓
RCL * *n*	X
%	f %
G y^x	h y^x
G log	f LOG
G 10^x	g 10^x
G SCI	DSP
G $\sqrt{x}$	f $\sqrt{x}$
G →R	f R←
G SIN^{-1}	g SIN^{-1}
G COS^{-1}	g COS^{-1}
G TAN^{-1}	g TAN^{-1}
G *n*!	h N!
G $\bar{x},s$	g s f $\bar{x}$ (but see note on Σ+ below)
G →D.MS	g →H.MS } Provided mode is DEG
G D.MS→	f H← } Provided mode is DEG
G Δ%	g %CH
G DEG	h DEG
G RAD	h RAD
G GRD	h GRD

Change (HP-45)	to (HP-67)
[G] [LASTX]	[h] [LSTX]
[G] [π]	[h] [π]
[G] [$\Sigma-$]	[h] [$\Sigma-$]

Algorithms employing [$\Sigma+$] usually have to be changed to work on the HP-67. In particular, [RCL]s in an algorithm containing [$\Sigma+$] need to have their register numbers changed. Such algorithms should be rewritten to take advantage of the increased capability of [$\Sigma+$] on the HP-67. See Table A.1.1.

HP-91

The HP-91 is a portable RPN calculator with a printer. It has more capabilities than the HP-45; only [RCL] arithmetic is lacking. Essentially all algorithms designated for the HP-45 will work on the HP-91 with the following changes in notation:

Change (HP-45)	to (HP-91)
[→P]	[R→P]
[G] [y^x]	[y^x]
[G] [$\sqrt{x}$]	[$\sqrt{x}$]
[G] [→R]	[P→R]
[G] [$\bar{x}$,s]	[G] [s] [G] [$\bar{x}$] (but see note on [$\Sigma+$] below)
[→D.MS]	[→H.MS] } Provided mode is DEG
[D.MS→]	[H.MS→] }
[G] [DEG]	Set switch to corresponding mode.
[G] [RAD]	
[G] [GRD]	
[RCL] [•] n	X

Algorithms employing [$\Sigma+$] usually must be changed to work on the HP-91. In particular, [RCL]s in an algorithm containing [$\Sigma+$] require a change in register numbers. Such algorithms should be rewritten to take advantage of the increased capability of [$\Sigma+$] on the HP-91. See Table A.1.1.

HP-97

The HP-97 is essentially an HP-67 with a portable printer similar to the HP-91. Although the HP-97 and HP-67 have almost identical capabilities, the keyboard and the keystrokes are different. Most algorithms designated for the HP-45 will work on the HP-97 in *nonprogrammed* form using the following tabulation. Converting to HP-97 programs is also possible.

Change (HP-45)	to (HP-97)
G	f * (exceptions below)
RCL • n	X
G y^x	y^x
G SCI	SCI
G $1/x$	$1/x$
G →R	→R
G $\bar{x}, s$	f s f $\bar{x}$ (but see note on Σ+ below)
→D.MS	→H.MS (Provided mode is DEG)
D.MS→	H.MS→ (Provided mode is DEG)
Δ%	%CH

Algorithms employing Σ+ usually have to be changed to work on the HP-97. In particular, RCL s in an algorithm containing Σ+ need to have their register numbers changed. Such algorithms should be rewritten to take advantage of the increased capability of Σ+ on the HP-97. See Table A.1.1.

A.1.3 Powers of 10

Append the keystrokes from the following tabulation to convert prefixed units into the corresponding units with no prefix:

exa	E	EEX	18
peta	P	EEX	15
tera	T	EEX	12
giga	G	EEX	9
mega	M	EEX	6
kilo	k	EEX	3
hecto	h	EEX	2
deca	da	EEX	1

*Most HP-97s have an "f" on the gold key.

deci	d	EEX	CHS	1
centi	c	EEX	CHS	2
milli	m	EEX	CHS	3
micro	μ	EEX	CHS	6
nano	n	EEX	CHS	9
pico	p	EEX	CHS	12
femto	f	EEX	CHS	15
atto	a	EEX	CHS	18

For example, to key 1665 MHz in hertz, press 1665 [EEX] 6.

A.2 NUMERICAL (TABULAR) INTERPOLATION AND DIFFERENTIATION

Algorithms of numerical interpolation and differentiation obtain approximations for the value or the derivative of a function specified by a table of values with equally spaced arguments,

$$h = x_1 - x_0 = x_2 - x_1 = \text{etc.}$$

And

$f_0 = f(x_0)$ is the tabulated value at x_0,

$f_1 = f(x_1)$ is the tabulated value at x_1, etc.

Linear interpolation (x should be between x_0 and x_1).
HP-45 or HP-35:
x_0 [↑] [↑] x [−] [$x \gtrless y$] x_1 [−] [÷] [↑] [↑] f_1 [×] [$x \gtrless y$] 1 [−] f_0 [×] [−] (see $\cong f(x)$).

Lagrange three-point interpolation (x can be anywhere between x_{-1} and x_1 but should be closer to x_0 than to x_{-1} or x_1).
HP-45:
x_0 [↑] [↑] x [−] [$x \gtrless y$] x_1 [−] [÷] [↑] [↑] [↑] 1 [−] f_{-1} [×] [$x \gtrless y$] 1 [+] f_1 [×] [+] [×] 2 [÷] [$x \gtrless y$] [x^2] 1 [−] f_0 [×] [−] (see $\cong f(x)$).

Lagrange four-point interpolation (x should be between x_0 and x_1).
HP-45:

x_0 [↑] [↑] x [−] [x≷y] x_1 [−] [÷] [↑] [↑] [↑] 1 [+] f_2 [×] [x≷y] 2 [−]
f_{-1} [×] [−] [x≷y] 1 [−] [×] [×] 6 [÷] [STO] 4 [CLX] f_1 [×] [x≷y] 1
[−] f_0 [×] [−] [R↓] [×] [−] 2 [+] [×] 2 [÷] [RCL] 4 [+] (see $\cong f(x)$).

Lagrange five-point interpolation (x can be anywhere between x_{-1} and x_1 but should be closer to x_0 than to x_{-1} or x_1).
HP-45:

x_0 [↑] [↑] x [−] [x≷y] x_1 [−] [÷] [↑] [↑] [↑] 1 [−] f_{-1} [×] [x≷y] 1 [+]
f_1 [×] [+] [x≷y] [x^2] 4 [−] [×] [×] [STO] 4 [CLX] 2 [−] f_{-2} [×]
[x≷y] 2 [+] f_2 [×] [+] [x≷y] [x^2] 1 [−] [×] [×] 4 [÷] [RCL] 4 [−] 6 [÷]
[R↓] [×] 1 [−] [x≷y] [x^2] 4 [−] [×] 4 [÷] f_0 [×] [+] (see $\cong f(x)$).

Lagrange six-point interpolation (x should be between x_0 and x_1).
HP-45:

x_0 [↑] [↑] x [−] [x≷y] x_1 [−] [÷] [↑] [↑] [↑] 2 [−] f_{-1} [×] [x≷y] 1 [+]
f_2 [×] [−] [x≷y] 1 [−] [×] [×] 2 [÷] [STO] 4 [CLX] f_1 [×] [x≷y] 1 [−]
f_0 [×] [−] [x≷y] 1 [+] [×] [x≷y] 2 [−] [×] [RCL] 4 [+] [x≷y] 2 [+]
[×] [x≷y] 3 [−] [×] [STO] 4 [CLX] 2 [+] f_3 [×] [x≷y] 3 [−] f_{-2} [×]
[−] [x≷y] 2 [−] [×] [×] [x≷y] [x^2] 1 [−] [×] 10 [÷] [RCL] 4 [+]
12 [÷] (see $\cong f(x)$).

For an HP-35 (any of the above), replace [x^2] by [↑] [×] and delete 4 after [STO] and [RCL].

Linear differentiation (x should be between x_0 and x_1).
HP-45 or HP-35:

f_1 [↑] f_0 [−] x_1 [↑] x_0 [−] [÷] (see $\cong f'(x)$).

Lagrange three-point differentiation (x can be anywhere between x_{-1} and x_1 but should be closer to x_0 than to x_{-1} or x_1).
HP-45:

x_0 [↑] [↑] x [−] [x≷y] x_1 [−] [STO] 4 [÷] [↑] [↑] [↑] [+] f_0 [×] [x≷y]
.5 [−] f_{-1} [×] [−] [x≷y] .5 [+] f_1 [×] [−] [RCL] 4 [÷]
(see $\cong f'(x)$).

For an HP-35, delete 4 after [STO] and [RCL].

Lagrange four-point differentiation (x should be between x_0 and x_1).
HP-45:

x_0 [↑] [↑] x [−] [x≷y] x_1 [−] [STO] 1 [÷] [↑] [↑] [↑] 3 [×] 6 [−] [×] 2
[+] f_{-1} [×] [STO] 2 [CLX] 3 [×] [×] 1 [−] f_2 [×] [STO] [−] 2 [CLX] 3
[STO] [÷] 2 [×] 4 [−] [×] 1 [−] f_0 [×] [STO] [−] 2 [CLX] 3 [×] 2 [−]
[×] 2 [−] f_1 [×] [RCL] 2 [+] 2 [÷] [RCL] 1 [÷] (see $\cong f'(x)$).

Lagrange five-point differentiation (x can be anywhere between x_{-1} and x_1 but should be closer to x_0 than to x_{-1} or x_1).
HP-45:

x_0 $\boxed{\uparrow}$ $\boxed{\uparrow}$ x $\boxed{-}$ $\boxed{x\gtrless y}$ x_1 $\boxed{-}$ $\boxed{\text{STO}}$ 1 $\boxed{\div}$ $\boxed{\uparrow}$ $\boxed{\uparrow}$ $\boxed{\uparrow}$ 2 $\boxed{\times}$ 3 $\boxed{-}$ $\boxed{\times}$ 1 $\boxed{-}$ $\boxed{\times}$ 1 $\boxed{+}$ 2 $\boxed{\div}$ f_{-2} $\boxed{\times}$ $\boxed{\text{STO}}$ 2 $\boxed{CLX}$ 2 $\boxed{\times}$ 3 $\boxed{+}$ $\boxed{\times}$ 1 $\boxed{-}$ $\boxed{\times}$ 1 $\boxed{-}$ 2 $\boxed{\div}$ f_2 $\boxed{\times}$ $\boxed{\text{STO}}$ $\boxed{+}$ 2 $\boxed{CLX}$ 4 $\boxed{\times}$ 3 $\boxed{-}$ $\boxed{\times}$ 8 $\boxed{-}$ $\boxed{\times}$ 4 $\boxed{+}$ f_{-1} $\boxed{\times}$ $\boxed{\text{STO}}$ $\boxed{-}$ 2 $\boxed{CLX}$ 4 $\boxed{\times}$ 3 $\boxed{+}$ $\boxed{\times}$ 8 $\boxed{-}$ $\boxed{\times}$ 4 $\boxed{-}$ f_1 $\boxed{\times}$ $\boxed{\text{RCL}}$ 2 $\boxed{-}$ 3 $\boxed{\div}$ $\boxed{\text{R}\downarrow}$ $\boxed{\times}$ 2 $\boxed{\times}$ 5 $\boxed{-}$ $\boxed{\times}$ f_0 $\boxed{\times}$ $\boxed{-}$ 2 $\boxed{\div}$ $\boxed{\text{RCL}}$ 1 $\boxed{\div}$ (see $\cong f'(x)$).

TEST CASE: See Sections 5.2 and 5.3.
REFERENCES: HMF 25.2, 25.3, p. x.

A.3 QUADRATURE (NUMERICAL INTEGRATION)

A.3.1 Tabular Quadrature

Tabular quadrature algorithms approximate the integral of a function specified by a table of values with equally spaced arguments,

$$h = x_1 - x_0 = x_2 - x_1 = \text{etc.}$$

And

$f_0 = f(x_0)$ is the tabulated value at x_0,

$f_1 = f(x_1)$ is the tabulated value at x_1, etc.

The integration interval begins at x_0 and ends at x_m, that is, the integral to be approximated is

$$\int_{x_0}^{x_m} f(x)\,dx.$$

Note that there are $m+1$ points.

Two points: Trapezoidal rule (HMF 25.4.1).
f_0 $\boxed{\uparrow}$ f_1 $\boxed{+}$ 2 $\boxed{\div}$ h $\boxed{\times}$ (see $\cong \int f(x)\,dx$).

Three points: Simpson's rule (HMF 25.4.5).
f_1 $\boxed{\uparrow}$ 4 $\boxed{\times}$ f_0 $\boxed{+}$ f_2 $\boxed{+}$ 3 $\boxed{\div}$ h $\boxed{\times}$ (see $\cong \int f(x)\,dx$).

Four points: $\frac{3}{8}$ rule (HMF 25.4.13).
f_1 $\boxed{\uparrow}$ f_2 $\boxed{+}$ 3 $\boxed{\times}$ f_0 $\boxed{+}$ f_3 $\boxed{+}$ 8 $\boxed{\div}$ 3 $\boxed{\times}$ h $\boxed{\times}$ (see $\cong \int f(x)\,dx$).

Five points: Bode's rule (HMF 25.4.14).

f_1 [↑] f_3 [+] 32 [×] f_0 [↑] f_4 [+] 7 [×] [+] f_2 [↑] 12 [×] [+] 45 [÷] 2 [×] h [×] (see $\cong \int f(x)\,dx$).

Six points: Newton-Cotes formula (HMF 25.4.15).

f_1 [↑] f_4 [+] 3 [×] f_2 [↑] f_3 [+] 2 [×] [+] 25 [×] f_0 [↑] f_5 [+] 19 [×] [+] 288 [÷] 5 [×] h [×] (see $\cong \int f(x)\,dx$).

Seven points: Newton-Cotes formula (HMF 25.4.16).

f_1 [↑] f_5 [+] 8 [×] f_2 [+] f_4 [+] 27 [×] f_3 [↑] 272 [×] [+] f_0 [↑] f_6 [+] 41 [×] [+] 140 [÷] h [×] (see $\cong \int f(x)\,dx$).

Eight points: Newton-Cotes formula (HMF 25.4.17).

f_1 [↑] f_6 [+] 73 [×] f_2 [↑] f_5 [+] 27 [×] [+] f_3 [↑] f_4 [+] 61 [×] [+] 49 [×] f_0 [↑] f_7 [+] 751 [×] [+] 17280 [÷] 7 [×] h [×] (see $\cong \int f(x)\,dx$).

Any number of points: Simpleton's formula (HMF 25.4.2).

f_0 [↑] f_m [+] 2 [÷]
LOOP: f_n [+] (for $n=1$ to $m-1$) :|
h [×] (see $\cong \int f(x)\,dx$).

TEST CASE: See Section 5.4.
REFERENCE: HMF 25.4 as indicated.

A.3.2 Gaussian Quadrature

Gaussian quadrature algorithms approximate the integral of a function specified by a formula or algorithm.

Three-point Gauss-Legendre formula

The integral to be approximated is $\int_{x_L}^{x_u} f(x)\,dx$. Subroutine [$\cdots f \cdots$] is any sequence of keystrokes that accepts x in X and leaves $f(x)$ in X.

HP-45:

x_L [↑] [↑] [↑] x_u [$x \gtrless y$] [−] [STO] 1 [+] [+] 2 [÷] [STO] 2 [$\cdots f \cdots$] 8 [×] [STO] 4 [RCL] 1 .15 [G] [$\sqrt{x}$] [×] [STO] 3 [RCL] 2 [+] [$\cdots f \cdots$] [STO] 5 [RCL] 2 [RCL] 3 [−] [$\cdots f \cdots$] [RCL] 5 [+] 5 [×] [RCL] 4 [+] [RCL] 1 [×] 18 [÷] (see $\cong \int f(x)\,dx$).

TEST CASE: $x_L=0$, $x_u=1$, $f(x)=e^x$, [$\cdots f \cdots$] = [e^x], get 1.718281005. The precise answer is $e-1=1.718281828$.

ANOTHER TEST CASE: $x_L=0$, $x_u=1$, $f(x)=\sqrt{x}$, [···f···] = [G] [√x], get 0.6691796339. The precise answer is $\frac{2}{3}=$ 0.6666666667.

REFERENCES: HMF 25.4.30; HMF Table 25.4; and QG3 in IBM SSP, p. 299.

Five-point Lobatto-Radau formula

The integral to be approximated is $\int_{x_L}^{x_u} f(x)\,dx$. [···f···] is any sequence of keystrokes that accepts x in X and leaves $f(x)$ in X.

HP-45:

x_L [STO] 1 [···f···] [STO] 3 x_u [STO] 2 [···f···] [RCL] 3 [+] 9 [×] [STO] 3 [RCL] 1 [RCL] 2 [+] 2 [÷] [STO] 5 [···f···] 64 [×] [STO] [+] 3 [RCL] 2 [RCL] 1 [−] [STO] 1 3 [↑] 28 [÷] [G] [√x] [×] [STO] 2 [RCL] 5 [+] [···f···] [STO] 4 [RCL] 5 [RCL] 2 [−] [···f···] [RCL] 4 [+] 49 [×] [RCL] 3 [+] [RCL] 1 [×] 180 [÷] (see $\cong \int f(x)\,dx$).

TEST CASE: $x_L=0$, $x_u=1$, $f(x)=e^x$, [···f···] = [e^x], get 1.718281829. The precise answer is $e-1=$ 1.718281828.

ANOTHER TEST CASE: $x_L=0$, $x_u=1$, $f(x)=\sqrt{x}$, [···f···] = [G] [√x], get 0.6621414106. The precise answer is $\frac{2}{3}=$ 0.6666666667.

REFERENCES: HMF 25.4.32 and HMF Table 25.6.

n-point Gauss-Chebyshev formula

The integral to be approximated is

$$\int_{x_L}^{x_u} \frac{f(x)}{\sqrt{(x-x_L)(x_u-x)}}\,dx.$$

[···f···] is any sequence of keystrokes that accepts x in X and leaves $f(x)$ in X.

HP-45:

SETUP: x_L [↑] [↑] x_u [+] 2 [÷] [STO] 1 [$x\gtrless y$] [−] [STO] 2 [G] [π] n [÷] 2 [÷] [STO] 3 [G] [RAD] [G] [CLEAR] [Σ+]

LOOP: [↑] [+] 1 [−] [RCL] 3 [×] [COS] [RCL] 2 [×] [RCL] 1 [+] [···f···] [Σ+] (Check: Is display $n+1$? If so, quit.) :|

CODA: [RCL] [Σ+] [RCL] 3 [×] 2 [×] (see $\cong \int\cdots$).

TEST CASE: $x_L = 0$, $x_u = 1$, $f(x) = \sqrt{x(1-x)}$, $n = 3$, [···f···] = [↑] [x^2] [−] [G] [$\sqrt{x}$]; get 1.04719755. The precise answer is 1.

REFERENCES: HMF 25.4.39; Hamming (1962), p. 160.

Two-point Gauss-Laguerre formula

The integral to be approximated is $\int_0^\infty f(x)e^{-x}dx$. [···f···] is any sequence of keystrokes that accepts x in X, Y, Z, and T, leaves $f(x)$ in X, and retains x in Y (Z and T are irrelevant).

HP-45:

2 [↑] [G] [$\sqrt{x}$] [+] [↑] [↑] [↑] [···f···] [$x \gtrless y$] 4 [$x \gtrless y$] [−] [↑] [↑] [R↓] [R↓] [×] [STO] 4 [CLX] [−] [···f···] [$x \gtrless y$] 4 [$x \gtrless y$] [ ] [×] [RCL] 4 [+] 4 [÷] (see $\cong\int\cdots$).

HP-35:

2 [↑] [$\sqrt{x}$] [+] [↑] [↑] [↑] [···f···] [$x \gtrless y$] 4 [$x \gtrless y$] [−] [↑] [↑] [R↓] [R↓] [×] [STO] [CLX] [−] [···f···] [$x \gtrless y$] 4 [$x \gtrless y$] [−] [×] [RCL] [+] 4 [÷] (see $\cong\int\cdots$).

TEST CASE: $f(x)=\sqrt{x}$, HP-45 [···f···] = [G] [$\sqrt{x}$], HP-35 [···f···] = [$\sqrt{x}$], get 0.9238795328. The precise answer is $\sqrt{\pi}/2=0.8862269255$.

ANOTHER TEST CASE: $f(x)=x^2$, [···f···] = [×], get 2.000000001. The precise answer is 2 (the formula is exact for polynomials of degree three or less).

REFERENCES: HMF 25.4.45, HMF Table 25.9; and QL2 in IBM SSP, p. 303.

Gauss-Hermite formulas

The integral to be approximated is $\int_{-\infty}^{\infty} f(x)e^{-x^2}dx$. [···f···] is any sequence of keystrokes that accepts x in X, Y, Z, and T, leaves $f(x)$ in X, and retains x in Y (Z and T are irrelevant).

Three-point form

HP-45:

1.5 [G] [$\sqrt{x}$] [↑] [↑] [↑] [···f···] [STO] 4 [R↓] [CHS] [↑] [↑] [↑] [···f···] [RCL] 4 [+] 2 [÷] [STO] 4 [CLX] [↑] [↑] [↑] [···f···] 2 [×] [RCL] 4 [+] [G] [π] [G] [$\sqrt{x}$] [×] 3 [÷] (see $\cong\int\cdots$).

HP-35:

1.5 [$\sqrt{x}$] [↑] [↑] [↑] [···f···] [STO] [R↓] [CHS] [↑] [↑] [↑] [···f···]
[RCL] [+] 2 [÷] [STO] [CLX] [↑] [↑] [↑] [···f···] 2 [×] [RCL] [+]
[π] [$\sqrt{x}$] [×] 3 [÷] (see $\cong\int\cdots$).

Five-point form

HP-45:

2.5 [↑] [G] [$\sqrt{x}$] [+] [STO] 1 [G] [$\sqrt{x}$] [↑] [↑] [↑] [···f···] [STO] 2 [R↓]
[CHS] [↑] [↑] [↑] [···f···] [RCL] 2 [+] [RCL] 1 1.5 [−] [x^2] [÷]
[STO] 2 5 [RCL] 1 [−] [G] [$\sqrt{x}$] [↑] [↑] [↑] [···f···] [STO] 3 [R↓]
[CHS] [↑] [↑] [↑] [···f···] [RCL] 3 [+] [RCL] 1 3.5 [−] [x^2] [÷]
[RCL] 2 [+] 3 [×] 8 [÷] [STO] 2 [CLX] [↑] [↑] [↑] [···f···] 8 [×] 3 [÷]
[RCL] 2 [+] [G] [π] [G] [$\sqrt{x}$] [×] 5 [÷] (see $\cong\int\cdots$).

HP-35:

2.5 [↑] [$\sqrt{x}$] [+] [$\sqrt{x}$] [↑] [↑] [↑] [···f···] [STO] [R↓] [CHS] [↑] [↑]
[↑] [···f···] [RCL] [+] 2.5 [$\sqrt{x}$] 1 [+] [↑] [×] [÷] [STO] 2.5 [↑] [$\sqrt{x}$]
[−] [$\sqrt{x}$] [↑] [↑] [↑] [···f···] 2.5 [$\sqrt{x}$] 1 [−] [↑] [×] [÷] [RCL] [+]
[STO] [R↓] [CHS] [↑] [↑] [↑] [···f···] 2.5 [$\sqrt{x}$] 1 [−] [↑] [×] [÷]
[RCL] [+] 3 [×] 8 [÷] [STO] [CLX] [↑] [↑] [↑] [···f···] 8 [×] 3 [÷]
[RCL] [+] [π] [$\sqrt{x}$] [×] 5 [÷] (see $\cong\int\cdots$).

TEST CASE: $f(x)=\cos(2x)$, HP-45 [···f···] = ([G] [RAD]) [+] [COS], HP-35 [···f···] = [π] [÷] 360 [×] [COS]; get 0.7267617763 (three-point form) or 0.653223752 (five-point form). The precise answer is $\sqrt{\pi}/e=$ 0.6520493323.

REFERENCES: HMF 25.4.46, HMF Table 25.10; and QH3 and QH5 in IBM SSP, p. 308.

A.4. NUMERICAL INTEGRATION OF DIFFERENTIAL EQUATIONS

The algorithms discussed in this section approximately integrate differential equations of prescribed form from specified initial conditions (initial-value problems).

A.4.1 First-Order Ordinary Differential Equations

The differential equation to be integrated is

$$y'=f(x,y)$$

(a given function) beginning from $y(a)=b$, with step size h; that is,

$$x=a,\ a+h,\ a+2h,\cdots.$$

Subroutine [$\cdots f\cdots$] is any sequence of keystrokes that accepts y in X, x in Y, Z, and T, puts $f(x,y)$ into X, and retains x in Y (Z and T are irrelevant). Note that with the HP-35 versions, auto-enter will be sometimes enabled, sometimes disabled, on the call to [$\cdots f\cdots$].

Point-slope formula (HMF 25.5.1)

HP-45 SETUP: h [STO] 5 a [↑] [↑] [↑] b [STO] 7
LOOP: [$\cdots f\cdots$] [RCL] 5 [×] [RCL] 7 [+] (see y_{n+1}) [STO] 7 [$x\gtrless y$] [RCL] 5 [+] (see x_{n+1}) [↑] [↑] [R↓] [R↓] [R↓] :|.

HP-35 SETUP: a [↑] [↑] [↑] b [STO]
LOOP: [$\cdots f\cdots$] h [×] [RCL] [+] (see y_{n+1}) [STO] [$x\gtrless y$] h [+] (see x_{n+1}) [↑] [↑] [R↓] [R↓] [R↓] :|.

First-order bastard formula

$$y_{n+1}\cong y_n+hf\left[x_n+\frac{h}{2},\left(\frac{3y_n-y_{n-1}}{2}\right)\right].$$

HP-45 SETUP: b [STO] 7 [STO] 8 h [STO] 5 2 [÷] a [$x\gtrless y$] [−]
LOOP: [RCL] 5 [+] (see $x_n-h/2$) [$x\gtrless y$] [↑] [↑] [+] [+] [RCL] 8 [−] 2 [÷] [$\cdots f\cdots$] [RCL] 5 [×] [RCL] 7 [STO] 8 [+] (see y_{n+1}) [STO] 7 [$x\gtrless y$] :|.

Second-order Runge-Kutta (HMF 25.5.6)

HP-35 SETUP: a [↑] [↑] [↑] b [STO] (and precalculate $h/2$)
LOOP: [$\cdots f\cdots$] $(h/2)$ [×] [↑] [↑] [RCL] [+] [STO] [+] [$x\gtrless y$] h [+] (see x_{n+1}) [↑] [↑] [R↓] [R↓] [R↓] [$\cdots f\cdots$] $(h/2)$ [×] [RCL] [+] (see y_{n+1}) [STO] :|.

HP-45 SETUP: h [STO] 5 2 [÷] [STO] 6 a [↑] [↑] [↑] b [STO] 7
LOOP: [$\cdots f\cdots$] [RCL] 6 [×] [↑] [↑] [RCL] 7 [+] [STO] 7 [+] [$x\gtrless y$] [RCL] 5 [+] (see x_{n+1}) [↑] [↑] [R↓] [R↓] [R↓] [$\cdots f\cdots$] [RCL] 6 [×] [RCL] 7 [+] (see y_{n+1}) [STO] 7 :|.

Second-order Runge-Kutta (HMF 25.5.7)

HP-35 SETUP: a [↑] [↑] [↑] b [STO] (and precalculate $h/2$)
LOOP: [$\cdots f\cdots$] $(h/2)$ [×] [RCL] [+] [$x\gtrless y$] $(h/2)$ [+] [↑] [↑] [R↓] [R↓] [R↓] [$\cdots f\cdots$] h [×] [RCL] [+] (see y_{n+1}) [STO] [$x\gtrless y$] $(h/2)$ [+] (see x_{n+1}) [↑] [↑] [R↓] [R↓] [R↓] :|.

HP-45 SETUP: h [STO] 5 2 [÷] [STO] 6 a [↑] [↑] [↑] b [STO] 7

LOOP: [$\cdots f \cdots$] [RCL] 6 [×] [RCL] 7 [+] [$x \gtrless y$] [RCL] 6 [+] [↑] [↑] [R↓] [R↓] [R↓] [$\cdots f \cdots$] [RCL] 5 [×] [RCL] 7 [+] (see y_{n+1}) [STO] 7 [$x \gtrless y$] [RCL] 6 [+] (see x_{n+1}) [↑] [↑] [R↓] [R↓] [R↓] :|.

Third-order Runge-Kutta (HMF 25.5.8)

HP-45 SETUP: h [STO] 5 2 [÷] [STO] 6 a [↑] [↑] [↑] b [STO] 7

LOOP: [$\cdots f \cdots$] [RCL] 6 [×] [STO] 1 [RCL] 7 [+] [$x \gtrless y$] [RCL] 6 [+] [↑] [↑] [R↓] [R↓] [R↓] [$\cdots f \cdots$] [RCL] 5 [×] 2 [×] [STO] 2 [RCL] 1 2 [×] [−] [RCL] 7 [+] [$x \gtrless y$] [RCL] 6 [+] (see x_{n+1}) [↑] [↑] [R↓] [R↓] [R↓] [$\cdots f \cdots$] [RCL] 6 [×] [RCL] 2 [+] [RCL] 1 [+] 3 [÷] [RCL] 7 [+] (see y_{n+1}) [STO] 7 :|.

Third-order Runge-Kutta (HMF 25.5.9)

HP-45 SETUP: h [STO] 5 3 [÷] [STO] 6 a [↑] [↑] [↑] b [STO] 7

LOOP: [$\cdots f \cdots$] [RCL] 6 [×] [STO] 1 [RCL] 7 [+] [$x \gtrless y$] [RCL] 6 [+] [↑] [↑] [R↓] [R↓] [R↓] [$\cdots f \cdots$] [RCL] 6 [×] 2 [×] [RCL] 7 [+] [$x \gtrless y$] [RCL] 6 [+] [↑] [↑] [R↓] [R↓] [R↓] [$\cdots f \cdots$] [RCL] 5 [×] [RCL] 1 [+] 3 [×] 4 [÷] [RCL] 7 [+] (see y_{n+1}) [STO] 7 [$x \gtrless y$] [RCL] 6 [+] (see x_{n+1}) [↑] [↑] [R↓] [R↓] [R↓] :|.

Fourth-order Runge-Kutta (HMF 25.5.10)

HP-45 SETUP: h [STO] 5 2 [÷] [STO] 6 a [↑] [↑] [↑] b [STO] 7

LOOP: [$\cdots f \cdots$] [RCL] 6 [×] [STO] 1 [RCL] 7 [+] [$x \gtrless y$] [RCL] 6 [+] [↑] [↑] [R↓] [R↓] [R↓] [$\cdots f \cdots$] [RCL] 5 [×] [STO] 2 2 [÷] [RCL] 7 [+] [$\cdots f \cdots$] [RCL] 5 [×] [STO] 3 [RCL] 7 [+] [$x \gtrless y$] [RCL] 6 [+] (see x_{n+1}) [↑] [↑] [R↓] [R↓] [R↓] [$\cdots f \cdots$] [RCL] 6 [×] [RCL] 3 [+] [RCL] 2 [+] [RCL] 1 [+] 3 [÷] [RCL] 7 [+] (see y_{n+1}) [STO] 7 :|.

Fourth-order Runge-Kutta (HMF 25.5.11)

HP-45 SETUP: h [STO] 5 3 [÷] [STO] 6 a [↑] [↑] [↑] b [STO] 7

LOOP: [$\cdots f \cdots$] [RCL] 5 [×] [STO] 1 3 [÷] [RCL] 7 [+] [$x \gtrless y$] [RCL] 6 [+] [↑] [↑] [R↓] [R↓] [R↓] [$\cdots f \cdots$] [RCL] 5 [×] [STO] 2 [RCL] 1 3 [÷] [−] [RCL] 7 [+] [$x \gtrless y$] [RCL] 6 [+] [↑] [↑] [R↓] [R↓] [R↓] [$\cdots f \cdots$] [RCL] 5 [×] [STO] 3 [RCL] 2 [−] [RCL] 1 [+] [RCL] 7 [+] [$x \gtrless y$] [RCL] 6 [+] (see x_{n+1}) [↑] [↑] [R↓] [R↓] [R↓] [$\cdots f \cdots$] [RCL] 5 [×] [RCL] 1 [+] [RCL] 2 [RCL] 3 [+] 3 [×] [+] 8 [÷] [RCL] 7 [+] (see y_{n+1}) [STO] 7 :|.

Gill's modified fourth-order Runge-Kutta (HMF 25.5.12)

HP-45 SETUP: h [STO] 5 2 [÷] [STO] 6 a [↑] [↑] [↑] b [STO] 7

LOOP: [···f···] [RCL] 6 [×] [STO] 1 [RCL] 7 [+] [x≷y] [RCL] 6
[+] [↑] [↑] [R↓] [R↓] [R↓] [···f···] [RCL] 6 [×] [STO] 2
[RCL] 1 [−] 2 [G] [√x] 1 [x≷y] [−] [×] [RCL] 2 [+] [RCL]
7 [+] [···f···] [RCL] 6 [×] [STO] 3 [RCL] 2 [−] 2 [G] [√x]
[×] [RCL] 3 2 [×] [+] [RCL] 7 [+] [x≷y] [RCL] 6 [+]
(see x_{n+1}) [↑] [↑] [R↓] [R↓] [R↓] [···f···] [RCL] 6 [×]
[RCL] 1 [+] 2 [↑] [G] [√x] [−] [RCL] 2 [×] [+] 2 [↑] [G]
[√x] [+] [RCL] 3 [×] [+] 3 [÷] [RCL] 7 [+] (see y_{n+1})
[STO] 7 :|.

TEST CASE: See Section 5.6.

REFERENCES: HMF 25.5 as indicated.

A.4.2 Second-Order Ordinary Differential Equations

The differential equation to be integrated is

$$y'' = f(x,y),$$

(a given function) beginning from $y(a)=b$, $y'(a)=c$, with step size h, that is,

$$x = a,\ a+h,\ a+2h, \cdots .$$

[···f···] is any sequence of keystrokes that accepts x in X, y in Y, and leaves $f(x,y)$ in X.

Simplified third-order Runge-Kutta (modified from HMF 25.5.22)

HP-45 SETUP: h [STO] 5 2 [÷] [STO] 6 c [STO] 8 b [STO] 7 a [STO] 4
[···f···] [RCL] 6 [×] [STO] 1 2 [÷] [RCL] 8

LOOP: [+] [RCL] 6 [×] [RCL] 7 [+] [RCL] 6 [RCL] 4 [+] [···f···]
[RCL] 5 [×] [STO] 2 [RCL] 1 [+] 3 [÷] [RCL] 8 [+] [RCL]
5 [×] [RCL] 7 [+] (see y_{n+1}) [STO] 7 [RCL] 4 [RCL] 5
[+] (see x_{n+1}) [STO] 4 [···f···] [RCL] 6 [×] [RCL] 1
[x≷y] [STO] 1 [+] [RCL] 2 2 [×] [+] 3 [÷] [RCL] 8 [+]
(see y'_{n+1}) [STO] 8 [RCL] 1 2 [÷] :|.

Third-order Runge-Kutta (HMF 25.5.22)

HP-45 SETUP: h [STO] 5 2 [÷] [STO] 6 c [STO] 8 b [STO] 7 a [STO] 4

LOOP: [···f···] [RCL] 6 [×] [STO] 1 2 [÷] [RCL] 8 [+] [RCL] 6
[×] [RCL] 7 [+] [RCL] 6 [RCL] 4 [+] [···f···] [RCL] 5 [×]
[STO] 2 2 [÷] [RCL] 8 [+] [RCL] 5 [×] [RCL] 7 [+] [RCL]
4 [RCL] 5 [+] (see x_{n+1}) [STO] 4 [···f···] [RCL] 6 [×]
[RCL] 1 [+] [RCL] 2 2 [×] [+] 3 [÷] [RCL] 8 [+] (see
y'_{n+1}) [RCL] 8 [$x \gtrless y$] [STO] 8 [$x \gtrless y$] [RCL] 1 [RCL] 2 [+]
3 [÷] [+] [RCL] 5 [×] [RCL] 7 [+] (see y_{n+1}) [STO] 7
[RCL] 4 :|.

Another differential equation to be integrated is

$$y'' = f(x,y,y'),$$

from $y(a)=b$, $y'(a)=c$ with step size h. The subroutine [···f···] is any sequence of keystrokes that accepts x in X, y in Y, and y' in Z, and puts $f(x,y,y')$ into X; R_4 and R_8 are usable.

Fourth-order Runge-Kutta (HMF 25.5.20)

HP-45 SETUP: h [↑] 2 [÷] [STO] 5 c [STO] 8 b [STO] 7 a [STO] 6

LOOP: [···f···] [RCL] 5 [×] [STO] 1 [RCL] 8 [+] [↑] [↑] [RCL] 1
2 [÷] [−] [RCL] 5 [×] [RCL] 7 [+] [RCL] 6 [RCL] 5 [+]
[STO] 6 [···f···] [RCL] 5 [×] [STO] 2 [RCL] 8 [+] [RCL]
1 2 [÷] [RCL] 8 [+] [RCL] 5 [×] [RCL] 7 [+] [RCL] 6
[···f···] [RCL] 5 [×] [STO] 3 2 [×] [RCL] 8 [+] [↑] [↑]
[RCL] 8 [+] [RCL] 5 [×] [RCL] 7 [+] [RCL] 6 [RCL] 5 [+]
(see x_{n+1}) [STO] 6 [···f···] [RCL] 5 [×] [RCL] 1 [+]
[RCL] 3 [RCL] 2 [+] 2 [×] [+] 3 [÷] [RCL] 8 [+] (see
y'_{n+1}) [RCL] 3 [RCL] 2 [+] [RCL] 1 [+] 3 [÷] [RCL] 8 [+] 2
[×] [RCL] 5 [×] [RCL] 7 [+] (see y_{n+1}) [$x \gtrless y$] [STO] 8
[$x \gtrless y$] [STO] 7 [RCL] 6 :|.

TEST CASE: See Section 5.6.
REFERENCES: HMF 25.5 as indicated.

A.4.3 System of Two First-Order Ordinary Differential Equations

The differential equations to be integrated are

$$y' = f(x,y,z),$$

$$z' = g(x,y,z),$$

from $y(a)=b$, $z(a)=c$ with step size h. Subroutines [$\cdots f \cdots$] and [$\cdots g \cdots$] are any sequences of keystrokes that accept x in X, y in Y, and z in Z, and put f or g respectively into X; R_4 and R_9 are usable.

Second-order Runge-Kutta (HMF 25.5.17)

HP-45 SETUP: h [STO] 5 c [STO] 8 b [STO] 7 a [STO] 6

LOOP: [$\cdots f \cdots$] [RCL] 5 [×] [STO] 1 [RCL] 8 [RCL] 7 [RCL] 6 [$\cdots g \cdots$] [RCL] 5 [×] [STO] 2 [RCL] 8 [+] [RCL] 1 [RCL] 7 [+] [RCL] 5 [RCL] 6 [+] (see x_{n+1}) [STO] 6 [$\cdots f \cdots$] [RCL] 5 [×] [STO] 3 [RCL] 2 [RCL] 8 [+] [RCL] 1 [RCL] 7 [+] [RCL] 6 [$\cdots g \cdots$] [RCL] 5 [×] [RCL] 2 [+] 2 [÷] [RCL] 8 [+] (see z_{n+1}) [STO] 8 [RCL] 1 [RCL] 3 [+] 2 [÷] [RCL] 7 [+] (see y_{n+1}) [STO] 7 [RCL] 6 :|.

Modified second-order Runge-Kutta (based on HMF 25.5.17)

HP-45 SETUP: h [STO] 5 c [STO] 8 b [STO] 7 a [STO] 6

LOOP: [$\cdots f \cdots$] [RCL] 5 [×] [STO] 1 [RCL] 8 [RCL] 7 [RCL] 6 [$\cdots g \cdots$] [RCL] 5 [×] [STO] 2 [RCL] 8 [+] [RCL] 7 [RCL] 1 [+] [RCL] 6 [RCL] 5 [+] (see x_{n+1}) [STO] 6 [$\cdots f \cdots$] [RCL] 5 [×] [RCL] 1 [+] 2 [÷] [RCL] 7 [+] (see y_{n+1}) [STO] 7 [RCL] 8 [RCL] 2 [+] [$x \gtrless y$] [RCL] 6 [$\cdots g \cdots$] [RCL] 5 [×] [RCL] 2 [+] 2 [÷] [RCL] 8 [+] (see z_{n+1}) [STO] 8 [RCL] 7 [RCL] 6 :|.

TEST CASE: See Section 5.6.

REFERENCES: HMF 25.5 as indicated.

A.5 SPECIAL FUNCTIONS

This section contains algorithms dealing with mathematical functions and operations that are not available on the keys of (most) calculators. Some of these special functions can be expressed exactly in terms of simple functions. For more difficult cases, one needs an approximation based on a truncated series, an iterative algorithm, or some special technique depending on the properties of the function.

A.5.1 Hyperbolic Functions and Gudermannians

gd (Gudermannian) *in degrees*

HP-45: x [e^x] [G] [TAN^{-1}] 2 [×] 90 [−].

HP-35: x [e^x] [ARC] [TAN] 2 [×] 90 [−].

gd^{-1} (inverse Gudermannian) (enter with *degrees*)

x [↑] [TAN] [$x \gtrless y$] [COS] [$1/x$] [+] [ln].

sinh(x)

x [e^x] [↑] [$1/x$] [−] 2 [÷].

cosh(x)

x [e^x] [↑] [$1/x$] [+] 2 [÷].

tanh(x)

HP-45: x [e^x] [G] [TAN^{-1}] 2 [×] [COS] [CHS].
HP-35: x [e^x] [ARC] [TAN] 2 [×] [COS] [CHS].

$\sinh^{-1}(x)$

HP-45: x [↑] [x^2] 1 [+] [G] [$\sqrt{x}$] [+] [ln].
HP-35: x [↑] [↑] [×] 1 [+] [$\sqrt{x}$] [+] [ln].

$\cosh^{-1}(x)$

HP-45: x [↑] [x^2] 1 [−] [G] [$\sqrt{x}$] [+] [ln].
HP-35: x [↑] [↑] [×] 1 [−] [$\sqrt{x}$] [+] [ln].

$\tanh^{-1}(x)$

HP-45: x [G] [SIN^{-1}] 2 [÷] 45 [+] [TAN] [ln].
HP-35: x [ARC] [SIN] 2 [÷] 45 [+] [TAN] [ln].

TEST CASES: gd(0.5) = 27.°5238, sinh(0.5) = 0.5211, cosh(0.5) = 1.1276, tanh(0.5) = 0.4621.

REFERENCES: CTC SMT, pp. 337–349; HMF 4.3.117, 4.5, and 4.6.

A.5.2 Roots of a Quadratic

The quadratic equation

$$Ax^2 + Bx + C = 0$$

has roots x_1 and x_2 (real) *or* $a \pm ib$ (complex).

HP-45:

B [↑] 2 [÷] A [STO] 4 [÷] [CHS] (see *a*) [↑] [x^2] C [RCL] 4 [÷] [−]

Check display at this point:

If negative, roots are complex: [CHS] [G] [$\sqrt{x}$] (see b).

If positive, roots are real: [G] [$\sqrt{x}$] [+] (see x_1) [G] [LASTX] 2 [×] [−] (see x_2).

HP-35:

B [↑] 2 [÷] A [STO] [÷] [CHS] (see a) [↑] [↑] [×] C [RCL] [÷] [−]

Check display at this point:

If negative, roots are complex: [CHS] [$\sqrt{x}$] (see b).

If positive, roots are real: [$\sqrt{x}$] [STO] [+] (see x_1) 2 [RCL] [×] [−] (see x_2).

TEST CASE: $x^2-18x+77=0$ ($A=1$, $B=-18$, $C=77$); get $x_1=11$, $x_2=7$.

$$x=\frac{-B\pm\sqrt{B^2-4AC}}{2A}.$$

A.5.3 Real Roots of a Cubic by Iteration

A cubic equation

$$x^3+Ax^2+Bx+C=0$$

has either three real roots (which may coincide) or a real root and a pair of conjugate complex roots. The following iterative procedure finds the real root nearest an initial guess x_0. Probably a graphical method is best to get x_0.

HP-45 SETUP: A [STO] 1 2 [×] [STO] 2 B [STO] 3 C [STO] 4 x_0

LOOP: [↑] [↑] [↑] [+] [RCL] 1 [+] [×] [×] [RCL] 4 [−] [R↓] [R↓] 3 [×] [RCL] 2 [+] [×] [RCL] 3 [+] [÷] (see x_i) : |.

Then the real root can be divided out, leaving a quadratic to solve.

TEST CASE: $x^3+x^2-2x-2=0$ ($A=1, B=-2, C=-2$), $x_i=2$, 1.571428571, 1.430276168, 1.414408047, 1.414213592, 1.414213562 ($=\sqrt{2}$).

REFERENCE: Section 3.2. See also *HP-45 Applications Book*, p. 74.

A.5.4 Random Numbers

The algorithm below generates a series of pseudo-random numbers approximately uniformly distributed over the range 0 to 1 and showing seven figures. If random digits are wanted instead, use the left-most part of the numbers (the right-most digits are not as random).

INITIALIZE: 987 [↑] [↑] [↑] .7654321
HP-45 ONLY: [FIX] 7
LOOP: [×] (note *integer part* of display (three figures or less) key it back in) [−] (see a random number) :|.
ALTERNATIVE LOOP: [×] [↑] [↑] .5 [−] [EEX] 9 [+] [EEX] 9 [−] [−] (see a random number) :|.

The first few numbers of this series are: 0.4814827, 0.2234249, 0.5203763,.... The series repeats after 5×10^5 numbers.

Other series of random numbers can be obtained by changing 0.7654321 to any other number between 0 and 1, but the right-most (seventh) digit should be 1, 3, 7, or 9. The multiplier (987) also is not unique.

REFERENCE: IBM GC20-8011. See also Knuth (1969), Chapter 3.

A.5.5 Fourier Series Summation

The Fourier series is

$$F(x)=\frac{a_0}{2}+\sum_{n=1}^{m} a_n\cos\left(\frac{n\pi x}{L}\right)\pm b_n\sin\left(\frac{n\pi x}{L}\right).$$

For the + sign before b_n and if neither the a_n nor b_n are all zero:

HP-45 SETUP: [G] [CLEAR] [G] [RAD] x [G] [π] [×] L [÷] [↑] [↑] a_0 [↑] 2 [÷] [Σ+] (mentally set n to 1)
LOOP: [×] 1 [G] [→R] a_n [×] [$x \gtrless y$] b_n [×] [+] [Σ+] (see *next n*) :|
CODA: [RCL] [Σ+] (see $F(x)$).

Or if the b_n are all zero, change the loop to

LOOP: [×] [COS] a_n [×] [Σ+] (see *next n*) :|

Or if the a_n are all zero, change the loop to

LOOP: [×] [SIN] b_n [×] [Σ+] (see *next n*) :|

HP-21 SETUP: (switch to RAD mode) x [B] [π] [×] L [÷] [↑] [↑] a_0 [↑] 2 [÷] [STO] (mentally set n to 1)

LOOP: [CLX] n [×] 1 [B] [→R] a_n [×] [$x \gtrless y$] b_n [×] [+] [B] [M+] (mentally add 1 to n) :|

CODA: [RCL] (see $F(x)$).

Or if the b_n are all zero, change the loop to

LOOP: [CLX] n [×] [COS] a_n [×] [B] [M+] (mentally add 1 to n) :|

Or if the a_n are all zero, change the loop to

LOOP: [CLX] n [×] [SIN] b_n [×] [B] [M+] (mentally add 1 to n) :|

HP-35 SETUP: x [↑] 180 [×] L [÷] [↑] [↑] a_0 [↑] 2 [÷] [STO] (mentally set n to 1)

LOOP: n [×] [↑] [COS] a_n [×] [$x \gtrless y$] [SIN] b_n [×] [+] [RCL] [+] [STO] (mentally add 1 to n) :|

CODA: [RCL] (see $F(x)$).

Or if the b_n are all zero, change the loop to

LOOP: n [×] [COS] a_n [×] [RCL] [+] [STO] (mentally add 1 to n) :|

Or if the a_n are all zero, change the loop to

LOOP: n [×] [SIN] b_n [×] [RCL] [+] [STO] (mentally add 1 to n) :|

This HP-35 algorithm utilizes the special feature of the HP-35 that [STO] disables auto-enter; do not convert to other calculators.

If the sign before b_n is negative, replace [+] in the loops by [−].

TEST CASE: $a_0=2$, $a_1=1$, $a_2=2$, $b_1=1$, $b_2=2$, $m=2$, + sign on b_n, $x=2$, $L=4$; get 0.

REFERENCES: HMF 25.2.53; Bracewell (1965), Chapter 10; CRC SMT, p. 474; CRC HTM, p. 584; Lanczos (1956), Chapter 4.

A.5.6 Chebyshev Polynomials

Chebyshev polynomials are related to sines and cosines, and these relations can be used to evaluate the polynomials.

HP-45:

$T_n(x)$ $(-1 \leqslant x \leqslant 1)$
x [G] [COS⁻¹] n [×] [COS].

$T_n^*(x)$ $(0 \leqslant x \leqslant 1)$
x [↑] [+] 1 [−] [G] [COS⁻¹] n [×] [COS].

$U_n(x)$ $(-1 < x < 1)$
x [G] [COS⁻¹] [↑] [↑] [↑] n [×] [+] [SIN] [$x \gtrless y$] [SIN] [÷].

$U_n^*(x)$ $(0 < x < 1)$
x [↑] [+] 1 [−] [G] [COS⁻¹] [↑] [↑] [↑] n [×] [+] [SIN] [$x \gtrless y$] [SIN] [÷].

$C_n(x)$ $(-2 \leqslant x \leqslant 2)$
x [↑] 2 [÷] [G] [COS⁻¹] n [×] [COS] 2 [×].

$S_n(x)$ $(-2 < x < 2)$
x [↑] 2 [÷] [G] [COS⁻¹] [↑] [↑] [↑] n [×] [+] [SIN] [$x \gtrless y$] [SIN] [÷].

For an HP-35, replace [G] [COS⁻¹] by [ARC] [COS].
To sum

$$\sum_{n=0}^{m} a_n T_n(x); \quad -1 < x < 1,$$

HP-45 SETUP: [G] [CLEAR] x [G] [COS⁻¹] [↑] [↑] [↑] a_0 [Σ+]
LOOP: (see n) [×] [COS] a_n [×] [Σ+] :|
CODA: [RCL] [Σ+].
HP-35 SETUP: x [ARC] [COS] [↑] [↑] [↑] a_0 (mentally set $n = 1$)
LOOP: [$x \gtrless y$] n [×] [COS] a_n [×] [+] (mentally add 1 to n) :|
(see sum when $n = m + 1$).

The following HP-25 program computes Chebyshev polynomials from their recurrence relations, usually gives somewhat more precise answers than the trigonometric algorithms above, and allows x to be outside the normal range.

Line	Code	Key Entry
00	—	—
01	24 00	RCL 0
02	23 06	STO 6

Line	Code	Key Entry
03	22	R↓
04	24 01	RCL 1
05	61	×
06	24 02	RCL 2
07	41	−
08	23 07	STO 7
09	24 03	RCL 3
10	61	×
11	24 04	RCL 4
12	21	$x \gtrless y$
13	01	1
14	23 41 06	STO − 6
15	24 06	RCL 6
16	15 41	g $x<0$
17	13 26	GTO 26
18	41	−
19	22	R↓
20	21	$x \gtrless y$
21	24 07	RCL 7
22	61	×
23	21	$x \gtrless y$
24	41	−
25	13 12	GTO 12
26	22	R↓
27	22	R↓
28	22	R↓
29	13 00	GTO 00

To run this program, first set r_0 through r_4 from the following tabulation, that is,

[f] [PRGM] n [STO] 0 r_1 [STO] 1 r_2 [STO] 2 r_3 [STO] 3 r_4 [STO] 4.

	r_0	r_1	r_2	r_3	r_4
$T_n(x)$	n	2	0	0.5	1
$T_n^*(x)$	n	4	2	0.5	1
$U_n(x)$	n	2	0	0	1
$U_n^*(x)$	n	4	2	0	1
$C_n(x)$	n	1	0	1	2
$S_n(x)$	n	1	0	0	1

Then:

x [R/S] (see $f_n(x)$) :|,

where f is the selected Chebyshev polynomial. Repeat as often as needed.

TEST CASE: $T_6(0.2) = -0.354752$, $T_6^*(0.2) = 0.752192$, $U_6(0.2) = -0.163904$, $U_6^*(0.2) = 0.257984$, $C_6(0.2) = -1.649536$, $S_6(0.2) = -0.767936$.
If $a_0 = 0.5$, $a_1 = 1$, $a_2 = 2$, then

$$\sum_{n=0}^{2} a_n T_n(0.5) = 0.$$

REFERENCES: HMF 22.3, 22.5, 22.7.

A.5.7 Elliptic Integrals

The complete elliptic integral of the first kind is

$$K(k) = \int_0^{\pi/2} \frac{d\theta}{\sqrt{1 - k^2 \sin^2\theta}},$$

and the complete elliptic integral of the second kind is

$$E(k) = \int_0^{\pi/2} \sqrt{1 - k^2 \sin^2\theta}\, d\theta,$$

where $0 < k < 1$. Beware; some authors define $K(m)$ and $E(m)$, or $K(\alpha)$ and $E(\alpha)$ where $k^2 = m = \sin^2\alpha$.

$K(k)$ (HMF 17.3.33)
HP-45:

k [x^2] 1 [$x \gtrless y$] [−] [↑] [↑] [↑] .0288729 [×] .1213478 [+] [×] .5 [+] [$x \gtrless y$] [$1/x$] [ln] [×] [R↓] [R↓] .0725296 [×] .1119723 [+] [×] [+] 1.3862944 [+] (see $K(k)$) (absolute error $\leqslant 3 \times 10^{-5}$).

$E(k)$ (HMF 17.3.35)
HP-45:

k [x^2] 1 [$x \gtrless y$] [−] [↑] [↑] [↑] .0412496 [×] .2452727 [+] [×] [$x \gtrless y$] [$1/x$] [ln] [×] [R↓] [R↓] .1077812 [×] .4630151 [+] [×] [+] 1 [+] (see $E(k)$) (absolute error $< 4 \times 10^{-5}$).

For an HP-35, replace [x^2] by [↑] [×].

The following HP-25 program computes both $K(k)$ and $E(k)$ using Gauss's formula for the arithmogeometrical mean. Define a sequence of number pairs (a_n, b_n) as follows:

$$a_{-1}=1+k,\quad b_{-1}=1-k,\quad a_{n+1}=(a_n+b_n)/2,\quad b_{n+1}=\sqrt{a_n b_n}\ .$$

Compute these for $n=0,1,2\cdots N$, until $a_n=b_n$ to an accuracy of 10^{-9}. Then

$$K(k)=\frac{\pi}{2a_N},$$

$$E(k)=\frac{K(k)\left[2-(a_0^2-b_0^2)-2(a_1^2-b_1^2)-4(a_2^2-b_2^2)-\cdots\right]}{2}.$$

Line	Code	Key Entry
00	—	—
01	23 00	STO 0
02	23 01	STO 1
03	04	4
04	15 22	g $1/x$
05	23 02	STO 2
06	01	1
07	23 03	STO 3
08	24 00	RCL 0
09	51	+
10	01	1
11	24 00	RCL 0
12	41	−
13	23 00	STO 0
14	21	$x\gtrless y$
15	23 61 00	STO×0
16	51	+
17	02	2
18	23 61 02	STO×2
19	71	÷
20	31	↑
21	15 02	g x^2
22	24 00	RCL 0
23	41	−
24	24 02	RCL 2
25	61	×
26	23 41 03	STO−3

Line	Code	Key Entry
27	34	CLX
28	09	9
29	51	+
30	09	9
31	41	−
32	24 00	RCL 0
33	14 02	f $\sqrt{x}$
34	09	9
35	51	+
36	09	9
37	41	−
38	14 61	f $x \neq y$
39	13 13	GTO 13
40	15 73	g π
41	21	$x \gtrless y$
42	71	÷
43	02	2
44	71	÷
45	23 61 03	STO×3
46	23 02	STO 2
47	13 00	GTO 00

Run this program by

[f] [PRGM] k [R/S] (Then $K(k)$ is in the display and R_2, and $E(k)$ is in R_3) [RCL] 3 (see $E(k)$).

TEST CASE: $K(0.3) = 1.60808$ or 1.60804862, $E(0.3) = 1.53487$ or 1.53483347. (The second value in each case is from the HP-25 program and is more precise.)

REFERENCE: HMF 17.3, 17.6.

The incomplete elliptic integral of the first kind is

$$F(\phi\backslash\alpha) = \int_0^{\phi} \frac{d\theta}{\sqrt{1-\sin^2\alpha\sin^2\theta}}$$

The following HP-25 program computes $F(\phi\backslash\alpha)$ using Landen's descending transformation. Define a sequence of number triples (ϕ_n, α_n, F_n) as follows: $\phi_0 = \phi$, $\alpha_0 = \alpha$, $F_0 = 1$, and

$$\tan(\phi_{n+1} - \phi_n) = \tan\phi_n \cos\alpha_n,$$

$$\sin\alpha_{n+1} = \frac{2}{1+\cos\alpha_n} - 1,$$

$$F_{n+1} = \frac{F_n}{1+\cos\alpha_n}.$$

Compute these for $n = 0, 1, 2 \cdots N$ until $\alpha_n = 0$ to an accuracy of 10^{-10}. Then

$$F(\phi \backslash \alpha) = F_N \phi_N,$$

with ϕ_N in radians. However, the program will operate correctly in any angular mode. The two angles must be in the range 0 to $\sin^{-1} 1$ inclusive.

Line	Code	Key Entry
00	—	—
01	31	↑
02	01	1
03	23 03	STO 3
04	32	CHS
05	15 05	g COS^{-1}
06	23 00	STO 0
07	24 01	RCL 1
08	24 01	RCL 1
09	24 02	RCL 2
10	14 05	f COS
11	31	↑
12	22	R↓
13	21	$x \gtrless y$
14	14 06	f TAN
15	61	×
16	15 06	g TAN^{-1}
17	21	$x \gtrless y$
18	51	+
19	14 73	f LAST*X*
20	24 00	RCL 0
21	71	÷
22	73	.

Line	Code	Key Entry
23	05	5
24	14 71	f $x=y$
25	34	CLX
26	51	+
27	14 01	f INT
28	24 00	RCL 0
29	61	×
30	51	+
31	21	$x \gtrless y$
32	02	2
33	21	$x \gtrless y$
34	01	1
35	51	+
36	23 71 03	STO ÷ 3
37	71	÷
38	01	1
39	41	−
40	15 04	g SIN^{-1}
41	15 61	g $x \neq 0$
42	13 10	GTO 10
43	34	CLX
44	24 00	RCL 0
45	71	÷
46	15 73	g π
47	61	×
48	23 61 03	STO × 3
49	24 03	RCL 3

Run this program by

[f] [PRGM] ϕ [STO] 1 α [STO] 2 [R/S] (see $F(\phi\backslash\alpha)$).

TEST CASES: $F(\pi/4\backslash\pi/4) = 0.82602$, $F(90°\backslash\sin^{-1} 0.3) = 1.60805$ $(= K(0.3))$.

REFERENCE: HMF 17.5.

These HP-25 programs and their descriptions were written by Norman M. Brenner (private communication) and are reproduced here with his permission.

A.5.8 Factorial and Gamma Functions

$$x! = \Gamma(x+1), \quad \Gamma(x) = (x-1)!$$

$2 \lesssim x$
HP-45:

x [↑] [↑] [↑] .4 [$x \gtrless y$] [÷] [+] 4 [×] [1/x] [+] 30 [×] [1/x] [+] 12 [×]
[1/x] [$x \gtrless y$] [−] [e^x] [$x \gtrless y$] 2 [×] [G] [:π:] [×] [G] [:$\sqrt{x}$:] [×] [$x \gtrless y$] [↑]
[G] [:y^x:] [×] (see $\cong x!$).

$1 \lesssim x$
HP-45:

x [↑] 3 [+] [↑] [↑] [↑] .4 [$x \gtrless y$] [÷] [+] 4 [×] [1/x] [+] 30 [×] [1/x]
[ı] 12 [×] [1/x] [$x \gtrless y$] [ ] [e^x] [$x \gtrless y$] 2 [:] [G] [:π:] [:] [G] [:$\sqrt{x}$:] [:]
[$x \gtrless y$] [↑] [G] [:y^x:] [×] [$x \gtrless y$] 1 [−] [÷] [$x \gtrless y$] 2 [−] [÷] (see $\cong x!$).

For an HP-35, replace [G] [:π:] by [π], [G] [:$\sqrt{x}$:] by [$\sqrt{x}$], and [G] [:y^x:] by [x^y] (sic) in either algorithm above.

The precision is about eight or nine significant figures for $x \lesssim 69.957$ (overflow).

TEST CASE: $5! \cong 119.9999996$, $(\frac{1}{2})! \cong 0.8862269253$ (the precise value is $\sqrt{\pi/4} = 0.8862269254$.)

REFERENCES: HMF 6.1.15, 6.1.17, 6.1.48.

This algorithm is based on an HP-25 program by Norman M. Brenner (private communication).

The following algorithm calculates the incomplete gamma function (HMF 6.5.3)

$$\Gamma(n,x) = \int_x^\infty e^{-t} t^{n-1}\, dt,$$

for integer n, and the chi-square probability function (HMF 26.4.2)

$$Q(\chi^2|\nu) = \left[2^{\nu/2}\Gamma\left(\frac{\nu}{2}\right)\right]^{-1} \int_{\chi^2}^\infty t^{\nu/2-1} e^{-t/2}\, dt,$$

for an even number of degrees of freedom ν. For Q, use $n=\nu/2$ and $x=\chi^2/2$.

FOR $n=1$: x [CHS] [e^x] (see $Q(2x|2)=\Gamma(1,x)$).
FOR $n\geqslant 2$: x [↑] [↑] [↑] (mentally set $j=n-1$)
LOOP: j [÷] 1 [+] (mentally subtract 1 from j; is $j=0$? If so, go to coda) [×] :|
CODA: [$x\gtrless y$] [e^x] [÷] (see $Q(2x|2n)$),
HP-45 ONLY: n[↑] 1 [−] [G] [$n!$] [×] (see $\Gamma(n,x)$).

This algorithm uses the sum of the Poisson distribution (HMF 26.4.21, 26.4.19)

$$Q(2x|2n)=\frac{\Gamma(n,x)}{(n-1)!}=e^{-x}\sum_{j=0}^{n-1}\frac{x^j}{j!}.$$

The sum in this formula is the truncated series for e^x (HMF 6.5.11). Other related functions are the chi-square distribution (HMF 26.4.1)

$$P(\chi^2|\nu)=1-Q(\chi^2|\nu)$$

and the exponential integrals (HMF 6.5.9)

$$E_n(x)=x^{n-1}\Gamma(1-n,x).$$

This expression is useful for finding $E_n(x)$ for $n<0$; see also Section A.5.11.

TEST CASE: $Q(6|6)=0.42319$, $\Gamma(3,3)=0.84638$.

This algorithm is based in part on an HP-25 program by Norman M. Brenner (private communication). See also the *HP-55 Statistics Programs*, pp. 16 and 50.

A.5.9 Bessel Functions

The ordinary ascending series can be used to calculate Bessel J functions for all x and n, but the series converges slowly for large x.

$J_n(x)$ (ascending series for x positive and n integer) HP-45:
INITIALIZE: [G] [CLEAR] x [↑] 2 [÷] [↑] [x^2] [CHS] [$x\gtrless y$] n [STO] 1 [G] [y^x] [RCL] 1 [G] [$n!$] [$1/x$] [×] [↑] [Σ+]

LOOP: [↑] [RCL] [+] 1 [×] [÷] [×] (Check: is term negligible? If so, go to coda.) [↑] [Σ+] :|

CODA: [RCL] [Σ+] (see $\simeq J_n(x)$).

For $n=0$ ($J_0(x)$), the foregoing procedure can be simplified to:

$J_0(x)$ (ascending series) HP-45:

INITIALIZE: [G] [CLEAR] x [↑] 2 [÷] [x^2] [CHS] [↑] [↑] 1 [Σ+] [↑]

LOOP: [x^2] [÷] [×] (Check: is term negligible? If so, go to coda.) [↑] [Σ+] :|

CODA: [RCL] [Σ+] (see $\simeq J_0(x)$).

The approximate number of times one must go through the loop to get four significant figures of precision is shown in the following tabulation:

x	Number of Loops
0.5	2
1	3
2	5
4	8
8	14
16	25

The numbers in this tabulation are approximately true for all n, however the series converges a bit faster for larger n.

Rather than the open-ended series just given, the following truncated series are usually preferable.

$J_0(x)$ (modified ascending series of five terms; gives absolute error $<10^{-4}$ for $0<x\lesssim 3$).

HP-45:

x [x^2] 4 [*CHS*] [÷] [↑] [↑] [↑] *26* [÷] 1 [+] [×] 16 [÷] 1 [+] [×] 9 [÷] 1 [+] [×] 4 [÷] 1 [+] [×] 1 [+] (see $\simeq J_0(x)$).

For an HP-35, replace [x^2] by [↑] [×]. For $I_0(x)$, use the algorithm above for $J_0(x)$ but delete [*CHS*] and change *26* to 24.

$J_1(x)$ (modified ascending series of five terms; gives absolute error $<10^{-4}$ for $0<x\lesssim 3.6$).

HP-45:

x $\boxed{\uparrow}$ 2 $\boxed{\div}$ $\boxed{\text{STO}}$ 4 $\boxed{x^2}$ $\boxed{CHS}$ $\boxed{\uparrow}$ $\boxed{\uparrow}$ $\boxed{\uparrow}$ *32* $\boxed{\div}$ 1 $\boxed{+}$ $\boxed{\times}$ 20 $\boxed{\div}$ 1 $\boxed{+}$ $\boxed{\times}$ 12 $\boxed{\div}$ 1 $\boxed{+}$ $\boxed{\times}$ 6 $\boxed{\div}$ 1 $\boxed{+}$ $\boxed{\times}$ 2 $\boxed{\div}$ 1 $\boxed{+}$ $\boxed{\text{RCL}}$ 4 $\boxed{\times}$ (see $\cong J_1(x)$).

For an HP-35, replace $\boxed{x^2}$ by $\boxed{\uparrow}$ $\boxed{\times}$ and delete 4 after $\boxed{\text{STO}}$ and $\boxed{\text{RCL}}$. For $I_1(x)$, use the algorithm above for $J_1(x)$ but delete $\boxed{CHS}$ and change *32* to 28.

$\boldsymbol{J_n(x)}$ (ascending series of five terms; gives absolute error $<10^{-4}$ for $0<x\lesssim3.3$, $n\geqslant2$).

HP-45:

x $\boxed{\uparrow}$ 2 $\boxed{\div}$ $\boxed{\text{STO}}$ 4 $\boxed{x^2}$ $\boxed{CHS}$ $\boxed{\uparrow}$ $\boxed{\uparrow}$ n $\boxed{\uparrow}$ 5 $\boxed{+}$ 5 $\boxed{\times}$ $\boxed{\div}$ 1 $\boxed{+}$ $\boxed{\times}$ n $\boxed{\uparrow}$ 4 $\boxed{+}$ 4 $\boxed{\times}$ $\boxed{\div}$ 1 $\boxed{+}$ $\boxed{\times}$ n $\boxed{\uparrow}$ 3 $\boxed{+}$ 3 $\boxed{\times}$ $\boxed{\div}$ 1 $\boxed{+}$ $\boxed{\times}$ n $\boxed{\uparrow}$ 2 $\boxed{+}$ 2 $\boxed{\times}$ $\boxed{\div}$ 1 $\boxed{+}$ $\boxed{\times}$ n $\boxed{\uparrow}$ 1 $\boxed{+}$ $\boxed{\div}$ 1 $\boxed{+}$ n $\boxed{\text{G}}$ $\boxed{n!}$ $\boxed{\div}$ $\boxed{\text{RCL}}$ 4 n $\boxed{\text{G}}$ $\boxed{y^x}$ $\boxed{\times}$ (see $\cong J_n(x)$).

For $I_n(x)$, use the algorithm above for $J_n(x)$ but delete $\boxed{CHS}$.

$\boldsymbol{J_0(x)}$ (truncated asymptotic series from CRC SMT, p. 534, gives absolute error $<10^{-4}$ for $2.7\lesssim x$).

HP-45:

x $\boxed{\text{STO}}$ 4 $\boxed{1/x}$ 8 $\boxed{\div}$ $\boxed{x^2}$ $\boxed{\text{CHS}}$ $\boxed{\uparrow}$ $\boxed{\uparrow}$ $\boxed{\uparrow}$ 679 $\boxed{\times}$ 1 $\boxed{+}$ $\boxed{\times}$ 326.7 $\boxed{\times}$ 1 $\boxed{+}$ $\boxed{\times}$ 1225 $\boxed{\times}$ 12 $\boxed{\div}$ 1 $\boxed{+}$ $\boxed{\times}$ 4.5 $\boxed{\times}$ 1 $\boxed{+}$ $\boxed{\text{R}\downarrow}$ $\boxed{\text{R}\downarrow}$ 198 $\boxed{\times}$ 1 $\boxed{+}$ $\boxed{\times}$ 37.5 $\boxed{\times}$ 1 $\boxed{+}$ $\boxed{\text{RCL}}$ 4 $\boxed{\div}$ 8 $\boxed{\text{CHS}}$ $\boxed{\div}$ $\boxed{\text{G}}$ $\boxed{\text{RAD}}$ $\boxed{\to\text{P}}$ 2 $\boxed{\text{RCL}}$ 4 $\boxed{\div}$ $\boxed{\text{G}}$ $\boxed{\pi}$ $\boxed{\div}$ $\boxed{\text{G}}$ $\boxed{\sqrt{x}}$ $\boxed{\times}$ $\boxed{x\gtrless y}$ $\boxed{\text{RCL}}$ 4 $\boxed{-}$ $\boxed{\text{G}}$ $\boxed{\pi}$ 4 $\boxed{\div}$ $\boxed{+}$ $\boxed{\text{SIN}}$ $\boxed{\times}$ (see $\sim J_0(x)$).

$\boldsymbol{J_1(x)}$ (truncated asymptotic series from CRC SMT, p. 534, gives absolute error $<10^{-4}$ for $2.9\lesssim x$).

HP-45:

x $\boxed{\text{STO}}$ 4 $\boxed{1/x}$ 8 $\boxed{\div}$ $\boxed{x^2}$ $\boxed{\text{CHS}}$ $\boxed{\uparrow}$ $\boxed{\uparrow}$ $\boxed{\uparrow}$ 300 $\boxed{\times}$ 1 $\boxed{+}$ $\boxed{\times}$ 78.75 $\boxed{\times}$ 1 $\boxed{+}$ $\boxed{\times}$ 7.5 $\boxed{\text{CHS}}$ $\boxed{\times}$ 1 $\boxed{+}$ $\boxed{\text{R}\downarrow}$ $\boxed{\text{R}\downarrow}$ 173 $\boxed{\times}$ 1 $\boxed{+}$ $\boxed{\times}$ 17.5 $\boxed{\times}$ 1 $\boxed{+}$ 3 $\boxed{\times}$ 8 $\boxed{\div}$ $\boxed{\text{RCL}}$ 4 $\boxed{\div}$ $\boxed{\text{G}}$ $\boxed{\text{RAD}}$ $\boxed{\to\text{P}}$ 2 $\boxed{\text{RCL}}$ 4 $\boxed{\div}$ $\boxed{\text{G}}$ $\boxed{\pi}$ $\boxed{\div}$ $\boxed{\text{G}}$ $\boxed{\sqrt{x}}$ $\boxed{\times}$ $\boxed{x\gtrless y}$ $\boxed{\text{RCL}}$ 4 $\boxed{-}$ $\boxed{\text{G}}$ $\boxed{\pi}$ 4 $\boxed{\div}$ $\boxed{+}$ $\boxed{\text{COS}}$ $\boxed{\times}$ (see $\sim J_1(x)$).

Figures A.5.1 and A.5.2 are error plots for these algorithms. The ascending series are (HMF 9.1.10)

$$J_n(x)=\left(\frac{x}{2}\right)^n\sum_{k=0}^{\infty}\frac{(-x^2/4)^k}{k!(n+k)!}$$

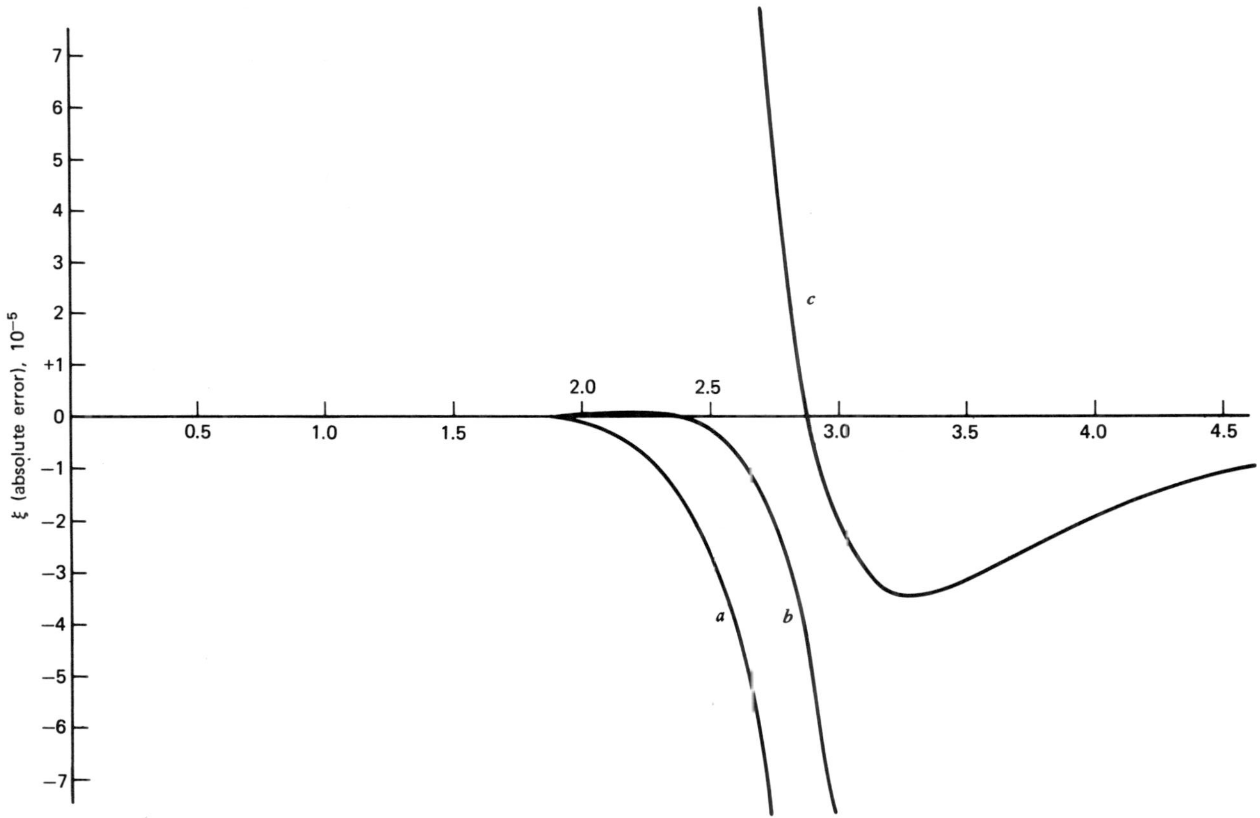

Figure A.5.1. An error plot for three formulas for the Bessel function $J_0(x)$. The abscissa is x and the ordinate is the difference between the value from the series and the precise value. The ascending series of five terms with the last constant 25 (a) and 26 (b); the latter is preferable. (c) The asymptotic series. The ideal point to change to the asymptotic series is about $x = 2.8$.

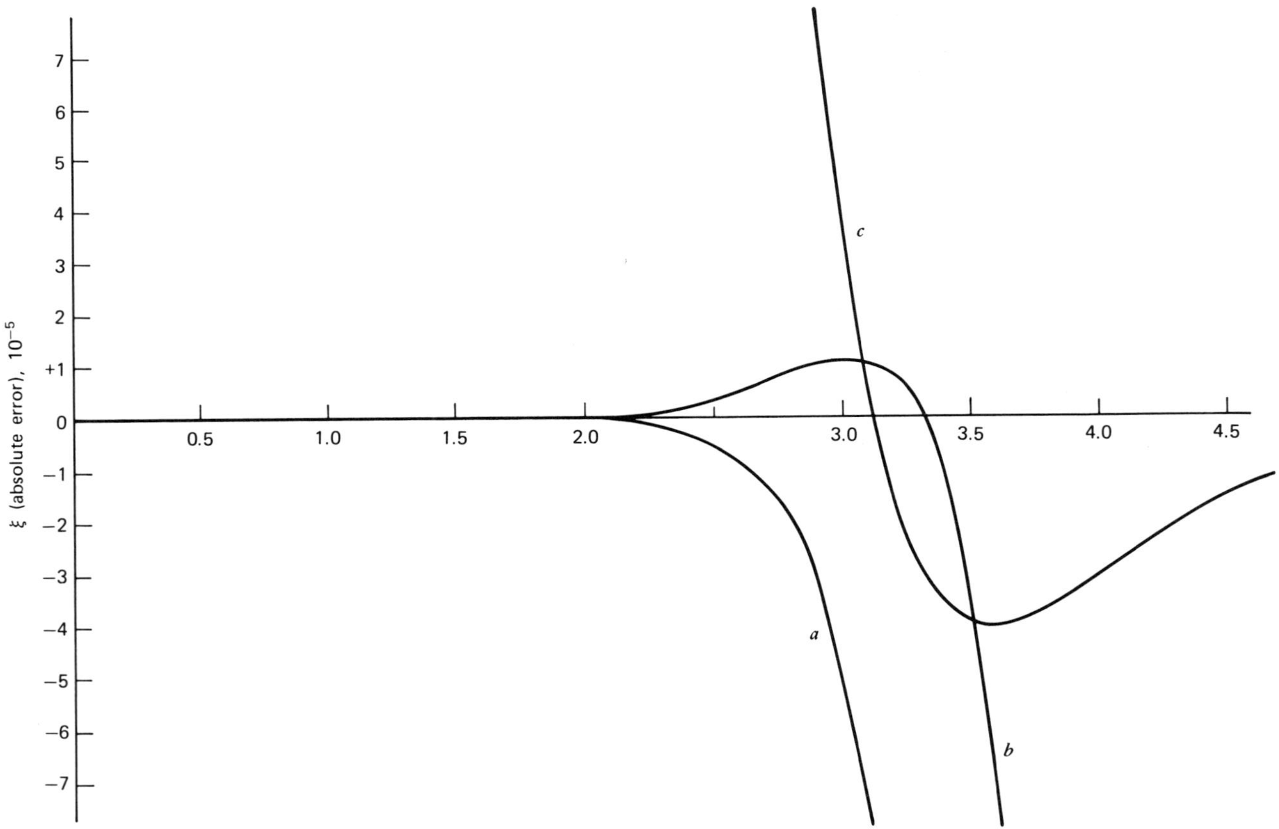

Figure A.5.2. Error plot similar to Figure A.5.1 for Bessel function $J_1(x)$. The ascending series of five terms with the last constant 30 (a), and 32 (b). (c) The asymptotic series.

and (HMF 9.6.10)

$$I_n(x) = \left(\frac{x}{2}\right)^n \sum_{k=0}^{\infty} \frac{(x^2/4)^k}{k!(n+k)!}.$$

The asymptotic series are (CRC SMT, p. 534)

$$J_0(x) \sim \left(\frac{2}{\pi x}\right)^{1/2} \left[P_0(x)\cos\left(x - \frac{\pi}{4}\right) - Q_0(x)\sin\left(x - \frac{\pi}{4}\right)\right],$$

$$J_1(x) \sim \left(\frac{2}{\pi x}\right)^{1/2} \left[P_1(x)\cos\left(x - \frac{3\pi}{4}\right) - Q_1(x)\sin\left(x - \frac{3\pi}{4}\right)\right],$$

where

$$P_0(x) \sim 1 - \frac{1^2 \cdot 3^2}{2!(8x)^2} + \frac{1^2 \cdot 3^2 \cdot 5^2 \cdot 7^2}{4!(8x)^4} - \frac{1^2 \cdot 3^2 \cdot 5^2 \cdot 7^2 \cdot 9^2 \cdot 11^2}{6!(8x)^6} + \cdots,$$

$$Q_0(x) \sim -\frac{1^2}{1!(8x)} + \frac{1^2 \cdot 3^2 \cdot 5^2}{3!(8x)^3} - \frac{1^2 \cdot 3^2 \cdot 5^2 \cdot 7^2 \cdot 9^2}{5!(8x)^5} + \cdots,$$

$$P_1(x) \sim 1 + \frac{1^2 \cdot 3 \cdot 5}{2!(8x)^2} - \frac{1^2 \cdot 3^2 \cdot 5^2 \cdot 7 \cdot 9}{4!(8x)^4} + \frac{1^2 \cdot 3^2 \cdot 5^2 \cdot 7^2 \cdot 9^2 \cdot 11 \cdot 13}{6!(8x)^6} - \cdots,$$

$$Q_1(x) \sim \frac{1 \cdot 3}{1!(8x)} - \frac{1^2 \cdot 3^2 \cdot 5 \cdot 7}{3!(8x)^3} + \frac{1^2 \cdot 3^2 \cdot 5^2 \cdot 7^2 \cdot 9 \cdot 11}{5!(8x)^5} - \cdots.$$

The point of diddling the constant in the last term of the truncated series is discussed in Section 2.4 and shown in Figures A.5.1 and A.5.2.

TEST CASES: $J_0(2.9) = -0.22436$ or -0.224315, $I_0(2.9) = 4.50269$, $J_1(2.9) = 0.37544$ or 0.3755, $I_1(2.9) = 3.61262$, $J_2(2.9) = 0.483221$, $I_2(2.9) = 2.011289$.

The first number in each case is from the modified ascending series, the second number for $J_0(2.9)$ and $J_1(2.9)$ is from the truncated asymptotic series.

A.5.10 Error Functions and Normal or Gaussian Distributions

One way to obtain numerical values for the error function

$$\text{erf}(x) = \frac{2}{\sqrt{\pi}} \int_0^x e^{-t^2}\, dt$$

is to use the ordinary ascending series.

HP-45 INITIALIZE: [G] [CLEAR] x [x^2] 2 [×] [↑] [CHS] [e^x] [G] [π] 2 [÷] [÷] [×] [G] [$\sqrt{x}$] [↑] [Σ+]

LOOP: [↑] [+] 1 [+] [÷] [×] (Check: Is display negligible? If so, go to coda.) [↑] [Σ+] :|

CODA: [RCL] [Σ+] (see $\cong$erf(x)).

Each cycle through the loop adds another term to the series. The approximate number of times one must go through the loop to get four significant figures of precision can be found in the following tabulation.

x	Number of Loops
0.1	1
0.2	2
0.4	3
1	6
2	13

An alternative method for erf(x) is a polynomial approximation such as the one suggested by Hastings and quoted in HMF 7.1.25,

$$\operatorname{erf}(x) \cong 1-\left(a_1 t + a_2 t^2 + a_3 t^3\right)e^{-x^2},$$

where $t=1/(1+px)$, $p=0.47047$, $a_1=0.3480242$, $a_2=-0.0958798$, $a_3=0.7478556$.

HP-45 INITIALIZE: .3480242 [STO] 1 .0958798 [CHS] [STO] 2 .7478556 [STO] 3 .47047 [STO] 4

OPERATION: x [STO] 5 [RCL] 4 [×] 1 [+] [$1/x$] [↑] [↑] [↑] [RCL] 3 [×] [RCL] 2 [+] [×] [RCL] 1 [+] [×] [RCL] 5 [x^2] [CHS] [e^x] [×] 1 [−] [CHS] (see $\cong$erf(x)) :|.

The operation can be repeated for various xs so long as the storage registers remain unchanged. This approximation gives about four significant figures. More precise polynomial approximations are also available.

The following truncated series are usually preferable to the open-ended series already given.

erf(x) (modified ascending series of seven terms; gives four or more significant figures for $0<x\lesssim 1.2$).

HP-45:

x [STO] 4 [x^2] [CHS] [↑] [↑] [↑] 90 [÷] 11 [$1/x$] [+] [×] 5 [÷] 9 [$1/x$] [+] [×] 4 [÷] 7 [$1/x$] [+] [×] 3 [÷] 5 [$1/x$] [+] [×] 2 [÷] 3 [$1/x$] [+] [×] 1 [+] [RCL] 4 [×] 2 [×] [G] [π] [G] [$\sqrt{x}$] [÷] (see $\cong$erf(x)).

erf(x) (modified continued fraction expansion of seven terms; gives four or more significant figures for $1 \lesssim x$).

HP-45:

x [↑] [↑] [↑] [+] 3 [$x \gtrless y$] [÷] [+] 6 [$x \gtrless y$] [÷] [+] [+] 5 [$x \gtrless y$] [÷] [+] 4 [$x \gtrless y$] [÷] [+] [+] 3 [$x \gtrless y$] [÷] [+] 2 [$x \gtrless y$] [÷] [+] [+] [1/x] [+] [$x \gtrless y$] [x^2] [e^x] [×] [G] [π] [G] [$\sqrt{x}$] [×] [1/x] 1 [$x \gtrless y$] [−] (see $\simeq$erf(x)).

For an HP-35, replace [x^2] by [↑] [×], [G] [π] by [π], and [G] [$\sqrt{x}$] by [$\sqrt{x}$]; and delete 4 after [STO] and [RCL].

The ascending series (from HMF 7.1.5) is

$$\operatorname{erf}(x) = \frac{2x}{\sqrt{\pi}} \sum_{n=0}^{\infty} \frac{(-x^2)^n}{n!(2n+1)},$$

and the continued fraction (from HMF 7.1.14) is

$$\operatorname{erf}(x) = 1 - \frac{C(x)}{\sqrt{\pi}\, e^{x^2}},$$

where

$$C(x) = \cfrac{1}{x + \cfrac{1/2}{x + \cfrac{1}{x + \cfrac{3/2}{x + \cfrac{2}{x + \cdots}}}}}.$$

Figure A.5.3 is an error plot for these algorithms. The point of diddling the last term of the truncated series is discussed in Section 2.4.

The normal or Gaussian distribution functions are related to the error function by

$$P(x) = \tfrac{1}{2} + \tfrac{1}{2}\operatorname{erf}(x/\sqrt{2}),$$

$$Q(x) = \tfrac{1}{2} - \tfrac{1}{2}\operatorname{erf}(x/\sqrt{2}),$$

$$A(x) = \operatorname{erf}(x/\sqrt{2}), \text{ all for } x \geqslant 0.$$

TEST CASE: erf(1.1) = 0.8802045 or 0.880206. (The first value is from the ascending series; the second from the continued fraction.)

REFERENCE: HMF 26.2.

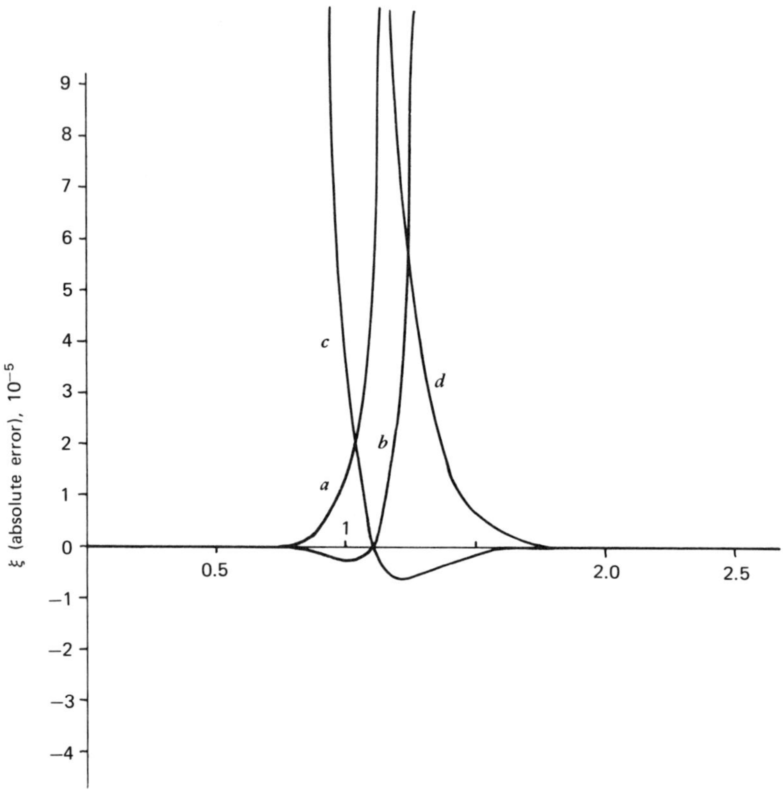

Figure A.5.3. An error plot for four formulas for the error function erf(x). The ascending series of seven terms with the last constant 78 (a); and 90 (b). The continued fraction of seven terms with the last constant 3 (c); 7 (d). The ideal point to change to the continued fraction is about $x=1.1$.

A.5.11 Exponential Integrals

The exponential integrals are

$$E_n(x)=x^{n-1}\int_x^{\infty}\frac{e^{-t}}{t^n}\,dt,$$

$$\mathrm{Ei}(x)=-E_1(-x)=\int_{-\infty}^{x}\frac{e^t}{t}\,dt.$$

$\boldsymbol{E_1(x)}$ (modified ascending series of eight terms, gives four or more significant figures for $0<x\lesssim 1.9$).

x [CHS] [↑] [↑] [↑] 76 [÷] 7 [1/x] [+] [×] 7 [÷] 6 [1/x] [+] [×] 6 [÷] 5 [1/x] [+] [×] 5 [÷] 4 [1/x] [+] [×] 4 [÷] 3 [1/x] [+] [×] 3 [÷] 2 [1/x] [+] [×] 2 [÷] 1 [+] [×] [x≷y] [CHS] [ln] [+] .5772156649 [+] [CHS] (see $\cong E_1(x)$).

For Ei(x), use series for $E_1(x)$ above, but delete [CHS] (in three places).

$E_1(x)$ (modified continued fraction expansion of eight terms, gives four or more significant figures for $1.9\lesssim x$).

x [↑] [↑] [↑] 2 [x≷y] [÷] 1 [+] 4 [x≷y] [÷] [+] 3 [x≷y] [÷] 1 [+] 3 [x≷y] [÷] [+] 2 [x≷y] [÷] 1 [+] 2 [x≷y] [÷] [+] [1/x] 1 [+] [1/x] [+] [x≷y] [e^x] [×] [1/x] (see $\cong E_1(x)$).

$E_1(x)$ (asymptotic expansion of seven terms, gives four or more significant figures for $12\lesssim x$).

x [CHS] [1/x] [↑] [↑] [↑] 7 [×] 1 [+] [×] 6 [×] 1 [+] [×] 5 [×] 1 [+] [×] 4 [×] 1 [+] [×] 3 [×] 1 [+] [×] 2 [×] 1 [+] [×] 1 [+] [×] [x≷y] [1/x] [e^x] [×] [CHS] (see $\sim E_1(x)$).

For Ei(x), use series for $E_1(x)$ above, but delete [CHS] (in two places).

For $E_n(x)$, $n>1$, first calculate $E_1(x)$, then use the recurrence relation:

SETUP: x [CHS] [↑] [↑] [↑] $E_1(x)$ (Note: after using the ascending series above to get $E_1(x)$, skip this setup; it has already been done.) (mentally set $n=1$)

LOOP: [×] [x≷y] [e^x] [+] n [÷] (see $E_{n+1}(x)$) (mentally add 1 to n) :|.

The ascending series is

$$E_1(x) = -\gamma - \ln x - \sum_{n=1}^{\infty} \frac{(-x)^n}{nn!},$$

where $\gamma = 0.5772156649\cdots$ is Euler's constant. The continued fraction expansion is

$$E_1(x)=\cfrac{e^{-x}}{x+\cfrac{1}{1+\cfrac{1}{x+\cfrac{2}{1+\cfrac{2}{x+\cfrac{3}{1+\cfrac{3}{x+\cfrac{4}{1+\cfrac{4}{x+\cdots}}}}}}}}}$$

The asymptotic expansion is

$$E_1(x) \sim \frac{e^{-x}}{x}\left\{1-\frac{1}{x}+\frac{2!}{x^2}-\frac{3!}{x^3}+\frac{4!}{x^4}-\frac{5!}{x^5}+\frac{6!}{x^6}-\frac{7!}{x^7}+\cdots\right\}.$$

Figure A.5.4 is an error plot. For $n<0$, see Section A.5.8.

TEST CASE: $E_1(1.9)=0.056203$ or 0.056208, $\mathrm{Ei}(1.9)=4.5935$. The first number is from the ascending series and the second number for $E_1(1.9)$ is from the continued fraction.

REFERENCE: HMF 5.1.

A.5.12 Modulo, gcd, lcm

Modulo function

To calculate m modulo n, where m and n are positive integers

n [↑] [↑] m [$x \gtrless y$] [÷] (note *integer part* of display, key it back in) [−] [×] (see m modulo n).

TEST CASE: 39 modulo 9 = 3.

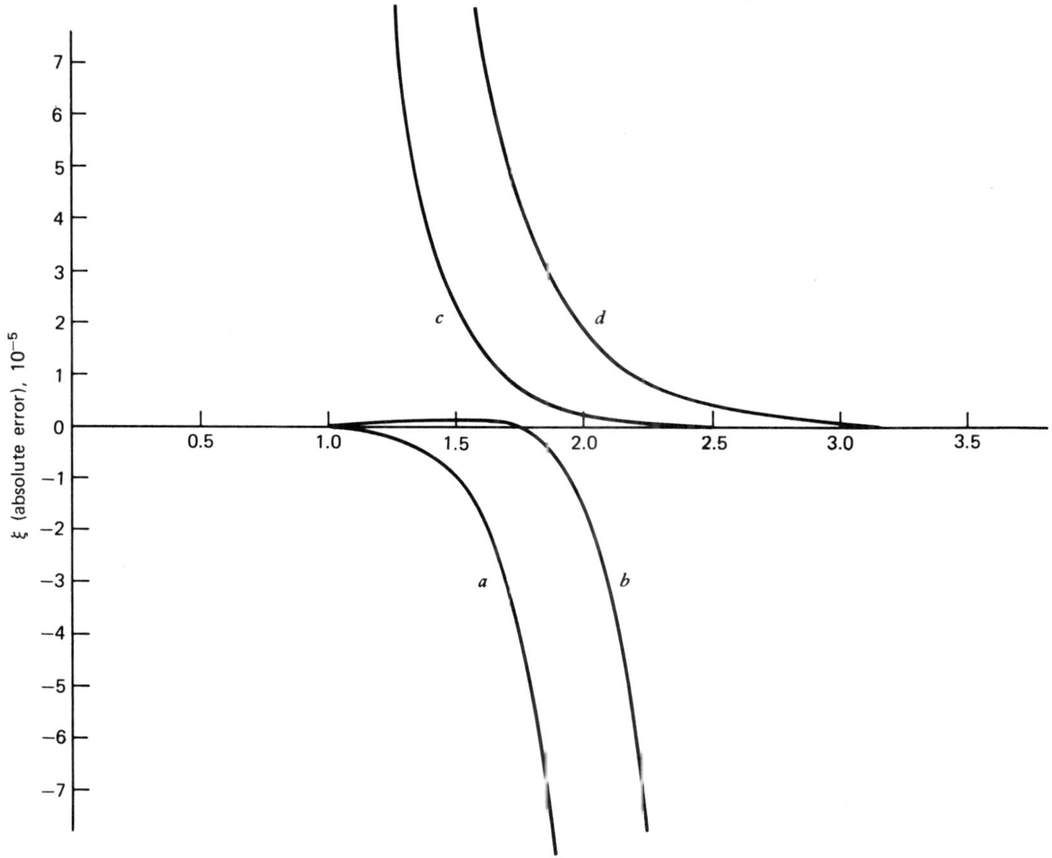

Figure A.5.4. An error plot for four formulas for the exponential integral $E_1(x)$. The ascending series of eight terms with the last constant 64 (a) and 75 (b) (the algorithm as printed has 76). The continued fraction of eight terms with the last constant 2 (c) and 4 (d).

Euclid's algorithm (c. 300 B.C.) for the greatest common divisor

To calculate the greatest common divisor (gcd) of m and n—positive integers,

SETUP: m [↑] [↑] [↑] n [$x \gtrless y$]
LOOP: [÷] (note *integer part* of display, key it back in) [−] [×] (Is display <1? If so, go to coda) [↑] [↑] [R↓] [R↓] :|
CODA: [R↓] (round to an integer if necessary; see gcd(m,n)).

Rounding errors necessitate using <1 rather than zero as the criterion for getting out of the loop. Choose $n > m$ to save one cycle through the loop. This "... may be called the granddaddy of all algorithms," says Donald Knuth (cited below).

Least common multiple

To calculate the least common multiple (lcm) of two positive integers m and n, first use the preceding algorithm to calculate the gcd, then

m [↑] n [×] (gcd(m,n)) [÷] (see lcm(m,n)).

TEST CASE: gcd(36,63) = 9, lcm(36,63) = 252.

REFERENCES: Euclid, Book 7, proposition 2; Knuth (1969), Vol. II, Section 4.5.2.

A.5.13 Permutations and Combinations

The number of permutations of n different things taken n at a time ${}_nP_n$ is

n [G] [$n!$],

and the number of permutations of n different things taken r at a time ${}_nP_r$ is

n [↑] [G] [$n!$] [$x \gtrless y$] r [−] [G] [$n!$] [÷].

The number of combinations of n different things taken r at a time ${}_nC_r$

is the same as the binomial coefficient $\binom{n}{r}$, and is

n [↑] [G] [$n!$] [$x \gtrless y$] r [↑] [G] [$n!$] [R↓] [−] [G] [$n!$] [÷] [$x \gtrless y$] [÷].

The preceding algorithms are for an HP-45. The following HP-25 program gives either ${}_nC_r$ or ${}_nP_r$.

Line	Code	Key Entry
00	—	—
01	23 01	STO 1
02	21	$x \gtrless y$
03	01	1
04	23 02	STO 2
05	51	+
06	23 03	STO 3
07	24 01	RCL 1
08	15 71	g $x=0$
09	13 30	GTO 30
10	71	÷
11	01	1
12	23 41 01	STO − 1
13	41	−
14	23 61 02	STO × 2
15	24 03	RCL 3
16	13 07	GTO 07
17	01	1
18	23 02	STO 2
19	41	−
20	23 01	STO 1
21	15 41	g $x<0$
22	13 30	GTO 30
23	21	$x \gtrless y$
24	23 61 02	STO × 2
25	01	1
26	23 41 01	STO − 1
27	41	−
28	24 01	RCL 1
29	13 21	GTO 21
30	24 02	RCL 2
31	13 00	GTO 00

Run this program by

$$n \; \boxed{\uparrow} \; r \; \boxed{\text{f}} \; \boxed{\text{PRGM}} \; \boxed{\text{R/S}} \qquad (\text{see } {}_nC_r),$$

or

$$n \; \boxed{\uparrow} \; r \; \boxed{\text{GTO}} \; 17 \; \boxed{\text{R/S}} \qquad (\text{see } {}_nP_r).$$

Both n and r are to be positive integers.

$$ {}_nC_r = \binom{n}{r} = \frac{n!}{(n-r)!r!},$$

$$ {}_nP_r = \frac{n!}{(n-r)!}.$$

TEST CASES: ${}_5C_3 = 10$, ${}_5P_3 = 60$.
REFERENCES: Lewart (1976); CRC SMT, p. 103.

A.5.14 Number in Base 10 to Number in Base b

The HP-25 program below converts a positive number N_{10} in base 10 into the equivalent number N_b in base b where $0 \leqslant b \leqslant 100$; b is an integer, but N_{10} need not be. When b is greater than 10, two display positions are necessary for each digit of N_b; the number must be partitioned both left and right from the decimal point. For example, 41106.12 in base 16 stands for 4B6.C in the usual notation.

Line	Code	Key Entry
00	—	—
01	23 07	STO 7
02	14 07	f ln
03	24 00	RCL 0
04	01	1
05	00	0
06	23 03	STO 3
07	14 51	f $x \geqslant y$
08	13 12	GTO 12
09	02	2
10	23 71 03	STO ÷ 3
11	14 03	f y^x
12	31	↑

Line	Code	Key Entry
13	22	R↓
14	15 22	g $1/x$
15	22	R↓
16	14 07	f ln
17	71	÷
18	14 01	f INT
19	14 03	f y^x
20	24 00	RCL 0
21	14 73	f LAST*X*
22	14 03	f y^x
23	23 71 07	STO ÷ 7
24	34	CL*X*
25	23 05	STO 5
26	34	CL*X*
27	33	EEX
28	32	CHS
29	08	8
30	24 07	RCL 7
31	51	+
32	13 38	GTO 38
33	22	R↓
34	61	×
35	24 00	RCL 0
36	24 07	RCL 7
37	61	×
38	23 07	STO 7
39	14 01	f INT
40	14 25	f Σ−
41	15 61	g $x \neq 0$
42	13 33	GTO 33
43	24 05	RCL 5
44	32	CHS
45	13 00	GTO 00

Run this program by

INITIALIZE: [f] [PRGM] [f] [FIX] 9 b [STO] 0

LOOP: N_{10} [R/S] (see N_b in the notation above) :|.

TEST CASES: $67.32 = 403.050114$ (as displayed for $b = 16$) $= 43.51\text{E}_{16}$. Or $3.141592654_{10} = 11.00100100_2$.

To convert N_b to N_{10}, see Section 2.4. This program and its description were written by Norman M. Brenner (private communication) and are reproduced here with his permission.

A.6 CURVE FITTING

The following algorithms obtain the parameters of a curve of specified form to go through or near given points. The criterion of goodness of fit is least squares or approximate least squares.

A.6.1 Straight-Line Fitting (Linear Regression)

The data, an indefinite number ($\geqslant 2$) of x_i, y_i pairs, are fitted in a least-squares sense to

$$y = a_0 + a_1 x,$$

where a_0 is the intercept and a_1 the slope. And σ is the RMS of the errors in this fit.

HP-45 INITIALIZE: [G] [CLEAR] [STO] 3 [STO] 4
ENTER DATA: x_i [↑] [↑] y_i [↑] [STO] [+] 3 [x^2] [STO] [+] 4 [R↓] [×] [$x \gtrless y$] [Σ+] (see i) :|
CODA: [RCL] 5 [RCL] 8 [×] [RCL] 3 [RCL] 7 [×] [−] [RCL] 5 [RCL] 6 [×] [RCL] 7 [x^2] [−] [÷] (see a_1) [↑] [↑] [RCL] 7 [×] [RCL] 3 [$x \gtrless y$] [−] [RCL] 5 [÷] (see a_0) [RCL] 3 [×] [$x \gtrless y$] [RCL] 8 [×] [+] [RCL] 4 [$x \gtrless y$] [−] [RCL] 5 [÷] [G] [$\sqrt{x}$] (see σ).

Then calculate the value of the fitted line at any x:

x [↑] a_1 [×] a_0 [+].

TEST CASE:

x_i	1	2	3	4
y_i	6.3	9.4	12.6	15.7

Get $a_1 = 3.14$, $a_0 = 3.15$, $\sigma = 0.022$.

REFERENCES: Elmore (1974); and Section 4.2.

A.6.2 Parabola Through Three Points

The y values are y_0 at $x=0$, y_- at $x=-1$, and y_+ at $x=1$; thus the x values are equally spaced and normalized. Then x_0 is the position of the peak (maximum or minimum y value) in these units, W is the full width to half-maximum (in these units), and A is the peak amplitude (maximum or minimum y value) in the same units as the ys.

The parabola fitted is

$$y=ax^2+bx+c,$$

and by inspection $c=y_0$. In case $a=0$, then x_0, A, and W are undefined.

HP-45:

(y_0) [↑] (y_-) [↑] [↑] (y_+) [$x \gtrless y$] [−] 2 [÷] (see b) [R↓] [−] [$x \gtrless y$] [R↓] [−] (see a) [STO] 4 [$x \gtrless y$] [↑] [R↓] [$x \gtrless y$] [÷] 2 [CHS] [÷] (see x_0) [$x \gtrless y$] [R↓] [×] 2 [÷] [$x \gtrless y$] [R↓] [+] (see A) [RCL] 4 [÷] 2 [CHS] [×] [G] [$\sqrt{x}$] (see W).

For an HP-35, change [G] [$\sqrt{x}$] to [$\sqrt{x}$] and delete 4 after [STO] and [RCL]. Then calculate the value of the parabola at any x:

x [↑] [↑] a [×] b [+] [×] c [+] (see y).

TEST CASE:

x_i	−1	0	1
y_i	1	3	2

Get $b=0.5$, $a=-1.5$, $x_0=\frac{1}{6}$, $A=3.04$, $W=2.01$.

A.6.3 Least-Squares Parabola

The data, an indefinite number ($\geqslant 3$) of x_i,y_i pairs, are fitted in a least-squares sense to

$$y=ax^2+bx+c.$$

In addition, x_0 is the position of the maximum or minimum y value, W is the full width to half-maximum, and A is the peak amplitude (i.e., the maximum or minimum y value). In case $a=0$, then x_0, A, and W are undefined.

HP-45 INITIALIZE: Turn calculator off, then on

ENTER DATA: x_i [↑] [↑] [↑] [×] [×] [STO] [+] 1 [×] [STO] [+] 2 [CLX] y_i [STO] [+] 3 [×] [STO] [+] 4 [×] [$x \gtrless y$] [Σ+] (see i) :|

CODA: [RCL] 5 [RCL] [×] 8 [RCL] 6 [RCL] [×] 3 [−] [RCL] 7 [x^2] [RCL] 5 [RCL] [×] 6 [−] [STO] 9 [×] [RCL] 5 [RCL] [×] 4 [RCL] 7 [RCL] [×] 3 [−] [RCL] 5 [RCL] [×] 1 [RCL] 7 [RCL] [×] 6 [−] [↑] [R↓] [×] [+] [RCL] 5 [RCL] [×] 2 [RCL] 6 [x^2] [−] [RCL] [×] 9 [R↓] [R↓] [x^2] [$x \gtrless y$] [R↓] [+] [÷] (see a) [$x \gtrless y$] [RCL] 7 [RCL] [×] 3 [RCL] 5 [RCL] [×] 4 [−] [R↓] [×] [$x \gtrless y$] [R↓] [+] [RCL] [÷] 9 (see b) [RCL] [×] 7 [$x \gtrless y$] [R↓] [$x \gtrless y$] [RCL] [×] 6 [+] [RCL] 3 [$x \gtrless y$] [−] [RCL] [÷] 5 (see c).

Then calculate the value of the parabola at any x:

x [↑] [↑] a [×] b [+] [×] c [+] (see y)

Then calculate x_0, A, and W if needed:

b [↑] [↑] 2 [CHS] [÷] a [STO] 9 [÷] (see x_0) [×] 2 [÷] c [+] (see A) [RCL] 9 [÷] 2 [CHS] [×] [G] [$\sqrt{x}$] (see W).

TEST CASE:

x_i	1	2	3	4	5
y_i	1	3	4	5	4

Get $a=-0.43$, $b=3.37$, $c=-2$, $x_0=3.93$, $A=4.63$, $W=4.65$.

REFERENCE: Section 4.2.

A.6.4 Polar Curve Fitting

The data, an indefinite number ($\geqslant 3$) of θ_i, Y_i pairs, are fitted in a least-squares sense to

$$Y = A\sin(\theta + B) + C,$$

or equivalently to

$$Y = A_x y + A_y x + C,$$

where $A_x = A\cos B,\ A_y = A\sin B,\ \ x = \cos\theta,\ \ y = \sin\theta.$

HP-45 INITIALIZE: Turn calculator off, then on (if θ is not to be in degrees, set angular mode).

ENTER DATA: θ_i [↑] 1 [G] [→R] [↑] [↑] [CLX] [−] [R↓] [×] [STO] [+] 1 [R↓] [$x \gtrless y$] [Σ+] (see i) Y_i [↑] [R↓] [STO] [+] 3 [×] [STO] [+] 2 [R↓] [×] [STO] [+] 4 :|

CODA: [RCL] 5 [RCL] 2 [×] [RCL] 3 [RCL] 8 [×] [−] [RCL] 7 [x^2] [RCL] 5 [RCL] 6 [×] [−] [STO] 9 [×] [RCL] 5 [RCL] 4 [×] [RCL] 7 [RCL] 3 [×] [−] [RCL] 5 [RCL] 1 [×] [RCL] 7 [RCL] [×] 8 [−] [↑] [R↓] [×] [+] [RCL] 5 [RCL] 6 [−] [RCL] 5 [×] [RCL] 8 [x^2] [−] [RCL] 9 [×] [R↓] [R↓] [R↓] [x^2] [+] [÷] (see A_x) [$x \gtrless y$] [RCL] 7 [RCL] 3 [×] [RCL] 5 [RCL] [×] 4 [−] [R↓] [×] [$x \gtrless y$] [R↓] [+] [RCL] 9 [÷] (see A_y) [RCL] 7 [×] [$x \gtrless y$] [R↓] [$x \gtrless y$] [RCL] 8 [×] [+] [RCL] 3 [$x \gtrless y$] [−] [RCL] 5 [÷] (see C).

Then calculate A and B if needed:

A_y [↑] A_x [→P] (see A) [$x \gtrless y$] (see B).

And calculate the value of the curve at any θ:

θ [↑] B [+] [SIN] A [×] C [+] (see Y).

TEST CASE:

i	1	2	3	4	5	6	7	8
θ_i	10°	40°	50°	100°	120°	190°	260°	300°
Y_i	4.06	3.07	2.72	1.30	1.04	1.94	4.32	4.96

Get $A_x = -1.49$, $A_y = 1.34$, $C = 3.00$; or $A = 2.00$, $B = 138°$.

REFERENCE: Section 4.2.

A.6.5 Power and Exponential Curve Fitting

The data, an indefinite number ($\geqslant 3$) of x_i, Y_i pairs, are fitted in an approximate least-squares sense to the power curve

$$Y = ax^b$$

or to the exponential or geometric curve

$$Y = ae^{bx}.$$

Power curve

HP-45 INITIALIZE: Turn calculator off, then on

ENTER DATA: Y_i [↑] [↑] [STO] [+] 1 [ln] [×] [STO] [+] 2 x_i [ln] [↑] [R↓] [×] [STO] [+] 3 [R↓] [×] [STO] [+] 4 [×] [STO] [+] 5 :|

CODA: [RCL] 1 [RCL] 3 [×] [RCL] 4 [RCL] 2 [×] [−] [RCL] 1 [RCL] 5 [×] [RCL] 4 [x^2] [−] [÷] (see b) [RCL] 4 [×] [RCL] 2 [$x \gtrless y$] [−] [RCL] 1 [÷] [e^x] (see a).

Then calculate the value of the power curve at any x:

x [↑] b [G] [y^x] a [×] (see Y).

Exponential curve

Use the algorithm above but omit [ln] after x_i. Then calculate the value of the exponential curve at any x,

x [↑] b [×] [e^x] a [×] (see Y).

The fit is only approximately least squares because the algorithm fits to $\ln Y$ but with weighting factors chosen so that the fit is approximately least squares in ΔY_i. For the exponential curve fit $Y_i > 0$, and for the power curve fit both $Y_i > 0$ and $x_i > 0$. These restrictions are necessary to work in log space, and this, in turn, is necessary to linearize the problem to make it solvable without iteration. Avoid points with Y_i very much smaller than the maximum Y_i.

TEST CASE:

x_i	1	2	3	4
Y_i	1	2	3	4

Get $a = 1.00$, $b = 1.00$ (power curve) or $a = 0.80$, $b = 0.42$ (exponential curve).

REFERENCE: Section 4.3.

A.6.6 Gaussian Through Three Points

The y values are y_0 at $x=0$, y_- at $x=-1$, y_+ at $x=1$; therefore the x values are equally spaced and normalized. They are fitted to

$$y = A\exp\left[\frac{-4\ln(2)(x-x_0)^2}{W^2}\right].$$

Then A is the peak amplitude (maximum y value), x_0 is the position of the peak in the normalized x units, and W is the full width to half-maximum in these units.

HP-45:

y_+ [↑] [↑] y_- [÷] [ln] [STO] 1 [$x \gtrless y$] y_0 [STO] 2 [÷] [ln] 2 [×] [−] [↑] [↑] [RCL] 1 2 [÷] [$x \gtrless y$] [÷] (see x_0) [x^2] 2 [÷] [×] [e^x] [RCL] 2 [×] (See A) [CLX] 2 [ln] 8 [×] [$x \gtrless y$] [÷] [G] [$\sqrt{x}$] (see W).

HP-35*:

y_+ [↑] [↑] y_- [÷] [ln] [STO] [$x \gtrless y$] y_0 [÷] [ln] 2 [×] [−] [↑] [↑] [RCL] 2 [÷] [$x \gtrless y$] [÷] (see x_0) [↑] [×] 2 [÷] [×] [e^x] y_0 [×] (see A) [CLX] 2 [ln] 8 [×] [$x \gtrless y$] [÷] [$\sqrt{x}$] (see W).

Then calculate the value of the Gaussian at any x,

HP-45: x [↑] x_0 [−] W [÷] [x^2] 2 [ln] [×] 4 [CHS] [×] [e^x] A [×] (see y).

HP-35: x [↑] x_0 [−] W [÷] [↑] [×] 2 [ln] [×] 4 [CHS] [×] [e^x] A [×] (see y).

TEST CASE:

x_i	−1	0	1
y_i	1	3	2

Get $x_0 = 0.23$, $A = 3.12$, $W = 1.92$.

*Sorry about having to key y_0 twice; I do not see any way to avoid it.

A.6.7 Gaussian Curve Fitting

The data, an indefinite number ($\geqslant 3$) of x_i, y_i pairs, are fitted in an approximate least-squares sense to

$$y = A\exp\left[\frac{-4\ln(2)(x-x_0)^2}{W^2}\right].$$

Then A is the peak amplitude (maximum y value), x_0 is the position of the peak, and W is the full width to half-maximum.

HP-45 INITIALIZE: Turn calculator off, then on

ENTER DATA: x_i [↑] [↑] [↑] y_i [STO] 9 [x^2] [×] [STO] [+] 1 [×] [STO] [+] 2 [×] [STO] [+] 3 [×] [STO] [+] 4 [CLX] [RCL] 9 [↑] [ln] [$x \gtrless y$] [x^2] [STO] [+] 5 [×] [STO] [+] 6 [×] [STO] [+] 7 [×] [STO] [+] 8 :|

CODA: [RCL] 5 [RCL] [×] 8 [RCL] 2 [RCL] [×] 6 [−] [RCL] 1 [x^2] [RCL] 5 [RCL] [×] 2 [−] [STO] 9 [×] [RCL] 5 [RCL] [×] 7 [RCL] 1 [RCL] [×] 6 [−] [RCL] 5 [RCL] [×] 3 [RCL] 1 [RCL] [×] 2 [−] [↑] [R↓] [×] [+] [RCL] 5 [RCL] [×] 4 [RCL] 2 [x^2] [−] [RCL] [×] 9 [R↓] [R↓] [x^2] [$x \gtrless y$] [R↓] [+] [÷] 2 [ln] 4 [CHS] [×] [$x \gtrless y$] [↑] [R↓] [÷] [G] [$\sqrt{x}$] (see W) [R↓] [×] [RCL] 5 [RCL] 7 [×] [−] [RCL] 1 [RCL] [×] 6 [+] [RCL] [÷] 9 [STO] 9 [$x \gtrless y$] [÷] 2 [CHS] [÷] (see x_0) [x^2] [×] [$x \gtrless y$] [RCL] [×] 2 [RCL] 9 [RCL] [×] 1 [+] [RCL] 6 [$x \gtrless y$] [−] [RCL] [÷] 5 [$x \gtrless y$] [−] [e^x] (see A).

Then calculate the value of the Gaussian at any x:

x [↑] x_0 [−] W [÷] [x^2] 2 [ln] [×] 4 [CHS] [×] [e^x] A [×] (see y).

The fit is only approximately least squares because the algorithm fits to $\ln y_i$ but with weighting factors chosen so that the fit is approximately least squares in Δy_i. Low-noise data and points only around the peak are best; avoid baseline. Negative y_i are forbidden, and y_i near the baseline (i.e., less than the noise level) produce unpredictable results. *Prescaling* the x_i may be necessary in some cases to reduce the accuracy requirements in the matrix inversion. As a rough rule, zero should be among the x_i; if it is not, subtract a constant from all the x_i and add the constant to x_0 after the fit.

TEST CASE:

x_i	1	2	3	4	5
y_i	1	3	4	5	4

Get $W=4.16$, $x_0=3.91$, $A=4.88$.

REFERENCE: Section 4.3.

A.7 DATES, TIMES, AND POINTING IN ASTRONOMY

But Aristarchus of Samos published a book of some hypotheses in which the premises lead to the result that the universe is many times larger than now believed. His hypotheses are that the fixed stars and the sun do not move, that the earth revolves around the sun along the circumference of a circle with the sun at the center of the orbit, and that the sphere of fixed stars, with the sun at the center, is so large that the circle on which he supposes the earth to revolve is in proportion to the distance of the fixed stars as the center [point] of a sphere to its radius.

—*from* The Sand-Reckoner, *Archimedes (c. 287–212 B.C.)*

A.7.1 Pointing Notes

Stars and other objects on the sky are known by their right ascension α and declination δ, usually for some standard epoch such as 1950. These coordinates change only very slowly because of precession, nutation, and annual aberration; the α,δ coordinate system is almost fixed on the sky. The hour angle (HA), declination coordinate system is fixed with respect to the earth, that is, with respect to the setting circles on an earthbound telescope. The hour angle is the (local) sidereal time minus the right ascension,

$$HA = LST - \alpha,$$

and this expression can be taken to define LST. Objects moving with respect to the sky, such as the sun, moon, and planets, have α,δ coordinates that change more or less rapidly with time; an ephemeris gives α and δ for such an object.

Beginning with α and δ for a given epoch, Section A.7.9 precesses these coordinates to date. If a more precise answer is needed, add nutation and annual aberration from Section A.7.10 (<47 arcseconds). Then calculate sidereal time (LST) from Section A.7.4 and use the expression above to calculate HA. If azimuth Az and elevation El are needed, they can now be calculated from Section A.7.6. For a more precise El, add atmospheric refraction from Section A.7.8. If only the rising and setting times are needed, use Section A.7.5 instead.

In addition to α, δ, two other coordinate systems fixed on the sky are sometimes useful: galactic latitude $\mathfrak{b}^{II}$ and longitude $\mathfrak{l}^{II}$ from Section A.7.12, and ecliptic latitude β and longitude L from Section A.7.13.

REFERENCES: AENA, ESE, Smart (1962), and TN1969-42.

A.7.2 Distances and Headings Between Points on the Earth

From λ_1, the longitude (west is +) and ϕ_1, the latitude (north is +) of station one on the earth, and λ_2, the longitude, and ϕ_2, the latitude of station two, this algorithm gives H, the initial heading (north reference clockwise azimuth) from station one toward station two, and the distance as indicated. This algorithm is approximate because it assumes a spherical earth.

HP-45:

(ϕ_2: DD.MMSS) [G] [D.MS→] 1 [G] [→R] (λ_1: DD.MMSS) [G] [D.MS→] (λ_2: DD.MMSS) [G] [D.MS→] [−] [$x \gtrless y$] [G] [→R] [$x \gtrless y$] [↑] [R↓] [R↓] [→P] [$x \gtrless y$] (ϕ_1: DD.MMSS) [G] [D.MS→] [−] [$x \gtrless y$] [G] [→R] [R↓] [$x \gtrless y$] [R↓] [→P] [$x \gtrless y$] (if <0; 360 [+]; see H in degrees) [R↓] [$x \gtrless y$] [→P] [$x \gtrless y$]

At this point choose one of the following options:

a. 60 [×] (see great-circle distance in nautical miles),
b. 69.05 [×] (see great-circle distance in statute miles),
c. 111.19 [×] (see great-circle distance in kilometers),
d. 2 [÷] [SIN] 12742 [×] (see straight-line distance in kilometers).

TEST CASE: $\lambda_1 = 71°05'$, $\phi_1 = 42°22'$ (Boston); $\lambda_2 = 70°40'$, $\phi_2 = -33°25'$ (Santiago de Chile); get $H = 179.^\circ 64$ (slightly *east* of south) and distance = 5233 statute miles.

This algorithm is based on a suggestion by George Rybicki (private communication). The spherical triangle solution is from Smart (1962), p.

13. For a modification of this algorithm to give pointing angles and slant ranges to an earth satellite, see Ball (1977).

A.7.3 Calendar

Day number from date

This algorithm calculates the day number (day of the year), given the month m (1 through 12) and the day of the month d (1 through 31). Leap years are evenly divisible by 4 (e.g., 1976); centennial years, however, are not leap years, except that centennial years evenly divisible by 400 are leap years.

m [↑] [↑] 1 [−] 31 [×] d [+] (if $m \leqslant 2$, see day number; otherwise go on) [$x \gtrless y$] .4 [×] [−] 1.8 [−] (if this is a leap year: 1 [+])

At this point choose either

HP-45: [FIX] 0 (see day number),

HP-35: [EEX] 9 [+] [EEX] 9 [−] (see day number).

TEST CASE: 1976 May 9, get day number = 130.

This algorithm is based on an HP-65 program by R. C. Vanderburgh (private communication).

Day of the week

This algorithm calculates the weekday (day of the week), where Sunday is 1, Monday is 2, and so on, given the current year Y (four digits, e.g., 1976) unless it is January or February, in which case Y is the previous year, M, the month number, unless it is January or February, in which case use 13 or 14 respectively (of the previous year) and d, the day of the month (1 through 31). This algorithm is for the New Style (Gregorian) calendar.

Y [↑] [↑] [EEX] 2 [÷] [↑] [↑] 2 [−] 4 [÷] 7 [EEX] 9 [+] [$x \gtrless y$] .5 [−] [−] [$x \gtrless y$] 5 [×] 2 [−] 4 [÷] [+] M [↑] 13 [×] 11 [+] 5 [÷] [+] d [+] [↑] [↑] 4 [−] 7 [÷] 7 [×] [−] (see weekday).

REFERENCES: *HP-45 Applications Book*, p. 213; Elmore (1976).

Day of the week, days elapsed, and phase of the moon

This HP-25 program computes the day of the week for a given date, the number of days between two dates, or the phase of the moon for a given date, for any dates since February 1, 1 B.C. That date is assigned day number one, and a corresponding number N is given to each succeeding day. All calendrical corrections are made by the program provided it is informed by the user whether the Old Style (Julian) or New Style (Gregorian) calendar was in use on the given date. The program needs to know the month number m (1 through 12), the day of the month d (1 through 31), the year y (e.g., 1976), and the style s (0=Old, 1=New). The day of the week D is then 0 for Sunday, 1 for Monday and so on. Note that this is *not* the same convention as in the previous algorithm. The phase of the moon P is 0 for the first quarter, 0.25 for full moon. 0.5 for third quarter and 0.75 for new moon.

Line	Code	Key Entry
00	—	—
01	24 01	RCL 1
02	02	2
03	41	−
04	01	1
05	02	2
06	71	÷
07	31	↑
08	34	CLX
09	14 51	f $x \geqslant y$
10	01	1
11	51	+
12	24 03	RCL 3
13	14 73	f LASTX
14	41	−
15	24 00	RCL 0
16	15 21	g %
17	14 01	f INT
18	15 71	g $x=0$
19	15 05	g $\cos^{-1}$
20	32	CHS
21	73	·
22	07	7
23	05	5
24	61	×
25	21	$x \gtrless y$
26	14 73	f LASTX
27	61	×

Line	Code	Key Entry
28	04	4
29	08	8
30	07	7
31	61	$\times$
32	14 01	f INT
33	51	+
34	21	$x \gtrless y$
35	03	3
36	06	6
37	07	7
38	61	$\times$
39	14 01	f INT
40	51	+
41	14 01	f INT
42	24 02	RCL 2
43	51	+
44	74	R/S
45	07	7
46	71	$\div$
47	15 01	g FRAC
48	07	7
49	61	$\times$

Run this program by

[g] [RAD] *s* [STO] 0 *m* [STO] 1 *d* [STO] 2 *y* [STO] 3 [f] [PRGM]
[R/S] (see *N*)

At this point choose one or the other of the following two options:

a. [R/S] (see *D*),
b. 29.53059 [÷] [g] [FRAC] (see *P*).

The Julian day number (*JD*) used by astronomers (beginning on January 1, 4713 B.C.*) is

$$JD = N + 1721089.$$

*Bishop Usher (1581–1656) calculated that the world began in 4004 B.C., but astronomers now know that he was incorrect by more than 700 years.

The Julian calendar was used until about 1582 in Roman Catholic countries such as France, Spain, and Italy, until 1752 in the British Empire including the American colonies, and until 1873 in Japan, 1912 in China, 1918 in Russia, 1923 in Greece, and 1927 in Turkey.

The basic formula used in this program is

$$N = d + \left[367\left(\frac{m-2}{12} + x \right) \right] + \left[\left[365.25(y-x) \right] - 0.75c \right],$$

where

$$x = \begin{cases} 1 & \text{if } \ m = 1 \text{ or } 2, \\ 0 & \text{if not,} \end{cases}$$

$$c = \begin{cases} 2 & \text{if } \ s = 0 \text{ (Old Style)}, \\ \left[(y-x)/100 \right] & \text{if } \ s = 1 \text{ (New Style)}, \end{cases}$$

and $[z]$ represents the truncation function f INT.

TEST CASES: July 4, 1776 (New Style) was Thursday, $D=4$. From May 27, 1948, until April 7, 1975, was 9872 days. George Washington was born on February 11, 1732 (OS) which is the same as February 22, 1732 (NS) because both dates give $N=632623$. The phase of the moon when George was born was $P=0.63$ —four days past third quarter.

This HP-25 program and its description were written by Norman M. Brenner (private communication) and reproduced here with his permission.

A.7.4 Sidereal Time

These algorithms compute the local (mean) sidereal time *LST* for any site on the earth. There are two versions, differing in precision. Day number means day of the year. Greenwich mean time *GMT* is the mean solar (civil) time at Greenwich, England, or equivalently the universal time *UT*. *GMT* differs from local (civil) time anywhere by an integer number of hours. *S* is from the tabulation below. Note that the site longitude λ is in hours, minutes, and seconds; west is +.

Precision $\pm 1^s$
HP-45:

(Day number) [↑] 24 [×] [↑] [↑](*GMT*: HH.MMSS) [G] [D.MS→] [+] 1.0027379 [×] [x≷y] [−] (λ: HH.MMSS) [G] [D.MS→] [−] (*S*) [+] (Check: If negative: 24 [+]; if >24: 24 [−]) [G] [→D.MS] (see *LST*: HH.MMSS).

Precision $\sim \pm 0^s.01$ (if the longitude is known that well).
HP-45:

(Day number) [↑] 24 [×] [↑] [↑] (*GMT*: HH.MMSS) [G] [D.MS→] [+] 1.002737909 [×] [x≷y] [−] (λ: HH.MMSS‸SS) [G] [D.MS→] [−] (*S*) [+] (Check: If negative: 24 [+]; if >24: 24 [−]) [↑] [G] [→D.MS] [↑] [↑] [R↓] [R↓] [G] [D.MS→] [−] .36 [×] [+] [FIX] 6 (see *LST*: HH.MMSS‸SS; if either MM or SS‸ is 99, read 59 instead).

Year, A.D.	*S* (hours)
1974	6.6183036
1975	6.6023892
1976	6.5864746
1977	6.6362698
1978	6.6203553
1979	6.6044408
1980	6.5885262
1981	6.6383215
1982	6.6224070
1983	6.6064925
1984	6.5905779

The Greenwich mean sidereal time *S* on January 0.0 of the indicated year is shown. The precision of this tabulation may deteriorate with time; check the AENA for the latest results.

TEST CASE: On May 9, 1976, (day number = 130), at the Harvard College Observatory in Harvard, Massachusetts, ($\lambda = 4^h 46^m 12^s.94$) at $GMT = 14^h$ (9:00 AM local time), the sidereal time was $LST = 0^h 23^m 48^s.55$.

REFERENCES: AENA; Smart (1962), Chapter VI.

A.7.5. Rising and Setting Times

This algorithm computes the local sidereal time (*LST*) of rising or setting (elevation = 0) for a celestial object specified by its right ascension α and declination δ. The latitude ϕ specifies the site on the earth. Because of refraction, the object will actually appear a few minutes sooner and disappear a few minutes later.

HP-45:

(ϕ: DD.MMSS) [G] [D.MS→] [TAN] (δ: DD.MMSS) [G] [D.MS→] [TAN] [CHS] [×] [G] [COS^{-1}] 15 [÷] [↑] [↑] [↑] (α: HH.MMSS) [G] [D.MS→] [$x \gtrless y$] [−] (if <0: 24 [+]) [↑] [G] [→D.MS] (see *LST* at rise in HH.MMSS) [R↓] [+] [+] (if >24: 24 [−]) [G] [→D.MS] (see *LST* at set in HH.MMSS).

TEST CASE: $\phi = 42°30'21.''71$. $\alpha = 17^h42^m28^s$, $\delta = -28°58'30''$; get *LST* rise $= 13^h44^m27^s$, *LST* set $= 21^h40^m28^s$.

REFERENCE: Smart (1962), pp. 46–48.

A.7.6 *Az*, *El*↔*HA*, δ

Hour angle *HA* and declination δ on the one hand, and azimuth *Az* and elevation *El* on the other, are two possible coordinate systems for objects on the sky, both referred to an observer on the earth: *El* is 90° minus the zenith angle and *Az* is north reference, clockwise. West *HA* is +. The latitude of the site on the earth is ϕ; north is +.

HP-45:

(δ: DD.MMSS) [G] [D.MS→] 1 [G] [→R] (*HA*: HH.MMSS) [G] [D.MS→] 6 [+] 15 [×] [$x \gtrless y$] [G] [→R] [↑] [R↓] [R↓] [→P] [$x \gtrless y$] (ϕ: DD.MMSS) [G] [D.MS→] [−] [$x \gtrless y$] [G] [→R] [R↓] [$x \gtrless y$] [R↓] [→P] [$x \gtrless y$] (if negative: 360 [+]; see *Az* in degrees) [R↓] [→P] [$x \gtrless y$] (see *El* in degrees).

(*El*, °) [↑] 1 [G] [→R] (*Az*, °) [$x \gtrless y$] [G] [→R] [$x \gtrless y$] [CHS] [↑] [R↓] [R↓] [→P] [$x \gtrless y$] (ϕ: DD.MMSS) [G] [D.MS→] [−] [$x \gtrless y$] [G] [→R] [R↓] [$x \gtrless y$] [R↓] [→P] [$x \gtrless y$] 15 [÷] [G] [→D.MS] (see *HA*: HH.MMSS) [R↓] [→P] [$x \gtrless y$] [G] [→D.MS] (see δ: DD.MMSS).

TEST CASE: $HA = 3^h04^m01^s$, $\delta = -13°11'10''$, $\phi = 42°22'$ corresponds to $Az = 228.°29$, $El = 20.°24$.

These algorithms are based on a suggestion by George Rybicki (private communication). The spherical triangle solution is from Smart (1962), p. 13.

A.7.7 Air Mass

The air mass (air path) a is the effective (normalized) thickness of the earth's atmosphere for an object seen at an elevation angle El. The curvature of the earth causes a to differ from cosecant (El): We also use Q, the ratio of the radius of the earth to the scale height in the atmosphere; a typical Q is 500 to 1000. There are several different formulas for a. One is

$$a=\sqrt{Q^2\sin^2 El+2Q+1}-Q\sin El.$$

HP-45:
(El, °) [SIN] Q [↑] [↑] [+] [R↓] [×] [↑] [R↓] [$x\gtrless y$] [R↓] [x^2] [+] 1 [+] [G] [$\sqrt{x}$] [$x\gtrless y$] [−] (see a).
a [1/x] [↑] [1/x] [↑] [1/x] [−] Q [↑] [+] [÷] [−] [G] [SIN^{-1}] (see El, °).

HP-35:
(E1, °) [SIN] Q [↑] [↑] [+] [R↓] [×] [↑] [R↓] [$x\gtrless y$] [R↓] [↑] [×] [+] 1 [+] [$\sqrt{x}$] [$x\gtrless y$] [−] (see a).
a [1/x] [↑] [1/x] [↑] [1/x] [−] Q [↑] [+] [÷] [−] [ARC] [SIN] (see El, °).

An alternative formula is

$$a^2=\frac{1/Q+1}{1/Q+\sin^2 El}.$$

HP-45: Q [1/x] [↑] [↑] 1 [+] [$x\gtrless y$] (El, °) [SIN] [x^2] [+] [÷] [G] [$\sqrt{x}$] (see a).
a [x^2] [1/x] [↑] [↑] 1 [−] Q [÷] [+] [G] [$\sqrt{x}$] [G] [SIN^{-1}] (see El, °).

HP-35:
Q [1/x] [↑] [↑] 1 [+] [$x\gtrless y$] (El, °) [SIN] [↑] [×] [+] [÷] [$\sqrt{x}$] (see a).
a [↑] [×] [1/x] [↑] [↑] 1 [−] Q [÷] [+] [$\sqrt{x}$] [ARC] [SIN] (see El, °).

TEST CASE: $El=9°$, $Q=600$, get $a=6.19$.

REFERENCES: Smart (1973), pp. 125 and 133; Marvin Litvak (private communication). See also Allen (1973), pp. 124–125.

A.7.8 Atmospheric Refraction

This algorithm calculates the total atmospheric refraction R given the elevation angle El, and, if standard conditions do not apply, the surface temperature T and pressure P.

(El, °) [TAN] [1/x] [↑] [↑] [×] .0668 [CHS] [×] 58.294 [+] [×] (see R in arcseconds at $T=50°$F, $P=30$ in. Hg; *or* go on:) (P, in. Hg) [×] 17 [×] (T, °F) [↑] 460 [+] [÷] (see R' in arcseconds).

The formulas are

$$R = 58''.294 \tan z - 0''.0668 \tan^3 z,$$

(at $P=30$ in. Hg and $T=50°$F), where z is the zenith angle $z=90°-El$. Then

$$R' = \frac{(17°\text{F/in. Hg})}{460°\text{F} + T} R.$$

These formulas deteriorate near the horizon (below $El=10°$, say). For a more precise R for $El<10°$, use

a [↑] El [COS] [×] 58.294 [×] (see R in arcseconds),

where a is the air mass from Section A.7.7 for $Q \cong 500$.

TEST CASE: $El=10°$, get $R=318$ arcseconds from the first formula or 320 arcseconds from the second.

REFERENCES: Smart (1962), especially p. 68; Allen (1973), p. 124.

A.7.9 Precession

The right ascension α declination δ coordinate system moves very slowly with respect to the fixed stars because of the precessional motion of the earth's axis. This motion is largely caused by the influence of the moon.

There are two algorithms for precession; the second, more complex version is needed only for very precise work over long periods of time. Add precessions from this algorithm to epoch coordinates to precess to date. "Day number" means day of the year.

These algorithms give precession only; nutation and annual aberration, which are *not* included, may amount to as much as 47 arcseconds. See Section A.7.10.

Precession from 1950 to date

HP-45:

(Year, e.g., 1974) [↑] 1950 [−] [EEX] 2 [÷] (Day number) [↑] 36525 [÷] [+] [↑] [↑] [↑] .0416 [CHS] [×] .426 [−] [×] 2004.255 [+] [×] (α: HH.MMSS) [G] [D.MS→] 15 [×] [$x \gtrless y$] [G] [→R] (see precession in δ in arcseconds) [CLX] (δ: DD.MMSS) [G] [D.MS→] [TAN] [×] [STO] 4 [CLX] .0371 [×] 1.395 [+] [×] 4609.896 [+] [×] [RCL] 4 [+] 15 [÷] (see precession in α in seconds of time).

The formulas used for this version are

$$\Delta\alpha = M + N \sin\alpha \tan\delta,$$

$$\Delta\delta = N \cos\alpha,$$

where

$$N = \theta = 2004.''255\,T - 0.''426\,T^2 - 0.''0416\,T^3,$$

$$M = \zeta_0 + z = 4609.''896\,T + 1.''395\,T^2 + 0.''0371\,T^3,$$

and T is the time since 1950 in centuries.

Precession with more precision

HP-45:

(Original epoch, e.g., 1950) [↑] [↑] 1900 [−] [EEX] 2 [÷] [STO] 1 [CLX] (Year, e.g., 1974) [$x \gtrless y$] [−] [EEX] 2 [÷] (Day number) [↑] 36525 [÷] [+] [↑] [↑] [↑] .042 [CHS] [×] .426 [−] [×] [RCL] 1 .853 [×] 2004.682 [−] [−] [×] 3600 [÷] [STO] 2 [CLX] .018 [×] .302 [+] [×] [RCL] 1 1.396 [×] 2304.25 [+] [+] [×] [↑] [↑] 3600 [÷] [STO] 3 [CLX] 2 [×] [$x \gtrless y$] [x^2] .791 [×] [+] 3600 [÷] [STO] 4 (α: HH.MMSS) [G] [D.MS→] 15 [×] [STO] [+] 3 [RCL] 2 2 [÷] [TAN] [RCL] 3 [COS] [×] (δ: DD.MMSS) [G] [D.MS→] [TAN] [+] [RCL] 2 [SIN] [×] [RCL] 3 [$x \gtrless y$] [G] [→R] 1 [$x \gtrless y$] [−] [→P] [R↓] [STO] 5 [RCL] 4 [+] 240 [×] (see precession in α in seconds of time) [RCL] 3 [RCL] 2 2 [÷] [TAN] [G] [→R] [$x \gtrless y$] [RCL] 5 2 [÷] [TAN] [×] [−] [G] [TAN^{-1}] 7200 [×] (see precession in δ in arcseconds).

The formulas used in this version are the so-called q formulas,

$$\Delta\alpha = B + \chi,$$

$$\tan\left(\frac{\Delta\delta}{2}\right) = \tan\left(\frac{\theta}{2}\right)\left(\cos A - \sin A \tan\left(\frac{B}{2}\right)\right),$$

where $\tan B = (q \sin A)/(1 - q\cos A)$,

$$q = \sin\theta\,(\tan\delta + \tan(\theta/2)\cos A),$$

$$A = \alpha + \zeta_0,$$

$$\chi = \zeta_0 + z = 2\zeta_0 + 0.''791\,T^2,$$

$$\zeta_0 = (2304.''250 + 1.''396\,T_0)T + 0.''302\,T^2 + 0.''018\,T^3,$$

$$\theta = (2004.''682 - 0.''853\,T_0)T - 0.''426\,T^2 - 0.''042\,T^3,$$

T_0 is the time from 1900 to the original epoch in centuries, and T is the time from the original epoch to the present in centuries. The constants in these formulas are not really arcseconds as labeled, but rather arcseconds per century, and so on. The answers in these formulas arè in arcseconds; the algorithm converts to degrees.

STORAGE ASSIGNMENTS: T_0 to 1, θ to 2, first ζ_0 then A to 3, χ to 4, B to 5.

TEST CASE: $\alpha = 7^h20^m53.^s$ $\delta = -25°40'24''$, epoch 1950 gives $\Delta\alpha = 65.^s13$, $\Delta\delta = -182.''61$, to precess to 1976 day number 132 using the first algorithm, or $\Delta\alpha = 65.^s13$, $\Delta\delta = -183.''78$ using the more precise algorithm.

REFERENCES: TN1969-42, pp. 13, 14; AENA; ESE, pp. 30, 31.

A.7.10 Nutation and Annual Aberration

The periodic (or short-term) components of the motion of the right ascension α and declination δ coordinate system are called nutation. Annual aberration is the apparent displacement of a celestial object due to the transverse component of the earth's motion about the sun. To point an earthbound telescope toward a celestial object, add nutation and annual aberration to the precessed α and δ, to give the *apparent coordinates* for a given date. The following algorithm is approximate, probably ± 2 arcseconds (see references below).

HP-45:

(Year, eg., 1974) [↑] 1900 [−] [EEX] 2 [÷] (Day number) [STO] 5 36525 [÷] [+] 1934.142 [CHS] [×] 259.183275 [+] [STO] 1 [COS] 9 [×] (α: HH.MMSS) [G] [D.MS→] 15 [×] [STO] 3 [SIN] [×] 23.4425 [STO] 2 [SIN] 17 [×] [RCL] 1 [SIN] [×] [RCL] 3 [COS] [×] [−] (see nutation in δ in arcseconds) [RCL] 3 (δ: DD.MMSS) [G] [D.MS→] [STO] 4 [TAN] [G] [→R] [RCL] 1 [COS] [×] .61 [CHS] [×] [$x \gtrless y$] [RCL] 2 [SIN] [×] [RCL] 2 [COS] [+] [RCL] 1 [SIN] [×] 1.15 [×] [−] (see nutation in α in seconds of time) .9856 [RCL] 5 [×] 282.51 [+] [STO] 6 1.36 [CHS] [RCL] 4 [COS] [÷] [G] [→R] [RCL] 2 [COS] [×] [RCL] 3 [COS] [×] [$x \gtrless y$] [RCL] 3 [SIN] [×] [+] (see annual aberration in α in seconds of time) [RCL] 2 [RCL] 6 [COS] [G] [→R] [RCL] 3 [RCL] 4 [SIN] [G] [→R] [R↓] [×] [$x \gtrless y$] [RCL] 4 [COS] [×] [−] [$x \gtrless y$] [RCL] 6 [SIN] [×] [−] 20.5 [×] (see annual aberration in δ in arcseconds).

TEST CASE: 1973, day number $=2$, $\alpha = 1^h03^m49^s$, $\delta = 12°19'42''$; get nutation $\Delta\delta = 6''.94$, $\Delta\alpha = 1^s.00$, and annual aberration $\Delta\delta = 2''.36$, $\Delta\alpha = 0^s.06$.

REFERENCES: TN1969-42, p. 14; ESE, especially p. 44.

A.7.11 Angular Separations Between Stars

This algorithm calculates the great-circle separation between two celestial objects specified by their right ascensions α_1 and α_2 and declinations δ_1 and δ_2.

HP-45:

(δ_1: DD.MMSS) [G] [D.MS→] 1 [G] [→R] [$x \gtrless y$] (δ_2: DD.MMSS) [G] [D.MS→] 1 [G] [→R] [R↓] [×] [R↓] [×] [$x \gtrless y$] [CLX] (α_1: HH.MMSS) [G] [D.MS→] (α_2: HH.MMSS) [G] [D.MS→] [−] 15 [×] [COS] [×] [+] [G] [COS^{-1}] 60 [×] (see separation in arcminutes).

TEST CASE: $\alpha_1 = 17^h42^m28^s$, $\delta_1 = -28°58'30''$, $\alpha_2 = 17^h44^m10^s.6$, $\delta_2 = -28°22'50''$, get separation $= 42'.17$.

$$\cos S = \sin\delta_1 \sin\delta_2 + \cos(\alpha_1 - \alpha_2)\cos\delta_1 \cos\delta_2.$$

A.7.12 $\alpha, \delta \leftrightarrow \mathfrak{l}^{II}, \mathfrak{b}^{II}$

Right ascension α and declination δ are astronomical coordinates defined with respect to the earth's equator. Galactic latitude $\mathfrak{b}^{II}$ and longitude $\mathfrak{l}^{II}$ are defined instead with respect to the galactic equator. The direction toward the galactic center is approximately $\mathfrak{l}^{II}=\mathfrak{b}^{II}=0$. The α,δ coordinates for these algorithms are epoch 1950.

HP-45:

(δ: DD.MMSS) [G] [D.MS→] 1 [G] [→R] (α: HH.MMSS) [G] [D.MS→] 15 [×] 77.75 [+] [x≷y] [G] [→R] [↑] [R↓] [R↓] [→P] [x≷y] 62.6 [−] [x≷y] [G] [→R] [R↓] [R↓] [R↓] [→P] [x≷y] 33 [+] (If negative: 360 [+]; see $\mathfrak{l}^{II}$ in degrees) [R↓] [→P] [x≷y] (see $\mathfrak{b}^{II}$ in degrees).
($\mathfrak{b}^{II}$, degrees) [↑] 1 [G] [→R] ($\mathfrak{l}^{II}$, degrees) [↑] 33 [−] [x≷y] [G] [→R] [↑] [R↓] [R↓] [→P] [x≷y] 62.6 [+] [x≷y] [G] [→R] [R↓] [R↓] [R↓] [→P] [x≷y] 77.75 [−] 15 [÷] (If negative: 24 [+]) [G] [→D.MS] (see α: HH.MMSS) [R↓] [→P] [x≷y] [G] [→D.MS] (see δ: DD.MMSS).

TEST CASE: $\alpha=07^h20^m56^s$, $\delta=-25°39'55''$ (1950), corresponds to $\mathfrak{l}^{II}=239.°35$, $\mathfrak{b}^{II}=-5.°06$.

REFERENCES: TN1969-42, p. 10; Allen (1973), p. 283.

These algorithms were written by George Rybicki (private communication) and are reproduced here, in slightly modified form, with his permission. The spherical triangle solution is from Smart (1962), p. 13.

A.7.13 $\alpha,\delta\leftrightarrow L,\beta$

These algorithms convert right ascension α and declination δ to and from ecliptic longitude L and latitude β. The ecliptic is the apparent track of the sun across the sky. The epoch for L,β is the same as for α,δ. If L and β are for Doppler velocities in Section A.8.2, use precessed α and δ.

HP-45:

(δ: DD.MMSS) [G] [D.MS→] 1 [G] [→R] (α: HH.MMSS) [G] [D.MS→] 15 [×] [x≷y] [G] [→R] [↑] [R↓] [R↓] [→P] [x≷y] 23.4423 [−] [x≷y] [G] [→R] [R↓] [R↓] [R↓] [→P] [x≷y] (see L in degrees) [R↓] [→P] [x≷y] (see β in degrees).
(β,°) [↑] 1 [G] [→R] (L,°) [x≷y] [G] [→R] [↑] [R↓] [R↓] [→P] [x≷y] 23.4423 [+] [x≷y] [G] [→R] [R↓] [R↓] [R↓] [→P] [x≷y] 15 [÷] (if <0: 24 [+]) [G] [→D.MS] (see α: HH.MMSS) [R↓] [→P] [x≷y] [G] [→D.MS] (see δ: DD.MMSS).

TEST CASE: $\alpha=07^h20^m55^s$, $\delta=-25°40'04''$, get $L=117.°31$, $\beta=-47.°21$.

REFERENCE: TN1969-42, p. 11.

These algorithms are based on a suggestion by George Rybicki (private communication). The spherical triangle solution is from Smart (1962), p. 13.

A.7.14 Sun Ephemeris

This algorithm calculates the ecliptic longitude Λ, and the right ascension α and declination δ of the center of the sun. These are geocentric apparent coordinates (do not precess) and are within about ± 1 arcminute. Day number means day of the year, and g' and ω' are from the tabulation below. Greenwich mean time (GMT) is the mean solar (civil) time at Greenwich, England, or equivalently universal time (UT). GMT differs from local (civil) time anywhere by an integer number of hours. GMT need not be very precise for this algorithm.

HP-45:

(GMT: HH.MM) [G] [D.MS→] 24 [÷] (Day number) [+] .9856 [×] (g') [−] [↑] [↑] [↑] [+] [SIN] .02 [×] [$x \gtrless y$] [SIN] 1.916 [×] [+] [+] (ω') [−] (see Λ in degrees) [↑] [SIN] 23.442 [$x \gtrless y$] [G] [→R] [$x \gtrless y$] [G] [SIN⁻¹] [G] [→D.MS] (see δ: DD.MMSS) [R↓] [$x \gtrless y$] [COS] [→P] [R↓] 15 [÷] (if <0: 24 [+]) [G] [→D.MS] (see α: HH.MMSS).

Year, A.D.	g' (°)	ω' (°)
1974	3.213	77.50
1975	3.469	77.48
1976	3.725	77.46
1977	2.995	77.45
1978	3.251	77.43
1979	3.507	77.41
1980	3.763	77.40
1981	3.033	77.38
1982	3.289	77.36
1983	3.545	77.34
1984	3.801	77.33

The angle g' is minus the mean anomaly on January 0.0 of the indicated year, and ω' is minus the longitude of perigee.

TEST CASE: 1976 day number = 134 (May 13), $GMT=0$; get $\Lambda = 52.^{\circ}3686$, $\alpha = 3^h19^m50^s$, $\delta = 18°21'51''$. The AENA (1976, p. 25) has $\Lambda = 52.^{\circ}3599$, $\alpha = 3^h19^m47.^s98$, $\delta = 18°21'38.''4$ (error$=0.'52$).

REFERENCES: AENA; Smart (1962), Chapter V.

A.8 DOPPLER VELOCITIES IN ASTRONOMY

A.8.1 Notes

Spectra of astonomical objects taken from the earth need to be corrected for the motion of the observer. The two principal contributions to the velocity of a site on the earth with respect to the sun are the velocity of the earth in its orbit around the sun (earth revolution) and the velocity of the site with respect to the earth's center (earth rotation). These are two velocity vectors changing as a function of time. What is needed is the Doppler velocity, that is, the component of the total velocity as projected onto the line of sight toward a specified celestial object. By convention, a positive velocity corresponds to increasing distance between source and observer; the Doppler velocity is the derivative of the distance.

Section A.8.2 calculates the earth revolution Doppler velocity V, which can amount to almost ± 30 km/s. Section A.8.3 calculates the earth rotation Doppler velocity V_{earth}, which can amount to almost ± 0.5 km/s. The largest neglected velocity is the motion of the earth's center with respect to the earth-moon barycenter, which can amount to about ± 0.015 km/s.

Optical astronomers traditionally use a velocity reference frame attached to the sun, and radio astronomers traditionally use the local standard of rest. Originally the local standard of rest was intended to be at the average velocity of all the stars within a few hundred parsecs of the sun. But now, by convention, the velocity of the sun with respect to the local standard of rest is taken to be 20 km/s toward $\alpha = 18^{\mathrm{h}}$, $\delta = 30°$, epoch 1900. The component V_{sun} of this solar velocity as projected onto the line of sight toward a specified celestial object is a constant for the object and can be calculated from Section A.8.4. Add V_{sun} to Doppler velocities with respect to the local standard of rest to obtain Doppler velocities with respect to the sun. Or subtract V_{sun} from Doppler velocities with respect to the sun to obtain Doppler velocities with respect to the local standard of rest.

The total Doppler velocity of a site on the earth with respect to the local standard of rest is the sum of these three velocities, namely, V from Section A.8.2 plus V_{earth} from Section A.8.3 plus V_{sun} from Section A.8.4. A "peculiar velocity" associated with the source itself may be added. Section A.8.5 then can be used to convert this total Doppler velocity to a Doppler frequency.

A.8.2 Earth Revolution Doppler Velocity

This algorithm calculates the component of the velocity of the earth in its orbit around the sun as projected onto the line of sight toward a celestial object specified by its ecliptic longitude L and latitude β; L and β should be apparent coordinates (precessed to date), and they can be calculated from Section A.7.13. Day number means day of the year, *GMT* is the Greenwich mean time, and g' and ω' are from the tabulation in Section A.7.14.

HP-45:

(*GMT*: HH.MM) [G] [D.MS→] 24 [÷] (Day number) [+] .9856 [×] (g') [−] [↑] [↑] [↑] [+] [SIN] .02 [×] [$x \gtrless y$] [SIN] 1.916 [×] [+] [+] (L, °) [↑] (ω') [+] [↑] [SIN] .01672 [×] [R↓] [−] [SIN] [$x \gtrless y$] [R↓] [−] (β, °) [COS] [×] 29.79 [×] (see V in kilometers per second).

HP-35: Key *GMT* in hours and change [G] [D.MS→] to [↑].

TEST CASE: 1976 day number = 135 (May 14), $GMT = 13^h$, $L = 320.^\circ 1$, $\beta = 0.^\circ 3$; get $V = -29.42$ km/s.

REFERENCE: TN1969-42, pp. 1–7.

A.8.3 Earth Rotation Doppler Velocity

This algorithm calculates the component of the velocity of a site on the earth with respect to the earth's center as projected onto the line of sight toward a celestial object specified by its hour angle *HA* and declination δ. There is a note on *HA* in Section A.7.1. The site on earth is specified by its latitude ϕ.

HP-45:

(ϕ: DD.MMSS) [G] [D.MS→] [COS] (δ: DD.MMSS) [G] [D.MS→] [COS] [×] (HA: HH.MMSS) [G] [D.MS→] 15 [×] [SIN] [×] .464 [×] (see V_{earth} in kilometers per second).

$$V_\rho = \frac{2\pi \times 6.368627 \times 10^3 \text{ km}}{(23^h 56^m 04.^s 09054) \times 3600 \text{ s/h}} = 0.464408 \text{ km/s}.$$

TEST CASE: $\phi = 42°30'21.''71$, $\delta = -28°58'30''$, $HA = -3^h$; get $V_{\text{earth}} = -0.21$ km/s.

REFERENCE: TN1969-42, p. 4.

A.8.4 Local Standard of Rest Doppler Velocity

This algorithm calculates the component of the (constant) velocity of the sun with respect to the local standard of rest as projected onto the line of sight toward a celestial object specified by its right ascension α and declination δ, epoch 1950. The sun's motion is taken to be 20 km/s toward $\alpha = 18^h$, $\delta = 30°$, epoch 1900.

HP-45:

(δ: DD.MMSS) [G] [D.MS→] 1 [G] [→R] 30 [COS] [×] 18.043 [↑] (α: HH.MMSS) [G] [D.MS→] [−] 15 [×] [COS] [×] 2 [×] [+] 10 [CHS] [×] (see V_{sun} in kilometers per second).

TEST CASE: $\alpha = 7^h 20^m 53^s$, $\delta = -25°40'24''$ (1950); get $V_{sun} = 19.04$ km/s.

REFERENCE: TN1969-42, p. 2.

A.8.5 Doppler Frequencies

This algorithm calculates the sky frequency f_{sky}, which is a line rest frequency f_0 as offset by a Doppler velocity v in kilometers per second, and f_{syn}, the frequency to be set into a frequency synthesizer to tune a receiver to f_{sky}. The receiver hardware is specified by the frequency multiplier N, usually an integer, and by the offset frequency f_{off}. The units of the fs are arbitrary but must all be the same.

f_0 [↑] [↑] v [×] 299792.5 [÷] [−] (see f_{sky})
f_{off} [+] N [÷] (see f_{syn}).

The appropriate formulas are

$$f_{sky} = f_0\left(1 - \frac{v}{c}\right),$$

where $c = 299792.5$ km/s is the speed of light. The minus sign occurs because, by convention, a positive v corresponds to increasing distance between source and observer. Then

$$f_{syn} = \frac{f_{sky} + f_{off}}{N}.$$

TEST CASE: $v = 11.3$ km/s, $f_0 = 1612231$ kHz, $f_{off} = -135000$ kHz, $N = 7$; get $f_{sky} = 1612170.231$ kHz and $f_{syn} = 211024.3187$ kHz.

See also Section A.8.6.

A.8.6 Relativistic Doppler Shifts

This algorithm converts a Doppler velocity v in kilometers per second into or from a wavelength offset $\Delta\lambda$ or a frequency offset Δf, where f_0 is the line rest frequency and λ_0 is the line rest wavelength. The formulas are correct if the velocity is radially toward or away from the observer.

HP-45:

(v, km/s) [↑] 299792.5 [÷] [↑] [↑] 1 [+] [x≷y] 1 [x≷y] [−] [÷] [G] [√x] 1 [−] (see $z=\Delta\lambda/\lambda_0$) ($\lambda_0$) [×] (see $\Delta\lambda$).
(v, km/s) [↑] 299792.5 [÷] [↑] [↑] 1 [x≷y] [−] [x≷y] 1 [+] [÷] [G] [√x] 1 [−] (see $q=\Delta f/f_0$) (f_0) [×] (see Δf).
($\Delta\lambda$) [↑] (λ_0) [÷] (see z) 1 [+] [G] [TAN⁻¹] 2 [×] [COS] [CHS] (see v/c) 299792.5 [×] (see v in kilometers per second).
(Δf) [↑] (f_0) [÷] (see q) 1 [+] [G] [TAN⁻¹] 2 [×] [COS] (see v/c) 299792.5 [×] (see v in kilometers per second).

For an HP-35, replace [G] [√x] by [√x], [x^2] by [↑] [×], and [G] [TAN⁻¹] by [ARC] [TAN].

z [↑] [↑] 1 [+] [CHS] [÷]	(see q) f_0 [×]	(see Δf).
q [↑] [↑] 1 [+] [CHS] [÷]	(see z) λ_0 [×]	(see $\Delta\lambda$).

$$\frac{\Delta\lambda}{\lambda_0}=z=\sqrt{\frac{1+v/c}{1-v/c}}-1,$$

$$\frac{\Delta f}{f_0}=q=\sqrt{\frac{1-v/c}{1+v/c}}-1,$$

$$\frac{v}{c}=\frac{(z+1)^2-1}{(z+1)^2+1}=\frac{1-(q+1)^2}{1+(q+1)^2}=\tanh(\ln(z+1))=-\tanh(\ln(q+1)).$$

TEST CASE: $v=10^5$ km/s, $v/c=0.3335640485$, $\lambda_0=18$ cm, $f_0=1665$ MHz; get $z=0.414580707$, $\Delta\lambda=7.462452726$ cm, $q=-0.2930767438$, $\Delta f=-487.9727784$ MHz.

REFERENCES: Lang (1974), Section 2.17; Born (1962), p. 300.

A.9 ASTRONOMY AND RADIO ASTRONOMY: SELECTED FORMULAS

When I heard the learn'd astronomer,
When the proofs, the figures, were ranged in columns
before me,
When I was shown the charts and diagrams, to add,
divide, and measure them,
When I sitting heard the astronomer where he lectured
with much applause in the lecture-room,
How soon unaccountable I became tired and sick,
Till rising and gliding out I wander'd off by myself,
In the mystical moist night air, and from time to time,
Look'd up in perfect silence at the stars.

—*Walt Whitman (1819–1892)*

Poets say science takes away from the beauty of the stars—mere globs of gas atoms. Nothing is "mere." I too can see the stars on a desert night, and feel them. But do I see less or more? The vastness of the heavens stretches my imagination—stuck on this carousel my little eye can catch one-million-year-old light. A vast pattern—of which I am a part—perhaps my stuff was belched from some forgotten star, as one is belching there. Or see them with the greater eye of Palomar, rushing all apart from some common starting point when they were perhaps all together. What is the pattern, or the meaning, or the *why*? It does not do harm to the mystery to know a little about it. For far more marvelous is the truth than any artists of the past imagined! Why do the poets of the present not speak of it? What men are poets who can speak of Jupiter if he were like a man, but if he is an immense spinning sphere of methane and ammonia must be silent?*

—*Richard P. Feynman (1918–   )*

A.9.1 RMS Noise: The Radiometer Equation

Perhaps the most important formula in radio astronomy, the radiometer (noise) equation, gives the expected dispersion or RMS noise in terms of system parameters and integration time. This RMS noise is ΔT, T_{sys} is the system noise temperature (usually in °K, but in any case ΔT and T_{sys} are in the same units), β is the resolution width or bandwidth in hertz, and τ is the total integration time in seconds. In its simplest form, appropriate for

*From Feynman, Leighton, and Sands (1963), Vol. I, p. 3–6.

an analog or multibit processor and with equal time spent on and off the signal (Dicke switching), we have

HP-45:
T_{sys} [↑] [+] β [↑] τ [×] [G] [$\sqrt{x}$] [÷] (see ΔT).

For a one-bit digital autocorrelator, we have approximately

HP-45:
T_{sys} [G] [π] [×] β [↑] τ [×] [G] [$\sqrt{x}$] [÷] (see ΔT).

For other cases, this ΔT should be multiplied by γ, a factor not too far from unity (see references). For an HP-35, replace [G] [$\sqrt{x}$] by [$\sqrt{x}$] and [G] [π] by [π].

TEST CASE: $T_{sys} = 89°K$, $\beta = 750$ Hz, $\tau = 3600$ s; get $\Delta T = 0.108°K$ (analog) or 0.170°K (one-bit).

REFERENCES: Ball (1976b); Ball (1975c), p. 206; Rice (1954), Section 3.9.

A.9.2 Beamwidths and Resolutions

The beamwidth of a radio telescope and the resolution of an optical telescope are given by the same formula, although the units are sometimes different. This algorithm calculates an approximate resolution or beamwidth θ (full width to half-maximum) given D, the diameter of the (circular) telescope, f the observing frequency, or λ the observing wavelength. For an optical telescope use λ around 5500 Å.

HP-45:
(f, MHz) [1/x] 67625 [×] (D, ft) [÷] [G] [→D.MS] (see θ: D.MMSS).

HP-35 or HP-45:
(f, MHz) [1/x] 67625 [×] (D, ft) [÷] (see θ in degrees),

or

(λ, Å) [↑] 9.7448 [EEX] [CHS] 4 [×] (D, in.) [÷] (see θ in arcseconds).

TEST CASE: $f=1665$ MHz, $D=84$ ft; get $\theta=29'01''$; or $\lambda=6000$ Å, $D=8$ in., get $\theta=0''.73$.

$$\theta=1.2\ \lambda/D=1.2\ c/(\nu D)\ \text{(converted from radians)},$$

$$1.2\times 2.997925\times 10^{10}\ \text{cm/s}\times\frac{1\ \text{in.}}{2.54\ \text{cm}}\times\frac{1\ \text{ft}}{12\ \text{in.}}\times\frac{\text{M}}{10^6}\times\frac{180°}{\pi\ \text{(rads)}}$$

$$=67625.37\ \frac{\text{Mft}\ °}{\text{(rad)s}},$$

$$1.2\times\frac{10^{-8}\ \text{cm}}{\text{Å}}\times\frac{1\ \text{in.}}{2.54\ \text{cm}}\times\frac{180°}{\pi\ \text{(rads)}}\times\frac{3600''}{1°}=9.7448\times 10^{-4}\frac{\text{in.}\ ''}{\text{Å (rad)}}$$

A.9.3 Beamwidths and Source Widths from Scans

If a celestial source small in angular size is scanned with a telescope, the scan gives information about the beamwidth. If, alternatively, the source can be taken to be Gaussian shaped and more or less comparable in angular size to the beamwidth, the source width can be calculated from the scan if the beamwidth is known. In these algorithms, θ_A is the beamwidth of the antenna, θ_S is the width of the celestial source, and θ_R is the response width of a scan through the source. The θs are all full widths to half-maximum, are all in the same units, and are for a given cut or direction on the sky. Then α is the right ascension, δ the declination, *Az* the azimuth, and *El* the elevation. The formulas follow from the assumption of Gaussian shapes for both source and antenna eam.

I Point source ($\theta_S=0$)
  A. Scan in α (or drift curve) (HP-45 only)
     (δ: DD.MMSS) [G] [D.MS→] [COS] (θ_R) [×] (see θ_A),
  or (δ, °) [COS] (θ_R) [×] (see θ_A).
  B. Scan in azimuth
     (*El*, °) [COS] (θ_R) [×] (see θ_A).
  C. Scan in δ or elevation: $\theta_A=\theta_R$.

II. Arbitrary (Gaussian-shaped) source
  A. Scan in α (or drift curve) (HP-45 only)
     (δ: DD.MMSS) [G] [D.MS→] [COS] (θ_R) [×] [x^2] (θ_A) [x^2] [−]
     [G] [$\sqrt{x}$] (see θ_S),

or (HP-35 only)
(δ, °) [COS] (θ_R) [×] [↑] [×] (θ_A) [↑] [×] [−] [$\sqrt{x}$] (see θ_S).

B. Scan in azimuth
(El, °) [COS] (θ_R) [×] [x^2] (θ_A) [x^2] [−] [G] [$\sqrt{x}$] (see θ_S),

or (HP-35 only)
(El, °) [COS] (θ_R) [×] [↑] [×] (θ_A) [↑] [×] [−] [$\sqrt{x}$] (see θ_S).

C. Scan in δ or elevation (HP-45 only)
(θ_R) [x^2] (θ_A) [x^2] [−] [G] [$\sqrt{x}$] (see θ_S),

or (HP-35 only)
(θ_R) [↑] [×] (θ_A) [↑] [×] [−] [$\sqrt{x}$] (see θ_S).

TEST CASE: (drift curves) $\theta_S = 0$, $\delta = 14°$, $\theta_R = 40'.2$;
get $\theta_A = 39'$, or $\theta_R = 51'$; get $\theta_S = 30'.46$.

A.9.4 Antenna Efficiencies versus Surface Tolerance

These algorithms relate the aperture efficiencies (η_1 at wavelength λ_1 and η_2 at wavelength λ_2) to ε the RMS surface tolerance (departure from a paraboloid); λ_1, λ_2, and ε are all in the same units. The formulas follow from the assumption of noiselike departures from a paraboloidal surface.

Given ε and η_1, find η_2

HP-45:
λ_1 [x^2] [$1/x$] λ_2 [x^2] [$1/x$] [−] 4 [G] [π] [×] ε [×] [x^2] [×] [e^x] η_1 [×] (see η_2).

HP-35:
λ_1 [↑] [×] [$1/x$] λ_2 [↑] [×] [$1/x$] [−] 4 [π] [×] ε [×] [↑] [×] [×] [e^x] η_1 [×] (see η_2).

Given η_1 at λ_1 and η_2 at λ_2, find ε

HP-45:
η_1 [↑] η_2 [÷] [ln] λ_2 [x^2] [$1/x$] λ_1 [x^2] [$1/x$] [−] [÷] [G] [$\sqrt{x}$] 4 [÷] [G] [π] [÷] (see ε).

HP-35:
η_1 [↑] η_2 [÷] [ln] λ_2 [↑] [×] [$1/x$] λ_1 [↑] [×] [$1/x$] [−] [÷] [$\sqrt{x}$] 4 [÷] [π] [÷] (see ε).

TEST CASE: $\eta_1=51\%$, $\lambda_1=2$ cm, $\eta_2=42\%$, $\lambda_2=1$ cm, $\varepsilon=0.04$ cm.

REFERENCE: Ruze (1952).

These algorithms were written by Hays Penfield (private communication) and are reproduced here, in slightly modified form, with his permission.

A.9.5 Brightness (Specific Intensity)↔Brightness Temperature

The brightness temperature is the temperature to which a black body would need to be heated to show the observed brightness or specific intensity. Brightness temperature can be a function of wavelength and need not equal any other temperature.

In these formulas I is the specific intensity in flux units per steradian (1 flux unit (fu) = 1 jansky = 10^{-26} joule/m^2), f is the frequency in megahertz, and T is the brightness temperature in degrees Kelvin. There are two formulas depending on the range of f and T: use the Planck law if $f/T > 10$ MHz/°K (because the Rayleigh-Jeans approximation is no good in this range) or use the Rayleigh-Jeans approximation if $f/T < 0.1$ MHz/°K (because the calculator runs out of significant figures in calculating the difference in the denominator of the Planck law).

Rayleigh-Jeans approximation
HP-45:

(f, MHz) [x^2] .03072297 [×] (T, °K) [×] (see I in fu/sr),

or

(I, fu/sr) [↑] .03072297 [÷] (f, MHz) [x^2] [÷] (see T in °K).

For an HP-35, replace [x^2] by [↑] [×].

Full Planck law
HP-45 or HP-35:

(f, MHz) [↑] [↑] [↑] [×] [×] 1.474527 [EEX] [CHS] 6 [×] [$x \gtrless y$] 4.799428 [EEX] [CHS] 5 [×] (T, °K) [÷] [e^x] 1 [−] [÷] (see I in flux units per steradian).

or

(f, MHz) [↑] [↑] [↑] [×] [×] 1.474527 [EEX] [CHS] 6 [×] (I, fu/sr) [÷] 1 [+] [ln] [÷] 4.799428 [EEX] [CHS] 5 [×] (see T in degrees Kelvin).

TEST CASE: $f=5000$ MHz, $T=1953$°K, $I \cong 1.5\times10^9$ fu/sr.

Rayleigh-Jeans approximation is

$$I=\frac{2kTf^2}{c^2}.$$

The full Planck law is

$$I = \frac{2hf^3}{c^2} \frac{1}{e^{hf/(kT)} - 1},$$

where
$$\frac{2h}{c^2} = 1.474527 \times 10^{-6}\ \text{fu/MHz}^3,$$

$$\frac{h}{k} = 4.799428 \times 10^{-5}\ \text{°K/MHz},$$

$$\frac{2k}{c^2} = 3.072297 \times 10^{-2}\ \text{fu/(°K MHz}^2\text{)}.$$

A.9.6 Kinetic Temperature↔Linewidths

In the simplest case, the width of an observed spectral line is due to a combination of turbulence (mass motions with a scale size smaller than the angular resolution) and kinetic temperature. If other contributions to the linewidth can be neglected, and if measurements are made with two different molecules or atoms (two different masses), the two contributions to the linewidth can be separated to give the kinetic temperature T_k (in degrees Kelvin) and the RMS turbulent velocity Δv_T (in kilometers per second). The equations follow from the assumptions that the two molecules or atoms are intermixed (and so share the same turbulence and kinetic temperature) and that the optical depth is small in both lines. The turbulence is characterized by Gaussian random motion. If other effects contribute to the linewidth, T_k from the algorithm is an upper limit, or Δv is a lower limit.

In the algorithm below, Δv or Δv_1 and Δv_2 are the full widths to half-maximum in kilometers per second of the corresponding lines; and M or M_1 and M_2 are the masses in atomic mass units (amu) of the corresponding emitters.

Case I: Only kinetic temperature affects the linewidths.
HP-45:

(Δv, km/s) [x^2] (M, amu) [×] 21.69 [×] (see T_k in degrees Kelvin),

or

(T_k, °K) [↑] 21.69 [÷](M, amu) [÷] [G] [$\sqrt{x}$] (see Δv in kilometers per second).

HP-35:

(Δv, km/s) [↑] [×] (M, amu) [×] 21.69 [×] (see T_k in degrees Kelvin),

or

(T_k, °K) [↑] 21.69 [÷] (M, amu) [÷] [$\sqrt{x}$] (see Δv in kilometers per second).

Case II: Kinetic temperature and turbulence affect the linewidths.

HP-45:

(Δv_1, km/s) [x^2] (Δv_2, km/s) [x^2] [−] 21.69 [×] (M_1, amu) [1/x] (M_2, amu) [1/x] [−] [÷] (see T_k in degrees Kelvin),

or

(M_1, amu) [↑] (M_2, amu) [÷] [↑] [↑] (Δv_1, km/s) [x^2] [×] (Δv_2, km/s) [x^2] [−] [$x \gtrless y$] 1 [−] [÷] [G] [$\sqrt{x}$] (see Δv_T in kilometers per second).

HP-35:

(Δv_1, km/s) [↑] [×] (Δv_2, km/s) [↑] [×] [−] 21.69 [×] (M_1, amu) [1/x] (M_2, amu) [1/x] [−] [÷] (see T_k in degrees Kelvin),

or

(M_1, amu) [↑] (M_2, amu) [÷] [↑] [↑] (Δv_1, km/s) [↑] [×] [×] (Δv_2, km/s) [↑] [×] [−] [$x \gtrless y$] 1 [−] [÷] [$\sqrt{x}$] (see Δv_T in kilometers per second).

TEST CASES: $M = 17$ amu (OH), $T_k = 299$°K, $\Delta v = 0.9$ km/s; or $M_1 = 17$ amu (OH), $M_2 = 13$ amu (CH), $\Delta v_1 = 0.9$ km/s, $\Delta v_2 = 1.0$ km/s, $T_k = 228$°K, and $\Delta v_T = 0.44$ km/s.

The formulas are

$$T_K = \frac{M(\Delta v)^2}{8k\ln 2},$$

$$\frac{1}{8k\ln 2} = 21.68986 \frac{°\text{K}}{\text{amu(km/s)}^2},$$

or

$$T_K = \frac{\Delta v_1{}^2 - \Delta v_2{}^2}{8k\ln 2[1/M_1 - 1/M_2]},$$

$$\Delta v_T{}^2 = \frac{M_2 \Delta v_2{}^2 - M_1 \Delta v_1{}^2}{M_2 - M_1}.$$

Note that the error limits are sometimes very large.

REFERENCE: Lang (1974), Section 2.18.

A.9.7 The Saha (Equilibrium Ionization) Equation

The Saha equation relates the free electron density N_e in number per cubic centimeter, the temperature T in degrees Kelvin, the ionization potential χ_r (from the ground state to r) in electron volts, and the partition functions U_r and U_{r+1} for the r and $r+1$ stages of ionization, to N_r and N_{r+1}, the number densities, in atoms per cubic centimeter, in the corresponding stages of ionization; U_r is approximately the statistical weight of the ground state, and $U_{r+1}/U_r \sim \frac{1}{2}$ very approximately.

HP-45:

(T, °K) [↑] [1/x] (χ_r, eV) [×] 11605 [CHS] [×] [e^x] [$x \gtrless y$] 1.5 [G] [y^x] [×] (U_{r+1}/U_r) [×] 4.829 [EEX] 15 [×] (see $N_e N_{r+1}/N_r$, /cm^3) (N_e, /cm^3) [÷] (see N_{r+1}/N_r).

For an HP-35, replace [G] [y^x] by [$x \gtrless y$] [x^y].

TEST CASE: $T = 3000$°K, $\chi_r = 11.26$ eV (carbon), $U_{r+1}/U_r = 0.65$, $N_e = 10/\text{cm}^3$; get $N_{r+1}/N_r = 6.25$ (i.e., most of the carbon is ionized).

The Saha equation is an example of a large class of equations relating excitations, ionizations, and chemical populations, all derivable by the methods of chemical equilibrium.

REFERENCES: Slater (1939), Chapter XX; Ambartsumyan (1958), Section 5.2; Allen (1973), Section 15; Lang (1974), Section 3.3.1.4.

A.9.8 Recombination Line Rest Frequencies

Recombination lines are misnamed. The physical process is the emission of a photon by a highly excited (but not usually ionized) atom—either hydrogen or hydrogenic in the sense that the nucleus and any other electrons can be approximated as a point charge. This process gives rise to a series of lines whose frequencies f can be calculated by this algorithm. Recombination, by contrast, means the reuniting of an electron and an ion to form a neutral atom, resulting in a bound-free continuum or band of frequencies rather than lines. The inappropriate nomenclature arose because in astronomical HII (ionized hydrogen) regions, recombination to a highly excited level, followed by a cascade by steps to the ground state, is a frequently occurring process, and the lines resulting from the cascade are observable toward HII regions by radio astronomers.

In this algorithm, n is the principal quantum number, Δn is the change in n and is 1 for an α line, 2 for a β line, 3 for a γ line, and so on; M is the total mass of the emitter in atomic mass units on the chemical scale. To use the physical scale of atomic mass units instead, change the constant 5.48593 to 5.48742.

HP-45 INITIALIZE: 5.48593 [CHS] [EEX] [CHS] 4 [↑] (M, amu) [÷] 1 [+] 3289842311 [×]

CALCULATE: (n) [↑] [x^2] [$1/x$] [$x \gtrless y$] (Δn) [+] [x^2] [$1/x$] [−] [×] (see f in MHz) [CLX] :|.

For an HP-35, replace [x^2] by [↑] [×].

TEST CASE: H109β, that is $M = 1.008$ amu, $n = 109$, $\Delta n = 2$, is at $f = 9883.082$ MHz.

REFERENCE: Lilley and Palmer (1968).

A.10 ELECTRICAL ENGINEERING: SELECTED FORMULAS

A.10.1 Noise Figure↔System Temperature

The system noise temperature and the noise figure are alternative ways of expressing the noise performance of an amplifier. The noise temperature T_{sys} (usually in degrees Kelvin) is the temperature to which a resistor would have to be heated to produce the same (input) noise level with a noiseless amplifier as is produced by the real amplifier. The noise figure F (usually in decibels, dB) is the ratio of the noise actually present to what would be present with a noiseless amplifier with a resistor at 290°K connected to its input. The resistance of this resistor in either case is whatever is specified for the amplifier. System noise temperature is usually the more useful quantity for an amplifier connected to a source that is colder than room temperature (e.g., a centimeter-wavelength radio astronomy antenna) and noise figure is usually the more useful quantity for an amplifier connected to a source at room temperature (e.g., a high-fidelity audio preamplifier). In a cascade of amplifiers, usually only the noise of the first amplifier needs to be considered because the noise of the second amplifier is divided by the gain of the first amplifier to refer to its input.

HP-45:

(T_{sys}, °K) [↑] 290 [÷] 1 [+] [G] [log] 10 [×] (see F in decibels).

For an HP-35, replace [G] [log] by [log]

HP-45:

(*F*, dB) [↑] 10 [÷] [G] [10^x] 1 [−] 290 [×] (see T_{sys} in degrees Kelvin).

HP-35:

(*F*, dB) [↑] 10 [÷] 10 [x^y] 1 [−] 290 [×] (see T_{sys} in degrees Kelvin).

TEST CASE: $F = 1.6$ dB corresponds to $T_{sys} = 129°K$.

$$F = \frac{T_{sys} + 290°K}{290°K},$$

$$F_{dB} = 10 \log_{10} F.$$

A.10.2 Noise Voltage and Noise Current from a Resistor

This algorithm calculates the open-circuit noise voltage e_n (RMS volts) or the short-circuit noise current i_n (RMS amperes) for an ideal resistor at room temperature (290°K); *B* is the noise bandwidth in hertz and *R* is the resistance in ohms (Ω).

HP-45:

(*B*, Hz) [↑] (*R*, Ω) [×] 1.6 [EEX] [CHS] 20 [×] [G] [$\sqrt{x}$] (see e_n in volts).

(*B*, Hz) [↑] (*R*, Ω) [÷] 1.6 [EEX] [CHS] 20 [×] [G] [$\sqrt{x}$] (see i_n in amperes).

For an HP-35, replace [G] [$\sqrt{x}$] by [$\sqrt{x}$].

TEST CASE: $R = 1$ kΩ, $B = 20$ kHz; get $e_n = 0.566$ μV or $i_n = 0.566$ nA.

The available power from the resistor is just

$$P_a = kTB = \frac{1}{R}\left(\frac{e_n}{2}\right)^2 = R\left(\frac{i_n}{2}\right)^2,$$

$$e_n^2 = 4kTBR,$$

$$i_n^2 = 4kTB/R,$$

where *T* is the temperature, taken to be 290°K, *k* is Boltzmann's constant,

1.380622×10^{-23} J/°K; thus $4kT = 1.6015215 \times 10^{-20}$ J. These are the Nyquist formulas, and they are intimately related to the Rayleigh-Jeans approximation to the Planck black body law (see Section A.9.5).

A.10.3 Noise Level in a Receiver

In a radio astronomy receiver or other chain of amplifiers, one needs to know the noise level at various points in the chain to avoid saturation and other problems. This algorithm calculates the noise power P_0 and the RMS noise voltage e_0 given G (the total system gain in decibels to the point where P_0 and e_0 are measured), T_{sys}, the total system noise temperature in degrees Kelvin referred to the input, B, the noise bandwidth in hertz, and R, the characteristic impedance (resistance) in ohms.

HP-45:

(G, dB) [↑] 10 [÷] [G] [10^x] 1.38 [EEX] [CHS] 23 [×] (T_{sys}, °K) [×] (B, Hz) [×] (see P_0 in watts) (R, Ω) [×] [G] [$\sqrt{x}$] (see e_0 in volts).

For an HP-35, replace [G] [10^x] by 10 [x^y] and [G] [$\sqrt{x}$] by [$\sqrt{x}$].

TEST CASE: $G = 135$ dB, $T_{sys} = 89$°K, $B = 20$ MHz, $R = 50$ Ω; get $P_0 = 0.777$ W, and $e_0 = 6.23$ V.

REFERENCE: See Section A.10.2.

A.10.4 Reactance and Impedance

These algorithms deal with various combinations of resistors (R in ohms), capacitors (C in farads), and inductors (L in henrys). Then f is a frequency in hertz, X is a (capacitive or inductive) reactance in ohms, Z is the magnitude of a (complex) impedance in ohms, and θ is its phase angle in degrees. For such quantities as microfarads (μF) millihenrys (mH), and megahertz (MHz), see Section A.1.3.

Capacitive reactance

HP-45:

(C, F) [G] [π] [×] 2 [×] (f, Hz) [×] [$1/x$] (see X in ohms),

or

(X, Ω) [G] [π] [×] 2 [×] (f, Hz) [×] [$1/x$] (see C in farads),

or

(X, Ω) [G] [π] [×] 2 [×] (C, F) [×] [$1/x$] (see f in hertz).

HP-35:

(*C*, F) [π] [×] 2 [×] (*f*, Hz) [×] [1/*x*] (see *X* in ohms),

or

(*X*, Ω) [π] [×] 2 [×] (*f*, Hz) [×] [1/*x*] (see *C* in farads),

or

(*X*, Ω) [π] [×] 2 [×] (*C*, F) [×] [1/*x*] (see *f* in hertz).

Inductive reactance

HP-45:

(*L*, H) [G] [π] [×] 2 [×] (*f*, Hz) [×] (see *X* in ohms),

or

(*X*, Ω) [G] [π] [÷] 2 [÷] (*f*, Hz) [÷] (see *L* in henrys),

or

(*X*, Ω) [G] [π] [÷] 2 [÷] (*L*, H) [÷] (see *f* in hertz).

HP-35:

(*L*, H) [π] [×] 2 [×] (*f*, Hz) [×] (see *X* in ohms),

or

(*X*, Ω) [π] [÷] 2 [÷] (*f*, Hz) [÷] (see *L* in henrys),

or

(*X*, Ω) [π] [÷] 2 [÷] (*L*, H) [÷] (see *f* in hertz).

***R* and *C* in series**

HP-45:

(*C*, F) [G] [π] [×] 2 [CHS] [×] (*f*, Hz) [×] [1/*x*]
(*R*, Ω) [→P] (see *Z* in ohms) [*x*≷*y*] (see *θ* in degrees).

***R* and *C* in parallel**

HP-45:

(*C*, F) [G] [π] [×] 2 [CHS] [×] (*f*, Hz) [×]
(*R*, Ω) [1/*x*] [→P] [1/*x*] (see *Z* in ohms) [*x*≷*y*] (see *θ* in degrees).

***R* and *L* in series**

HP-45:

(*L*, H) [G] [π] [×] 2 [×] (*f*, Hz) [×]
(*R*, Ω) [→P] (see *Z* in ohms) [*x*≷*y*] (see *θ* in degrees).

***R* and *L* in parallel**

HP-45:

(*L*, H) [G] [π] [×] 2 [×] (*f*, Hz) [×] [1/*x*] (*R*, Ω) [1/*x*] [→P] [1/*x*]
(see *Z* in ohms) [*x*≷*y*] (see *θ* in degrees).

***R*, *L*, and *C* in series**
HP-45:
(*f*, Hz) [G] [π] [×] 2 [×] [↑] [↑] (*L*, H) [×] [x≷y] (*C*, F) [×] [1/x] [−] (*R*, Ω) [→P] (see *Z* in ohms) [x≷y] (see *θ* in degrees).

***R*, *L*, and *C* in parallel**
HP-45:
(*f*, Hz) [G] [π] [×] 2 [×] [↑] [1/x] (*L*, H) [÷] [x≷y] (*C*, F) [×] [−] (*R*, Ω) [1/x] [→P] [1/x] (see *Z* in ohms) [x≷y] (see *θ* in degrees).

Resonance ($X_C = X_L$)
HP-45:
(*L*, H) [1/x] (*C*, F) [÷] [G] [√x] [G] [π] [÷] 2 [÷] (see *f* in hertz),
or
(*f*, Hz) [G] [π] [×] 2 [×] [x²] (*L*, H) [×] [1/x] (see *C* in farads),
or
(*f*, Hz) [G] [π] [×] 2 [×] [x²] (*C*, F) [×] [1/x] (see *L* in henrys).

HP-35:
(*L*, H) [1/x] (*C*, F) [÷] [√x] [π] [÷] 2 [÷] (see *f* in hertz),
or
(*f*, Hz) [π] [×] [↑] [+] [↑] [×] (*L*, H) [×] [1/x] (see *C* in farads),
or
(*f*, Hz) [π] [×] [↑] [+] [↑] [×] (*C*, F) [×] [1/x] (see *L* in henrys).

TEST CASE: $R = 1$ kΩ, $L = 10$ mH, and $C = 0.1$ μF, all in parallel at 5033 Hz give 1 kΩ at 0 degrees—resonance.

A.10.5 Standard Resistors

A resistance calculated from a formula does not usually coincide in value with a real resistor available from a vendor. Given *R*, an arbitrary resistance in ohms, these algorithms give R_s, the nearest (or at least a near) EIA–RETMA or military standard resistance value *as displayed*. Values of *M* are from the tabulation following the algorithms.

For 1 or 2% standard values or military preferred values
HP-45 SETUP: [G] [SCI] 2
LOOP: *R* [G] [log] *M* [×] [EEX] 9 [+] [EEX] 9 [−] *M* [÷] [G] [10ˣ] (see R_s) :|.

For 5, 10, or 20% EIA–RETMA standard values

HP-45 SETUP: [G] [SCI] 1

LOOP: R [G] [log] M [×] [EEX] 9 [+] [EEX] 9 [−] M [÷] [↑] [↑] .02 [+] 180 [×] [SIN] [x^2] 6 [×] [e^x] 2 [EEX] 4 [÷] [+] [EEX] [CHS] 3 [−] [G] [10^x] (see R_s) :|.

%	M	
1	96	(military standard values) (MSV)
2	48	
—	24	(military preferred values) (MPV)
5	24	(EIA–RETMA standard values)
10	12	(EIA–RETMA standard values)
20	6	(EIA–RETMA standard values)

TEST CASE: $R = 58\ \Omega$, get $R_s = 57.6\ \Omega$ (1%), $R_s = 59.0\ \Omega$ (2%), $R_s = 56.2\ \Omega$ (MPV), $R_s = 56\ \Omega$ (5%), $R_s = 56\ \Omega$ (10%), and $R_s = 68\ \Omega$ (20%).

A.10.6 American Wire Gauge

These algorithms relate g, the American Wire Gauge (B & S) number (but use -1 for 00, -2 for 000, -3 for 0000, etc.) to d, the diameter of the wire in mils (milli-inches), A, the cross-sectional area in circular mils (the area of a circle one mil in diameter), and r, the resistance of the wire in ohms per foot.

HP-45:

g [↑] 3 [+] 39 [÷] 92 [$x \gtrless y$] [G] [y^x] 460 [$x \gtrless y$] [÷] (see d in mils)
[x^2] (see A in circular mils).
A [G] [$\sqrt{x}$] 460 [$x \gtrless y$] [÷] [ln] 39 [×] 92 [ln] [÷] 3 [−] (see g).
460 [↑] d [÷] [ln] 92 [ln] [÷] 39 [×] 3 [−] (see g).

For an HP-35, replace [$x \gtrless y$] [G] [y^x] by [x^y], [x^2] by [↑] [×], and [G] [$\sqrt{x}$] by [$\sqrt{x}$].

10 [↑] A [÷] (see r in Ω/ft for annealed copper).
17 [↑] A [÷] (see r in Ω/ft for aluminum).

TEST CASE: $g = 14$ (number 14 wire); get $d = 64.08$ mils, $A = 4107$ circular mils, $r = 2.4 \times 10^{-3}\ \Omega$/ft for copper or $4.1 \times 10^{-3}\ \Omega$/ft for aluminum.

The formulas are

$$d = (460 \text{ mils})92^{-(3+g)/39},$$

$$A = d^2 \quad (\text{no } \pi/4 \text{ because of the funny units}).$$

As an approximation

$$d \cong 10^{(50-g)/20},$$

which is a formula for voltage decibels, and

$$A \cong 10^{(50-g)/10},$$

which is a formula for power decibels. These approximate formulas can be estimated without a calculator.

The formulas for r use 10 Ω circular mils/ft for copper and 17 Ω circular mils/ft for aluminum, which are approximations for 68°F.

REFERENCES: CRC HCP, pp. F-159 to F-161; *65 Notes*, V1N3P1.

A.10.7 Diodes and Resistors (Diodes as Nonlinear Circuit Elements)

Semiconductor diodes are usually characterized by the "diode equation"

$$I = I_0(T)\left\{\exp\left(\frac{qV}{kTm}\right) - 1\right\}.$$

In this equation, I is the current through the diode; V is the voltage across the diode; $I_0(T)$ is the "leakage current" as a function of T, the absolute temperature. Typically $I_0 \cong 3$ pA$=3\times10^{-12}$ A for silicon diodes and $T \cong 295$°K. And q is the charge of an electron, $q = 1.6021917\times10^{-19}$ coulomb (C); k is Boltzmann's constant, $k = 1.380622\times10^{-23}$ J/°K; and m is the "m factor" of the diode, $1 \leqslant m \leqslant 2$, depending on the construction and material of the diode. Usually m is nearer 1 than 2. Note that $q/(kT) \cong 39$/V and that $U = q/(mkT)$ is $20/\text{V} \lesssim U \lesssim 40/\text{V}$. The "nominal" forward voltage drop across the diode is $(-\ln I_0)/U$, which is about 0.68 V for silicon diodes. For reverse bias, $V<0$, a good approximation is just $I = -I_0(T)$; and for forward bias, $V>0$, the -1 is usually negligible so we can write

$$I \cong I_0 \exp(UV),$$

or

$$V \boxed{\uparrow} U \boxed{\times} \boxed{e^x} I_0 \boxed{\times} \qquad \text{(see } I\text{)},$$

and

$$V = \frac{\ln I - \ln I_0}{U},$$

or

$$I \boxed{\ln} I_0 \boxed{\ln} \boxed{-} U \boxed{\div} \qquad \text{(see } V\text{)}.$$

These equations, however, are not adequate in all cases to characterize real semiconductor diodes. Two principal problems are that m is not really constant but varies with I; and the bulk resistance, or an equivalent series resistance, may not be negligible. If either of these refinements is added to the model, finding I through the diode for a given applied V becomes a transcendental equation to solve by an iterative technique. A transcendental equation also results if one puts an external resistor R in series with the diode D and needs to know I for a given V; or if one puts an external resistor in parallel with the diode and needs to know V for a given I.

This section contains iterative solutions for the transcendental equations for four cases,

Case I: R and D in parallel.
Case II: R and D in series.
Case III: Two Rs and a D in series-parallel.
Case IV: Two Rs and a D in parallel-series.

as shown in Figure A.10.1. These iterative solutions are based on the g method with $g(x) = -f'(x)$, and this case is equivalent to the Newton-Raphson method. The diode itself is characterized by the equations above in their simplest form for forward bias, $V > 0$. All these algorithms are for the HP-45.

Case I: *R* and *D* in parallel (Figure A.10.1).
Case Ia: Given *V*, find *I*.
This is the trivial case; just use I from the equation for the diode alone, and add V/R:

$$V \boxed{\uparrow} \boxed{\uparrow} U \boxed{\times} \boxed{e^x} I_0 \boxed{\times} \boxed{x \gtrless y} R \boxed{\div} \boxed{+} \qquad \text{(see } I\text{)}.$$

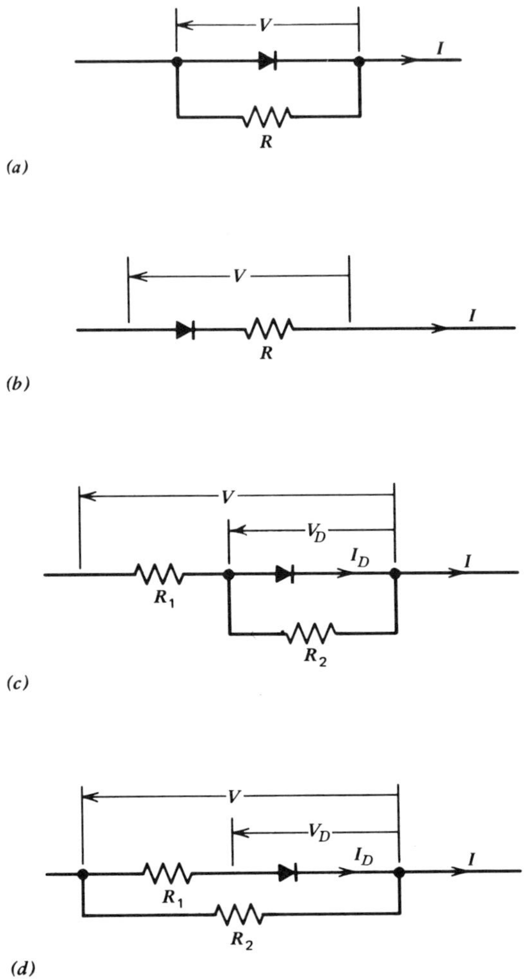

Figure A.10.1. Four circuits containing diodes and resistors. (*a*) Case I: *R* and *D* in parallel. (*b*) Case II: *R* and *D* in series. (*c*) Case III: two *R*s and a *D* in series-parallel. (*d*) Case IV: two *R*s and a *D* in parallel-series.

Case Ib: Given *I*, find *V*.
Case Ib, formula 1 (exponential form).

$$I = \frac{V}{R} + I_0 \exp(UV),$$

$$F(V) = \frac{IR + I_0 R (UV - 1) \exp(UV)}{I_0 R U \exp(UV) + 1}.$$

SETUP: R [↑] [↑] I [×] [STO] 3 [$x \gtrless y$] I_0 [×] [CHS] [STO] 2 U [STO] 1 [CHS] [×] [STO] 4 V_g

LOOP: [RCL] [×] 1 [e^x] [↑] [↑] [↑] [ln] [×] [−] [RCL] [×] 2 [RCL] [+] 3 [$x \gtrless y$] [RCL] [×] 4 1 [+] [÷] (see V_i) :|.

RULE: If $IR \lesssim (-\ln I_0)/U$, use this algorithm and guess $V_g \cong IR$.

TEST CASE: $I = 5$ mA $= 5 \times 10^{-3}$ A, $R = 100\ \Omega$, $U = 39/\mathrm{V}$, $I_0 = 3 \times 10^{-12}$ A; get $V_i = 0.5$, 0.480130171, 0.4720860921, 0.4712488631, 0.4712414768, 0.4712414762 V.

Case Ib, formula 2 (logarithmic form).

$$UV = \ln(IR - V) - \ln(I_0 R),$$

$$F(V) = \frac{(IR - V)\left[\ln(IV - V) - \ln(I_0 R)\right] + V}{1 + U(IR - V)}.$$

SETUP: R [↑] [↑] I [×] [STO] 1 [$x \gtrless y$] I_0 [×] [ln] [STO] 2 U [STO] 3 V_g

LOOP: [↑] [CHS] [RCL] [+] 1 [↑] [ln] [RCL] [−] 2 [$x \gtrless y$] [↑] [R↓] [×] [+] [$x \gtrless y$] [RCL] [×] 3 1 [+] [÷] (see V_i) :|.

RULE: If $IR \gtrsim (-\ln I_0)/U$, use this algorithm and guess $V_g \cong (-\ln I_0)/U$.

TEST CASE: $I = 10$ mA $= 10^{-2}$ A, $R = 100\ \Omega$, $U = 39/\mathrm{V}$, $I_0 = 3 \times 10^{-12}$ A; get $V_i = 0.68$, 0.543924139, 0.5422029833, 0.5422028108 V.

Case II: ***R* and *D* in series** (Figure A.10.1).

Case IIa: Given *I*, find *V*.

This is the trivial case; just use V from the equation for the diode alone, and add IR.

I [↑] [ln] I_0 [ln] [−] U [÷] [$x \gtrless y$] R [×] [+] (see V).

Case IIb: Given *V*, find *I*.

Case IIb, formula 1 (exponential form).

$$I = I_0 \exp(U(V - IR)),$$

$$F(I) = \frac{IUR + 1}{UR + \exp(IUR - UV)/I_0}.$$

SETUP: U [↑] [↑] R [×] [STO] 1 [$x \gtrless y$] V [×] [STO] 3 I_0 [1/x] [STO] 2 I_g

LOOP: [RCL] [×] 1 [↑] [RCL] [−] 3 [e^x] [RCL] [×] 2 [RCL] [+] 1 [$x \gtrless y$] 1 [+] [$x \gtrless y$] [÷] (see I_i) :|.

RULE: If $V \lesssim (-\ln I_0)/U$, use this algorithm and guess $I_g \cong I_0 \exp(UV)$; however this one does not converge very fast, especially for large R.

TEST CASE: $R = 100\ \Omega$, $V = 0.5$ V, $U = 39/\text{V}$, $I_0 = 3 \times 10^{-12}$ A; get $I_i = 8 \times 10^{-4}$, 1.39×10^{-4}, 2.64×10^{-4}, 2.87×10^{-4}, $2.875848819 \times 10^{-4}$, $2.875852387 \times 10^{-4}$ A.

Case IIb, formula 2 (logarithmic form).

$$I = \frac{V}{R} - \frac{\ln I - \ln I_0}{UR}.$$

$$F(I) = \frac{(\ln I_0 + 1 + UV - \ln I)I}{IUR + 1}.$$

SETUP: U [↑] [↑] R [×] [STO] 2 [$x \gtrless y$] V [×] I_0 [ln] [+] 1 [+] [STO] 1 I_g

LOOP: [↑] [↑] [ln] [RCL] 1 [$x \gtrless y$] [−] [×] [$x \gtrless y$] [RCL] [×] 2 1 [+] [÷] (see I_i) :|.

RULE: If $V \gtrsim (-\ln I_0)/U$, then use this algorithm and guess $I_g \cong (V + (\ln I_0)/U)/R$.

TEST CASE: $V = 1$ V, $R = 100\ \Omega$, $U = 39/\text{V}$, $I_0 = 3 \times 10^{-12}$ A; get $I_i = 3 \times 10^{-3}$, 4.55×10^{-3}, $4.577968412 \times 10^{-3}$, $4.577971889 \times 10^{-3}$ A.

Case III: Two *R*s and a *D* in series-parallel (Figure A.10.1).

$$I_D = I_0 \exp(UV_D),$$

$$I = I_D + \frac{V_D}{R_2},$$

$$V = V_D + IR_1,$$

$$V_D = \frac{\ln I_D - \ln I_0}{U}.$$

Case IIIa: Given I, find V.
$V=IR_1+$solution to Case Ib.

Case IIIb: Given V, find I.

Case IIIb, formula 1 (exponential form).

$$I=I_0\exp\left[U(V-IR_1)\right]+\frac{V-IR_1}{R_2},$$

$$F(I)=\frac{(IUR_1-UV)+(1+UV)+\exp(IUR_1-UV)V/(I_0R_2)}{UR_1+\exp(IUR_1-UV)(R_1+R_2)/(I_0R_2)}$$

(written in this peculiar form because it yields a shorter algorithm).

SETUP: U [↑] [↑] V [STO] 3 [×] [STO] 1 1 [+] [STO] 2 [$x\gtrless y$] R_2 [↑] [↑] I_0 [×] [STO] [÷] 3 [$1/x$] [STO] 5 [CLX] R_1 [↑] [R↓] [+] [STO] [×] 5 [R↓] [×] [STO] 4 [↑] [↑] I_g

LOOP: [×] [RCL] [−] 1 [↑] [↑] [e^x] [RCL] [×] 3 [+] [RCL] [+] 2 [$x\gtrless y$] [e^x] [RCL] [×] 5 [RCL] [+] 4 [÷] (see I_i) :|.

RULE: If $VR_2/(R_1+R_2)\lesssim(-\ln I_0)/U$, use this algorithm and guess $I_g\cong V/(R_1+R_2)$.

TEST CASE: $V=1.5$ V, $R_1=200$ Ω, $R_2=100$ Ω, $U=39/\text{V}$, $I_0=3\times10^{-12}$ A; get $I_i=5\times10^{-3}$, 5.089299525×10^{-3}, 5.116045258×10^{-3}, 5.117590627×10^{-3}, 5.117595119×10^{-3} A.

Case IIIb, formula 2 (logarithmic form).

$$I=\frac{V}{R_1}+\frac{\ln I_0-\ln(I(1+R_1/R_2)-V/R_1)}{UR_1},$$

$$F(I)=$$

$$\frac{UV+\ln I_0-\ln\left[(I(R_1+R_2)-V)/R_2\right]+I(R_1+R_2)/\left[I(R_1+R_2)-V\right]}{(R_1+R_2)/\left[I(R_1+R_2)-V\right]+UR_1}.$$

SETUP: I_0 [ln] U [STO] 3 V [↑] [R↓] [×] [+] 1 [+] [$x\gtrless y$] [ln] [−] R_2 [↑] [R↓] [ln] [+] [STO] 1 [R↓] [$x\gtrless y$] R_1 [STO] [×] 3 [+] [$x\gtrless y$] [÷] [STO] 2 [↑] [↑] I_g

LOOP: [×] 1 [−] [$1/x$] [↑] [↑] [ln] [+] [RCL] [+] 1 [$x\gtrless y$] [RCL] [×] 2 [RCL] [+] 3 [÷] (see I_i) :|.

RULE: If $VR_2/(R_1+R_2)\gtrsim(-\ln I_0)/U$, use this algorithm and guess $I_g\cong(V+(\ln I_0)/U)/R_1$.

TEST CASE: $V=2.5$ V, $R_1=200\ \Omega$, $R_2=100\ \Omega$, $U=39/\text{V}$, $I_0=3\times10^{-12}$ A; get $I_i=9.1\times10^{-3}$, 9.765883757×10^{-3}, 9.794523778×10^{-3}, 9.794547023×10^{-3} A.

Case IV: Two *R*s and a *D* in parallel-series (Figure A.10.1).

$$I_D=I\exp(UV_D),$$

$$I=I_D+\frac{V}{R_2},$$

$$V=V_D+I_DR_1,$$

$$V_D=\frac{\ln I_D-\ln I_0}{U}.$$

Case IVa: Given *V*, find *I*.

$$I=\frac{V}{R_2}+\text{solution to Case IIb.}$$

Case IVb: Given *I*, find *V*.

Case IVb, formula 1 (exponential form).

$$V=IR_2-I_0R_2\exp\left(UV-UIR_1+\frac{UVR_1}{R_2}\right),$$

$$F(V)=\frac{IR_2\exp(UIR_1-UV-UVR_1/R_2)-I_0R_2+UVI_0(R_1+R_2)}{\exp(UIR_1-UV-UVR_1/R_2)+UI_0(R_1+R_2)}.$$

SETUP: I_0 [1/x] [STO] 2 [CHS] [STO] 4 R_2 [STO] [÷] 4 R_1 [↑] [↑] I [STO] [×] 2 [×] U [x≷y] [R↓] [R↓] [x≷y] [÷] 1 [+] [R↓] [×] [STO] 1 1 [−] [STO] 3 [R↓] [×] [STO] 5 [CHS] [↑] [↑] V_g

LOOP: [×] [RCL] [+] 1 [↑] [↑] [e^x] [RCL] [×] 2 [RCL] [+] 3 [−] [x≷y] [e^x] [RCL] [×] 4 [RCL] [−] 5 [÷] (see V_i) :|.

RULE: If $IR_2 \lesssim (-\ln I_0)/U$, use this algorithm and guess $V_g \cong IR_2$.

TEST CASE: $I=5$ mA$=5\times10^{-3}$ A, $R_2=100\ \Omega$, $R_1=200\ \Omega$, $U=39/\text{V}$, $I_0=3\times10^{-12}$ A; get $V_i=0.5$, 0.492207442, 0.486832915, 0.485043932, 0.4849055156, 0.4849047923 V.

Case IVb, formula 2 (logarithmic form).

$$V=\frac{\ln(IR_2-V)-\ln(I_0R_2)}{U}+IR_1-\frac{VR_1}{R_2},$$

$$F(V)=\frac{\left[\ln(IR_2-V)-\ln(I_0R_2)+IR_1U-1\right](IR_2-V)+IR_2}{1+U(1+R_1/R_2)(IR_2-V)}.$$

SETUP: I [↑] [↑] R_1 [↑] [R↓] [×] U [STO] 3 [×] [STO] 1 [CLX] R_2 [↑] [R↓] [×] [STO] 2 [R↓] [+] [$x\gtrless y$] [÷] [STO] [×] 3 [G] [LASTX] I_0 [×] [ln] 1 [+] [STO] [−] 1 [R↓] [R↓] V_g

LOOP: [−] [↑] [↑] [ln] [RCL] [+] 1 [×] [RCL] [+] 2 [$x\gtrless y$] [RCL] [×] 3 1 [+] [÷] (see V_i) :|.

RULE: If $IR_2 \lesssim (-\ln I_0)/U$, use this algorithm and guess $V_g \cong IR_1R_2/(R_1+R_2)$ *or* $V_g \cong IR_1-(\ln I_0)/U$ whichever is *smaller*.

TEST CASE: $I=10$ mA$=10^{-2}$ A, $R_1=200\ \Omega$, $R_2=100\ \Omega$, $U=39/\text{V}$, $I_0=3\times10^{-12}$ A; get $V_i=0.67$, 0.840195155, 0.838496303, 0.8384958477 V.

RULE OF RULES: If it is borderline between formula 1 and formula 2 (any case), prefer formula 2 (logarithmic form).

A.11 TRIGONOMETRY AND COORDINATE SYSTEMS

A.11.1 Trigonometry Notes

There are essentially five cases to solve in plane trigonometry and a sixth case in spherical trigonometry. These cases can be classified by the given quantities: two sides and the included angle (SAS for side-angle-side), two angles and the included side (ASA), two sides and an opposite angle (SSA), two angles and an opposite side (AAS), and three sides (SSS). In spherical trigonometry one can also solve a triangle given three angles (AAA). In one case (SSA) there are usually two solutions—two triangles with the given parameters. In spherical trigonometry, the AAS case also usually has two solutions.

Section A.11.2 contains complete solutions (i.e., the three unknowns can be found) for the five cases of plane (oblique) triangles. Section A.11.3 contains incomplete solutions (i.e., only one or two unknowns can be found) for the six cases of spherical triangles. A complete spherical triangle solution then requires two of these algorithms, or two applications of an algorithm with different parameters.

A.11.2 Plane (Oblique) Triangles

Angles A, B, and C are in the units for which the calculator is set; a, b, and c are the opposite sides. The following algorithms are all for the HP-45.

Given a, b, C, find A, B, c: SAS.
C [↑] [↑] b [G] [→R] a [$x \gtrless y$] [−] [→P] (see c) [R↓] (see B) [+] [COS] [CHS] [G] [COS^{-1}] (see A).

Given a, b, c, find A, B, C: SSS.
a [↑] [↑] b [÷] [$x \gtrless y$] c [÷] [↑] [↑] [$1/x$] [+] [$x \gtrless y$] [R↓] [R↓] [R↓] [x^2] [÷] [↑] [R↓] [−] 2 [÷] [↑] [G] [COS^{-1}] (see B) [R↓] [R↓] [×] [G] [LASTX] [R↓] [$1/x$] [R↓] [R↓] [×] [−] [G] [COS^{-1}] (see A) [+] [COS] [CHS] [G] [COS^{-1}] (see C).

Given a, A, C, find B, b, c: AAS.
A [↑] [↑] C [+] [COS] [CHS] [G] [COS^{-1}] (see B)
a [G] [→R] [R↓] [R↓] [$x \gtrless y$] [R↓] 1 [G] [→R] [R↓] [$x \gtrless y$] [R↓] [÷] (see b) [×] [+] (see c).

Given a, B, C, find A, b, c: ASA.
C [↑] [↑] B [+] [COS] [CHS] [G] [COS^{-1}] (see A) 1 [G] [→R] [↑] [R↓] [R↓] [$x \gtrless y$] a [G] [→R] [R↓] [$x \gtrless y$] [÷] (see c) [×] [+] (see b).

Given B, b, c, find a, A, C: SSA.
B [↑] [↑] [SIN] c [×] b [÷] [G] [LASTX] [R↓] [G] [SIN^{-1}] (see C; if $C > B$, then $C' = 180° - C$ is also a possible solution. To get C': [COS] [CHS] [G] [COS^{-1}] (see C'), then go on) [+] [COS] [CHS] [G] [COS^{-1}] (see A or A') [SIN] [$x \gtrless y$] [SIN] [÷] [×] (see a or a').

Given a, b, c, find area.
a [↑] [↑] c [÷] [↑] [↑] [$1/x$] [+] [R↓] [÷] [×] [↑] [↑] [R↓] [R↓] b [x^2] [$x \gtrless y$] [÷] [−] 2 [÷] [G] [COS^{-1}] [SIN] [×] 2 [÷] (see area).

Given a, b, C, find area.
C [SIN] a [×] b [×] 2 [÷] (see area).

Given a, B, C, find area.
B [↑] [SIN] C [↑] [R↓] [SIN] [×] a [$x \gtrless y$] [R↓] [R↓] [+] [SIN] [÷] [×] [×] 2 [÷] (see area).

Given three vertices, (x_1,y_1), (x_2,y_2), (x_3,y_3), find area.
y_2 [↑] [↑] y_3 [↑] [R↓] [−] x_1 [×] [$x \gtrless y$] y_1 [−] [G] [LAST X] [R↓] [R↓] [R↓] [−] x_2 [×] [$x \gtrless y$] x_3 [×] [+] [−] 2 [÷] (see area).

TEST CASE: $A=30°$, $B=45°$, $C=105°$, $a=50$, $b=70.71$, $c=96.59$, area$=$ 1707.53.

REFERENCES: CRC SMT, p. 236; *HP-45 Applications Book*, pp. 190–198; *HP-21 Applications Book*, pp. 37–48; or any trigonometry book.

A.11.3 Spherical Triangles

Call A, B, and C the angles, and a, b, and c the opposite sides, all in the units for which the calculator is set, and in the range 0 to 180° or the equivalent; R is the radius of the sphere. The following algorithms are all for the HP-45.

Given a, b, C, find A, c: SAS.
a [↑] [↑] [SIN] C [$x \gtrless y$] [G] [→R] [R↓] [R↓] [R↓] [COS] [→P] [$x \gtrless y$] b [$x \gtrless y$] [−] [$x \gtrless y$] [G] [→R] [R↓] [→P] [R↓] (see A) [R↓] [R↓] [→P] [$x \gtrless y$] (see c).

Given A, B, c, find a, C: ASA.
A [↑] [↑] [SIN] c [$x \gtrless y$] [G] [→R] [R↓] [R↓] [R↓] [COS] [CHS] [→P] [$x \gtrless y$] B [−] [$x \gtrless y$] [G] [→R] [R↓] [→P] [R↓] (see a) [R↓] [R↓] [→P] [$x \gtrless y$] (see C).

Given a, b, B, find A, C: SSA.
a [↑] [SIN] b [↑] [R↓] [SIN] [÷] [R↓] [+] 2 [÷] [↑] [COS] [R↓] [−] [COS] [R↓] [R↓] [R↓] [÷] B [↑] [SIN] [R↓] [R↓] [R↓] [×] [G] [SIN^{-1}] (see A) ($A'=180°-A$ is another possible solution. To get A': [COS] [CHS] [G] [COS^{-1}] (see A'), then go on) [+] 2 [÷] [TAN] [÷] [G] [TAN^{-1}] 2 [×] (see C or C').

Given A, B, b, find a, c: AAS.

A [↑] [SIN] B [↑] [R↓] [SIN] [÷] [R↓] [−] 2 [÷] [↑] [COS] [R↓] [−] [COS] [R↓] [R↓] [R↓] [÷] b [↑] [SIN] [R↓] [R↓] [R↓] [×] [G] [SIN⁻¹] (see a) ($a' = 180° - a$ is another possible solution. To get a': [COS] [CHS] [G] [COS⁻¹] (see a'), then go on) [+] 2 [÷] [TAN] [×] [G] [TAN⁻¹] 2 [×] (see c or c').

Given a, b, c, find C: SSS.

a [↑] 1 [G] [→R] b [↑] [SIN] [R↓] [COS] [×] c [COS] [$x \gtrless y$] [−] [÷] [×] [$1/x$] [G] [COS⁻¹] (see C).

Given A, B, C, find c: AAA.

A [↑] 1 [G] [→R] B [↑] [SIN] [R↓] [COS] [×] C [COS] [+] [÷] [×] [$1/x$] [G] [COS⁻¹] (see c).

Given A, B, C; find solid angle Ω and area.

A [↑] B [+] C [+] 1 [CHS] [G] [COS⁻¹] [÷] 1 [−] [G] [π] [×] (see Ω in steradians) R [x^2] [×] (see area in (units of $R)^2$).

Right-angled spherical triangles

For the special case $C = 90°$, and assuming that all sides and angles are less than 180°, the spherical triangle formulas are simpler.

Given a, b ($C = 90°$), find A, B, c: S90S.

a [↑] [↑] [TAN] [$x \gtrless y$] [COS] b [COS] [×] [G] [COS⁻¹] (see c) [TAN] [÷] [G] [COS⁻¹] (see B) [SIN] [$x \gtrless y$] [COS] [×] [G] [COS⁻¹] (see A).

Given a, B, ($C = 90°$), find A, b, c: AS90.

B [↑] [COS] [$x \gtrless y$] [SIN] a [COS] [↑] [R↓] [×] [G] [COS⁻¹] (see A) [SIN] [÷] [↑] [G] [COS⁻¹] (see b) [R↓] [×] [G] [COS⁻¹] (see c).

Given a, c ($C = 90°$), find A, B, b: SS90.

a [↑] [TAN] c [TAN] [÷] [↑] [↑] [G] [COS⁻¹] (see B) [SIN] [R↓] [R↓] [R↓] [COS] [×] [G] [COS⁻¹] (see A) [SIN] [÷] [G] [COS⁻¹] (see b).

Given A, c ($C = 90°$), find B, a, b: 90AS.

c [↑] 1 [G] [→R] A [↑] [R↓] [COS] [÷] [→P] [R↓] (see b) [COS] [$x \gtrless y$] [SIN] [×] [G] [COS⁻¹] (see B) [R↓] [COS] [×] [G] [COS⁻¹] (see a).

Given A, a ($C=90°$), find B, b, c: A90S.

a [↑] [COS] [$x \gtrless y$] [SIN] A [SIN] [↑] [R↓] [÷] [G] [SIN^{-1}] (see c) [COS] [$x \gtrless y$] [÷] [G] [COS^{-1}] (see b) [COS] [×] [G] [COS^{-1}] (see B).

The alternative solution for this case is $B' = 180° - B$, $b' = 180° - b$, and $c' = 180° - c$.

Given A, B ($C=90°$), find a, b, c: AA90.

A [↑] 1 [G] [→R] B [↑] 1 [G] [→R] [R↓] [÷] [↑] [R↓] [G] [COS^{-1}] (see a) [R↓] [÷] [↑] [G] [COS^{-1}] (see b) [R↓] [×] [G] [COS^{-1}] (see c).

TEST CASE: $A=60°$, $B=100°$, $C=90°$, $a=59.°49$, $b=101.°57$, $c=95.°84$, $\Omega=1.22$ sr.

REFERENCES: A good discussion of spherical trigonometry is in Smart (1962), Chapter I. The "trick" in the SAS and ASA algorithms is from Figure 6 (p. 13) in Smart (1962) and the associated discussion. I thank George Rybicki (private communication) for his suggestions. See also CRC SMT, p. 238 or CRC HTM, p. 299.

A.11.4 Coordinate Translation and Rotation

This algorithm uses x and y, the coordinates of a point in an ordinary Cartesian reference frame, to calculate x' and y', the coordinates of the point in a new reference frame specified by x_0 and y_0, the origin of the new frame in the old coordinates, and by α, the angle of rotation in degrees (positive counterclockwise) of the new frame with respect to the old. See Section 2.3.2 and especially Figure 2.3.1.

HP-45:

y [↑] y_0 [−] x [↑] x_0 [−] [→P] [$x \gtrless y$] α [−] [$x \gtrless y$] [G] [→R] (see x') [$x \gtrless y$] (see y').

HP-35:

y [↑] y_0 [−] (see $y - y_0$; note sign) [↑] [×] x [↑] x_0 [−] [↑] [↑] [R↓] [×] [+] [$\sqrt{x}$] [STO] [÷] [ARC] [COS] (if $y - y_0 < 0$: [CHS] [↑]) α [−] [↑] [COS] [RCL] [×] (see x') [$x \gtrless y$] [SIN] [RCL] [×] (see y').

TEST CASE: $x=6$, $y=5$, $x_0=2$, $y_0=3$, $\alpha=18°$; get $x'=4.42$, $y'=0.67$.

REFERENCE: Section 2.3.2.

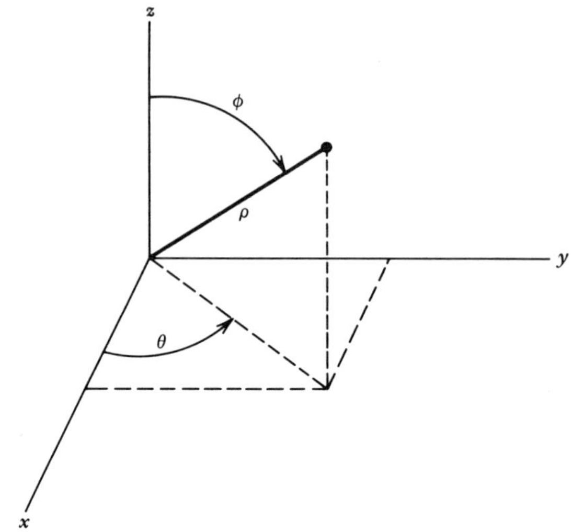

Figure A.11.1 Spherical and rectangular coordinates.

A.11.5 Spherical↔Rectangular Coordinates

HP-45:

y [↑] x [→P] z [→P] (see ρ) [R↓] (see ϕ) [R↓] (see θ).
ϕ [↑] ρ [G] [→R] (see z) [$x \gtrless y$] θ [$x \gtrless y$] [G] [→R] (see x) [$x \gtrless y$] (see y).

TEST CASE: $x = 3$, $y = 4$, $z = 12$, $\rho = 13$, $\phi = 22.^\circ 62$, $\theta = 53.^\circ 13$.

REFERENCES: CRC SMT, p. 385; CRC HTM, p. 513.

A.12 MONEY WITH INTEREST

The following algorithms deal with money invested or borrowed, with interest. The interest rate for a single interest period is i, expressed as a decimal fraction (i.e., 0.09 rather than 9%). Alternatively, key i in % followed by [EEX] [CHS] 2. Then P is the principal or present value (alias PV), S is the sum or future value (alias FV), and R is a single payment (alias PMT) in a series of n equal payments made at the end of each of the n interest periods. There are three cases: *SPin*, in which a payment or money transfer is made at the beginning (P) and end (S) of the time; *SRin*, in which n payments of R are made through the time and S at the end; and *PRin*, in which n payments of R are made through the time and

P at the beginning. The *SRin* and *PRin* cases are called ordinary annuities.

All these algorithms apply to interest compounded at the end of each of the n periods. For simple interest, set $n=1$ in the *SPin* case. For interest compounded continuously, replace i in the algorithms below by i', where

i [e^x] 1 [−] (see i'),

or

i' [↑] 1 [+] [ln] (see i).

Thus i' is the effective period rate if i is compounded continuously.

***SPin* case**

Given P, i, n, find S (single-payment compound amount factor).
- HP-45: i [↑] 1 [+] n [G] [y^x] P [×] (see S).
- HP-35: i [↑] 1 [+] n [x≷y] [x^y] P [×] (see S).

Given S, i, n, find P (single-payment present worth factor).
- HP-45: i [↑] 1 [+] n [G] [y^x] [1/x] S [×] (see P).
- HP-35: i [↑] 1 [+] n [x≷y] [x^y] [1/x] S [×] (see P).

Given S, P, n, find i.
- HP-45: S [↑] P [÷] n [1/x] [G] [y^x] 1 [−] (see i).
- HP-35: n [1/x] S [↑] P [÷] [x^y] 1 [−] (see i).

Given S, P, i, find n.
- HP-45 or HP-35: S [↑] P [÷] [ln] i [↑] 1 [+] [ln] [÷] (see n).

***SRin* case**

Given R, i, n, find S (equal-payment-series compound amount factor) (alias future value of an ordinary annuity).
- HP-45: i [↑] [↑] 1 [+] n [G] [y^x] 1 [−] [x≷y] [÷] R [×] (see S).
- HP-35: i [↑] [↑] 1 [+] n [x≷y] [x^y] 1 [−] [x≷y] [÷] R [×] (see S).

Given S, i, n, find R (equal-payment-series sinking fund factor).
- HP-45: i [↑] [↑] 1 [+] n [G] [y^x] 1 [−] [÷] S [×] (see R).
- HP-35: i [↑] [↑] 1 [+] n [x≷y] [x^y] 1 [−] [÷] S [×] (see R).

Given S, R, i, find n.
- HP-45 or HP-35: i [↑] [↑] S [×] R [÷] 1 [+] [ln] [x≷y] 1 [+] [ln] [÷] (see n).

(To solve for i, see below.)

PRin case

Given P, i, n, find R (equal-payment-series capital recovery factor).

HP-45: i [↑] [↑] [↑] 1 [+] n [G] [y^x] 1 [−] [÷] [+] P [×] (see R).

HP-35: i [↑] [↑] [↑] 1 [+] n [$x \gtrless y$] [x^y] 1 [−] [÷] [+] P [×] (see R).

Given R, i, n, find P (equal-payment-series present worth factor) (alias present value of an ordinary annuity).

HP-45: i [↑] [↑] [↑] 1 [+] n [G] [y^x] 1 [−] [÷] [+] [$1/x$] R [×] (see P).

HP-35: i [↑] [↑] [↑] 1 [+] n [$x \gtrless y$] [x^y] 1 [−] [÷] [+] [$1/x$] R [×] (see P).

Given P, R, i, find n.

HP-45 or HP-35: R [↑] P [÷] [↑] [↑] i [↑] [R↓] [−] [÷] [ln] [$x \gtrless y$] 1 [+] [ln] [÷] (see n).

(To solve for i, see below.)

The equation for i in the *SRin* and *PRin* cases is a transcendental equation and can be solved using some iterative technique. Each cycle through the loop below gives an improved approximation for the interest rate i. Three or four cycles usually give an approximation good to three or four significant figures. The .1 at the end of the setup is a first guess at i; if a better first guess is known, use it in place of the .1.

Given S, R, n, find i.

HP-45 SETUP: S [↑] R [÷] n [↑] [$1/x$] 1 [$x \gtrless y$] [−] [STO] 1 [CLX] [+] [÷] [STO] 2 [CLX] .1

LOOP: [×] [↑] [RCL] [×] 1 1 [+] [$x \gtrless y$] 1 [+] [RCL] 1 [G] [y^x] [−] [G] [LASTX] [RCL] [−] 2 [÷] (see i_i) :|.

HP-35 SETUP: S [↑] R [÷] [STO] .1

LOOP: [RCL] [×] [↑] [↑] 1 [+] n [$1/x$] 1 [$x \gtrless y$] [−] [↑] [R↓] [$x \gtrless y$] [x^y] [↑] [R↓] [R↓] [×] 1 [+] [$x \gtrless y$] [−] [$x \gtrless y$] [RCL] n [÷] [−] [÷] (see i_i) :|.

Given P, R, n, find i.

HP-45 SETUP: P [↑] R [÷] [STO] 1 n [STO] 2 1 [+] [STO] 3 .1

LOOP: [↑] [RCL] [×] 3 1 [+] [CHS] [$x \gtrless y$] 1 [+] [RCL] 3 [G] [y^x] [+] [G] [LASTX] [RCL] [×] 1 [RCL] [−] 2 [÷] (see i_i) :|.

HP-35 SETUP: P [↑] R [÷] [STO] .1 (and mentally calculate $n+1$ for later use)

LOOP: [↑] [↑] 1 [+] $(n+1)$ [↑] [R↓] [$x \gtrless y$] [x^y] [↑] [R↓] [R↓] [×] [−] 1 [−] [$x \gtrless y$] [RCL] [×] n [−] [÷] (see i_i) :|.

TEST CASE: $P=\$2000$, $R=\$311.64$, $n=10$, $i=0.09$, $S=\$4734.73$.

REFERENCES: Section 3.6; Thuesen (1950); CRC SMT, p. 634.

A.13 MISCELLANEOUS

A.13.1 Musical Scale

In common use today is the equal-tempered chromatic musical scale from the standard concert pitch of $A_4 = 440$ Hz. The 12 tones of each octave are equally spaced (tempered) in log space. The following algorithms convert from a note in standard notation to a frequency f in hertz, or from f to the (nearest) note; V is the octave number in the standard scheme; middle C begins octave number 4, and N is the note number from the following tabulation.

1	2	3	4	5	6	7	8	9	10	11	12
C	C#	D	E♭	E	F	F#	G	G#	A	B♭	B

HP-45: N [↑] 22 [−] 12 [÷] V [+] 2 [$x \gtrless y$] [G] [y^x] 55 [×] (see f in hertz).
HP-35: N [↑] 22 [−] 12 [÷] V [+] 2 [x^y] 55 [×] (see f in hertz).
HP-45 ONLY: [FIX] 6
EITHER: f [↑] 55 [÷] [ln] 2 [ln] [÷] 43 [↑] 24 [÷] [+] (see V as *integer part* of display) V (i.e., key V back in) [−] 12 [×] .5 [+] (see N; if N is not an integer, f falls in the crack between notes; the nearest note is N rounded to an integer.)

TEST CASE: A_4 ($N=10$, $V=4$); $f=440$ Hz.

$$\log_2\left(\frac{f}{55}\right) = V + \frac{N-22}{12}.$$

REFERENCE: CRC HCP, p. E-48.

A.13.2 Sailboat Speeds

The skipper of a sailboat beating to windward often needs to know S, the speed made good to windward, that is, the component of the sailboat's velocity directly to windward. From the boat, the skipper can measure θ, the angle (in degrees) of the apparent wind with respect to the bow of the boat; v_{wb}, the speed of the apparent wind with respect to the boat; and v_b, the speed of the boat through the water. Assume no currents and no leeway. Then the algorithm below calculates S and v_w, the speed of the wind over the water and ϕ, the angle (in degrees) between the boat's course and the true wind. The units of S and the vs are arbitrary but all the same.

HP-45:

θ [↑] v_{wb} [G] [→R] v_b [↑] [R↓] [−] [→P] (see v_w) [R↓] (see ϕ) [COS] [×] (see S).

TEST CASE: $\theta = 28°$, $v_{wb} = 17$ knots, $v_b = 9$ knots, get $v_w = 9.99$ knots, $\phi = 53.°02$, and $S = 5.41$ knots (which is a pretty hot sailboat!).

REFERENCE: Texas Instruments SR-50 instruction booklet, p. 39.

A.13.3 Stopwatch (Timer) on the HP-45

Some HP-45s have an arcane stopwatch or timer. The format of the display in stopwatch mode is

$$\text{HH.MM SS } \underbrace{{}_\wedge\text{SS}}_{\text{optional}}.$$

The timer is not very precise (typically $\pm 10\%$) because it is based on a free-running metal oxide semiconductor (MOS) clock. Presumably this is why Hewlett-Packard is officially unaware of the timer's existence. The timer in the HP-55 is very similar but much more precise.

A legal time number $< 14^{\text{h}}$ in HH.MMSS‸SS format in the display on entering stopwatch mode will be converted to timer display format. The timer runs only to 12 hours; the next hour after 12 is 1. The numbers originally in Y, Z, and T are preserved if the exit from stopwatch mode is by [·], but if the exit is by [↑], then

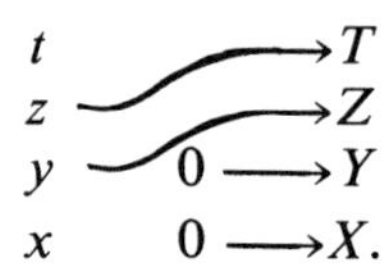

Into stopwatch mode: [RCL] [CHS] [7] [8]

(i.e., press [RCL], then [CHS], [7] and [8] all three simultaneously).

While in stopwatch mode:

Start timer: [CHS].

Stop timer and store time in LASTX (alias split 0): [CHS] or [Σ+].

Reset timer (i.e., clear display): [CLX].

Eliminate 1/100 second display: [EEX].

Restore 1/100 second display: [EEX].

Store a split: any number key 1 through 9 (timer running). (The format of the stored number is very peculiar, namely 0.HHMM0SS00E‸SS, unless converted by [·] below.)

Recall time when timer was stopped: 0.

Recall a split: any number key 1 through 9. (For both these recalls the timer must *not* be running and the split must *not* have been converted by [·] below.)

Out of stopwatch mode into calculator mode: [↑].

Out of stopwatch mode into calculator mode *and* convert display and all splits and LASTX into the format HH.MMSS‸SS: [·].

I thank Charles V. Briggs, III (private communication) for pointing out the stopwatch mode to me. See also *65 Notes*, V3N8P8.

A.13.4 Temperature Conversions

(T, °C) [↑] 1.8 [×] 32 [+] (see T, °F).
(T, °F) [↑] 32 [−] 1.8 [÷] (see T, °C).
(T, °C) [↑] 273.15 [+] (see T, °K).
(T, °K) [↑] 273.15 [−] (see T, °C).
(T, °F) [↑] 459.67 [+] 1.8 [÷] (see T, °K)
(T, °K) [↑] 1.8 [×] 459.67 [−] (see T, °F)
(T, °C) [↑] 273.15 [+] 1.8 [×] (see T, °R).
(T, °R) [↑] 1.8 [÷] 273.15 [−] (see T, °C).
(T, °K) [↑] 1.8 [×] (see T, °R).
(T, °R) [↑] 1.8 [÷] (see T, °K).
(T, °F) [↑] 459.67 [+] (see T, °R).
(T, °R) [↑] 459.67 [−] (see T, °F).

TEST CASE: $T = 68°F = 20°C = 293.15°K = 527.67°R$.

A.13.5 Weather

When two Englishmen meet, their first talk is of the weather.

—*Samuel Johnson (1709–1784)*

Everybody talks about the weather, but nobody does anything about it.

—*attributed to Mark Twain (1835–1910)*

The Tetens equation (as quoted in Haurwitz, 1941, p. 9), is a good approximation relating the equilibrium (or saturation) partial pressure e_m of water vapor, to the temperature T.

HP-45:

(T, °C) [↑] [↑] 236.87 [+] [÷] 7.49 [×] [G] [10^x] 4.579 [×] (see e_m in millimeters of mercury).

Change the preceding constant from 4.579 to 6.105, then see e_m in millibars.

The relative humidity f (expressed as a decimal fraction, not as a percentage) is the ratio of the amount of water vapor in the air to the amount of water vapor that the air could have if saturated at the same T. Thus f is also the ratio of the actual water vapor pressure to the saturation water vapor pressure at the same T. The dew point T_D is the temperature at which the actual water vapor pressure would equal the saturation vapor pressure, or f would equal unity. If the actual temperature reaches the dew point ($T \cong T_D$), fog forms ($f \gtrsim 1$).

HP-45:

(T, °C) [↑] [↑] 236.87 [STO] 4 [+] [÷] (f) [G] [log] 7.49 [÷] [+] [$1/x$] 1 [−] [RCL] 4 [÷] [$1/x$] (see T_D, °C).

(T_D, °C) [↑] [↑] 236.87 [STO] 4 [+] [÷] (T, °C) [↑] [↑] [RCL] 4 [+] [÷] [−] 7.49 [×] [G] [10^x] (see f).

To convert the preceding algorithms to the HP-35, change [G] [10^x] to 10 [x^y], [G] [log] to [log], and delete 4 after [STO] and [RCL].

A psychrometer is a device for measuring humidity employing two thermometers. The bulb of one thermometer is kept wet with water, and the air in contact with the wet bulb is at the dew point. But since this air was cooled *by addition of water*, the temperature T_w of the wet bulb in a psychrometer is not the same as T_D. The "psychrometer equation" (see, e.g., Humphreys, 1940, p. 15) gives f from T_w and T. There is also a weak dependence on P, the atmospheric pressure; for approximate values, use $P = 742.7$ mm Hg.

HP-45:

(T_w, °C) [↑] [↑] [↑] 236.87 [STO] 1 [+] [÷] 7.49 [×] [G] [10^x] [$x \gtrless y$] (T, °C) [STO] 2 [−] [↑] [↑] 870 [÷] 1 [−] [×] 6938 [÷] (P, mm Hg) [×] [−] [RCL] 2 [↑] [↑] [RCL] 1 [+] [÷] 7.49 [×] [G] [10^x] [÷] (see f).

HP-35:

(T_w, °C) [↑] [↑] [↑] 236.87 [+] [÷] 7.49 [×] 10 [x^y] [$x \gtrless y$] (T, °C) [STO] [−] [↑] [↑] 870 [÷] 1 [−] [×] 6938 [÷] (P, mm Hg) [×] [−] [RCL] [RCL] 236.87 [+] [÷] 7.49 [×] 10 [x^y] [÷] (see f).

The constants in all the preceding algorithms are for vapor over liquid water. For ice instead, change 236.87 to 270.8 and change 7.49 to 9.67. With this change, the psychrometer equation works with ice on the wet bulb.

TEST CASE: $T = 30°C$, $T_w = 23°C$, $T_D = 20°C$, $P = 750$ mm Hg, $f = 0.55$, water vapor pressure = 17.5 mm Hg.

The wind-chill temperature* is the equivalent temperature that would produce, on a calm day, the same heat loss from exposed human skin (at about 90°F) as is produced by the existing temperature and wind. The heat loss due to wind is approximately proportional to $\sqrt{v}$ or $v^{0.622}$ (see Conrad and Pollak, 1950, p. 195), where v is the wind speed. However the currently popular wind-chill charts take the heat loss to be proportional to $\log(v/v_0)$, where v_0 is a wind speed below which there is no additional heat loss due to the wind; v_0 is 4 to 5 mph.

In the algorithm below, ΔT is the depression of the effective temperature due to the wind; that is, ΔT is the difference between the actual temperature and the wind-chill temperature.

HP-45:

.634 [↑] [↑] (v, mph) [G] [log] [−] [×] (T, °F) [↑] 90 [−] [×] (see ΔT in Fahrenheit degrees).

For an HP-35, change [G] [log] to [log].

TEST CASE: $v = 20$ mph, $T = 15°F$, get $\Delta T = 32$ F°; thus it feels like $-17°F$.

*It is not a wind-chill *factor* because one does not multiply anything by it.

The temperature-humidity index (THI) is a measure of summer discomfort. Most people are uncomfortable for $THI > 80°F$ and somewhat uncomfortable for $75°F < THI < 80°F$.

(T_w, °F) [↑] (T, °F) [+] .4 [×] 15 [+] (see THI in degrees Fahrenheit).

TEST CASE: $T_w = 68°F$, $T = 77°F$; get $THI = 73°F$.

See Section A.13.4 to convert temperatures from Celsius to Fahrenheit, and vice versa.

REFERENCES: Humphreys (1940), Chapter 1; Conrad and Pollack (1950), pp. 195 ff; Haurwitz (1941), Chapter 1; CRC HCP, pages D-179 to D-181 and E-44 to E-45.

APPENDIX B
Answers to Exercises

To find the solution of a problem by our own means is a discovery. If the problem is not difficult, the discovery is not so momentous, but it is a discovery nevertheless. Having made some discovery, however modest, we should not fail to inquire whether there is something more behind it, we should not miss the possibilities opened up by the new result, we should try to use again the precedure used. Exploit your success! *Can you use the result, or the method, for some other problem?**

—*George Pólya (1887–   )*

Round numbers are always false.

—*Samuel Johnson (1709–1784)*

Each answer in this appendix has the same number as the corresponding exercise, but with a B prefix. Thus B.3.7.2 is the answer to exercise number 3.7.2, which is the second exercise in Section 3.7.

Even though one would normally round off, many of the numerical answers in this appendix are given to the full calculator accuracy as a check on computing techniques.

*From Pólya (1973), pp. 64–65.

B.1.5.1 (*a*) [$x \gtrless y$], [CHS], and [$1/x$] provided $x \neq 0$,
(*b*) [CLX], [CLR], [·], [EEX], [STO], [FIX], and any prefix or shift keys. On a calculator that needs a register number after [RCL], then [RCL] is also such a key. On a calculator with [STO] or [RCL] arithmetic, the arithmetic keys [+], [−], [×], and [÷] following [STO] or [RCL] are also such keys. There are other special cases, for example, when X contains zero.

B.1.5.2 (*a*) [+], (*b*) [×], (*c*) [+] [×], (*d*) [×] [×], (*e*) [×] [−].

B.1.5.3 (*a*) $2A^3$, (*b*) $A(A^2+1)$, (*c*) $A^2(A+1)$, (*d*) $A(2-A)$, (*e*) $A(1-2A)$, (*f*) $1/(1-A)$, (*g*) $\frac{1}{3}$.

B.1.5.4 $10^{-10} \times \pi \times 12740$ km$\times(10^6$ mm/km$)\cong 4$ mm, or $10^{-5}\times(10^6$ mm/km$)\cong 10$ mm.

B.1.5.5 $SF = 7.072$, $DD = 6.575$.

B.1.5.6 $(355/113-\pi)\times 12740$ km$\times(10^3$ m/km$)\cong 3$ m.

B.1.5.7 Using the Pythagorean theorem, the height of the buckled track at its center would be

$$h^2 = \left(2640 + \frac{0.5}{12}\ \text{ft}\right)^2 - (2640\ \text{ft})^2,$$

since there are 2640 feet in a half-mile and 0.5 inch expansion in the half-mile up to the peak of the buckle. Then $h = 14.8$ feet! Railroads avoid this problem by having expansion joints between segments of the rails, or by forcing the rails to flex in compression.

B.1.5.8 If the numbers are ϕ_n, then

$$\phi_n = 1/(\phi_n - n),$$

where n is an integer. Then

$$\phi_n^2 - n\phi_n = 1,$$

or

$$\phi_n = \frac{n + \sqrt{n^2+4}}{2}.$$

Only the plus sign on the radical will do (why?). Thus $\phi_1 =$ 1.618033989 (the golden ratio), $\phi_2 = 2.414213562$, $\phi_3 =$ 3.302775638, $\phi_4 = 4.236067977$, and so on.

The regular continued fraction expansion of these numbers is interesting; it is

$$\phi_1 = 1 + /1, 1, 1, \cdots /,$$

$$\phi_2 = 2 + /2, 2, 2, \cdots /,$$

$$\phi_3 = 3 + /3, 3, 3, \cdots /, \text{ etc.}$$

B.1.5.9 (*a*) 0.11 inch, (*b*) 0.030 inch, (*c*) 107.5 feet, (*d*) 3.3 inches, (*e*) 5500 miles (!), (*f*) approximately 865,000 miles/93,000,000 miles $\times 180°/\pi = 0°.53$, (*g*) approximately 2160 miles/239,000 miles $\times$ $180°/\pi = 0°.52$.
Because the last two numbers are so nearly equal, we are able to see both total and annular eclipses of the sun. The astronomical numbers are from Allen (1973).

B.1.5.10 (*a*) $2^{64} - 1 \cong 1.844674407 \times 10^{19}$ grains. (*b*) About 3,159,091 miles (13 times the distance to the moon).

B.1.5.11 $\ln(\$6.95/\$0.99)/200 \cong 1\%$.

B.1.5.12 From the "Sun Ephemeris" with day number = 117, $GMT = 7^h$ (there are about 5 hours between Boston and Greenwich, but this number does not matter much), get $\delta = 13°35'$ and then elevation $= 61°15'$. Thus the tower is (93 feet) tan(elevation) $\cong$ 169.5 feet tall. For any time other than local noon, calculate the sidereal time, hour angle, and elevation from the appropriate routines in Section A.7.

B.1.5.13 (*a*) About 4.7 s/mile. This is useful for estimating the distance to lightning by timing the delay to the associated thunder. (*b*) About 5.4 μs/mile or about one ft/ns.
(*c*) 231 cubic inches per gallon (exactly—why?). (*d*) Approximately 1000 fluid ounces per cubic foot.

B.2.7.1 (*a*) INITIALIZE: [CLR]
LOOP: x_i [+] [$x \gtrless y$]
y_i [+] [$x \gtrless y$] :|
CODA: (see Σx_i) [$x \gtrless y$] (see Σy_i).

(*b*) INITIALIZE: [CLR]
LOOP: x_i [+] [↑] [R↓] [R↓]
y_i [+] [↑] [R↓] [R↓]
z_i [+] [↑] [R↓] [R↓] :|
CODA: (see Σx_i) [R↓] (see Σy_i) [R↓] (see Σz_i).

B.2.7.2

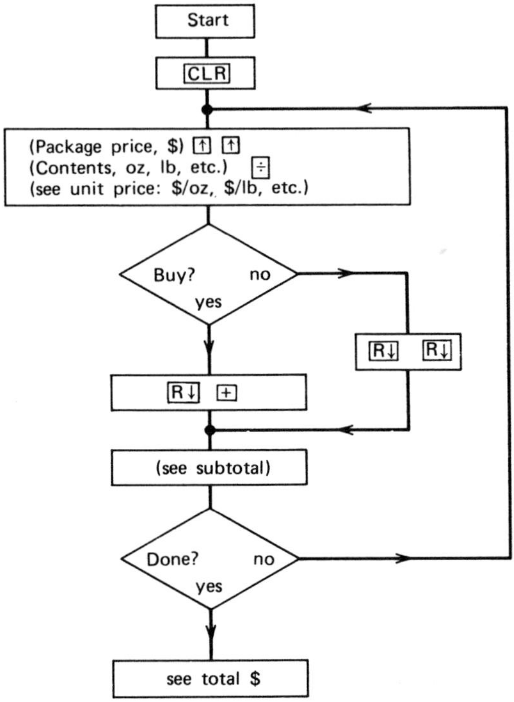

Figure B.2.1

B.2.7.3 $(a_1+ib_1)(a_2+ib_2)=u+iv$,
b_1 [↑] a_1 [→P] [$x \gtrless y$]
b_2 [↑] a_2 [→P] [R↓] [+] [↑] [R↓] [R↓] [×] [G] [→R] (see u)
[$x \gtrless y$] (see v).

B.2.7.4 $\dfrac{a_1+ib_1}{a_2+ib_2}=u+iv$,
b_1 [↑] a_1 [→P]
b_2 [↑] a_2 [→P] [R↓] [$x \gtrless y$] [R↓] [−] [↑] [R↓] [R↓] [÷] [G] [→R]
(see u) [$x \gtrless y$] (see v).

B.2.7.5 $\dfrac{1}{a+ib}=u+iv$,
b [CHS] [↑]
a [→P] [$1/x$] [G] [→R] (see u) [$x \gtrless y$] (see v).

B.2.7.6 $(a+ib)^2=u+iv$,
b [↑] a [→P] [$x \gtrless y$] 2 [×] [$x \gtrless y$] [x^2] [G] [→R] (see u) [$x \gtrless y$]
(see v).

B.2.7.7 The first one (*ABCD*) is, of course, no operation. The other four rearrangements each require eight keystrokes (and these are the only four rearrangements that do). Assume that the *last* keystroke is, say, [R↓], undo it, and look up the result in the table. For example, if we have *BADD*, we can get *ADDB* with a [R↓]. But since *BADD* is in the table, we have

ADDB [x≷y] [R↓] [R↓] [R↓] [↑] [R↓] [R↓] [R↓].
BADA [R↓] [x≷y] [CLX] [−] [R↓] [R↓] [R↓] [x≷y], or [R↓] [x≷y] [CLX] [−] [x≷y] [R↓] [x≷y] [R↓].
BADB [R↓] [R↓] [CLX] [−] [R↓] [x≷y] [R↓] [x≷y], or [R↓] [R↓] [CLX] [−] [x≷y] [R↓] [R↓] [R↓].
DADB [x≷y] [R↓] [R↓] [R↓] [↑] [R↓] [x≷y] [R↓].

There are other possibilities also.

B.2.7.8 Using almost any RPN calculator,

SETUP: 1 [↑]
LOOP: [↑] [↑] [R↓] [R↓] [+] (see the next number) :|.

Or, using the HP-45,

SETUP: 1 [↑] [↑] [+] (see the first number)
LOOP: [G] [LAST*X*] [x≷y] [+] (see the next number) :|.

Each Fibonacci number is the sum of the preceding two. The sequence is 2, 3, 5, 8, 13, 21, 34, 55, 89, 144, 233, 377 (at a year).

REFERENCES: *HP-35 Math Pac*, p. 79; *HP-45 Applications Book*, p. 113; Knuth (1973), Vol. I, p. 78.

B.2.7.9

(octal)		(decimal)
43172	=	18042
4.3172	=	4.4047851563
431.72	=	281.90625

(Do it in two parts and add them up.)

(hexadecimal)		(decimal)
AE19C2	=	11409858
FDA7.CB5	=	64935.7941894531

(binary)		(decimal)
1101011011101	=	6877
1011001.1110101101	=	89.9189453125

B.2.7.10 (*a*) t [↑] [↑] a [×] (see v) [×] 2 [÷] (see s).
(*b*) $s = 282.24$ ft, $v = 134.4$ ft/s.
These answers are imprecise because air resistance cannot be neglected after 2 or 3 seconds.

B.2.7.11 Definition:

HP-45 SETUP: [G] [CLEAR] [Σ+]
LOOP: [1/x] [Σ+] (Check: If display is $n+1$, go to coda.) :|
CODA: [RCL] [Σ+] (see H_n).

Approximation: n [1/x] [↑] [↑] [↑] 120 [÷] [×] 12 [1/x] [−] [×] 2 [1/x] [+] [×] .5772156649 [+] [x≷y] [ln] [−] (see $\cong H_n$).

n	"Exact" H_n	$\cong H_n$	$-\log_{10}\lvert\xi\rvert$
1	1.	1.002215665	2.65
2	1.5	1.500050345	4.30
3	1.833333333	1.833338242	5.31
4	2.083333333	2.083334245	6.04
5	2.283333333	2.283333577	6.61
6	2.45	2.450000083	7.08
7	2.592857143	2.592857176	7.48
8	2.717857143	2.717857158	7.82
9	2.828968254	2.828968261	8.15
10	2.928968254	2.928968258	8.40

REFERENCES: Knuth (1973), Vol. I, pp. 73–74; *HP-45 Applications Book*, p. 121.

B.2.7.12 HP-35 or HP-21:

i [↑] [↑] [↑] r [+] [STO] [−] [+], [↑] [SIN] [RCL] [SIN] [÷] [↑] [×] [x≷y] [TAN] [RCL] [TAN] [÷] [↑] [×] [+] 2 [÷] (see R).

Or for an HP-21 only, replace the first part of this algorithm, up to the comma, with

i [STO] r [B] [M+] [−] · · · .

B.2.7.13 HP-45 SETUP: [G] [CLEAR] x [G] [COS⁻¹] [↑] [↑] [↑] [CLX] [Σ+]
LOOP: [×] [COS] [RCL] 5 n [G] [y^x] [÷] [Σ+] (Check: Is display 7, if so, go to coda.) :|
CODA: [RCL] [Σ+] (see sum).

The sequence n [G] [y^x] in the loop can be omitted for $n=1$, replaced by [x^2] for $n=2$, and replaced by [x^2] [x^2] for $n=4$. With $x=0.5$ and $n=1$ get 0.05833333333 (the precise value of the infinite series is 0), and $n=2$, get 0.2804166670 (the precise value is 0.2741556780), and $n=4$ get 0.4560228011 (the precise value is 0.4559787700).

REFERENCES: Section A.5.6; and HMF 27.8.6.

B.2.7.14 HP-35:

x [↑] [+] y [STO] [÷] [↑] [↑] [↑] [×] 1 [+] [$\sqrt{x}$] [↑] [R↓] [+] [ln] [$x \gtrless y$] [÷] [+] [RCL] [×] (see s).

B.2.7.15 HP-21:

(Select RAD mode) R [↑] [↑] [↑] d [$x \gtrless y$] [÷] [B] [$\cos^{-1}$] [↑] [+] [↑] [SIN] [−] [×] [×] 2 [÷] (see A).

B.2.7.16 I do not know of a strictly shorter algorithm (nine steps total), but at the expense of more total steps (14), the following algorithm has fewer steps *in the loop* (four rather than five), and thus takes slightly less *time* for large arguments.

[STO] 1 1 [CHS] [↑] [↑] [↑] [RCL] 1 [+] [g] [$x=0$] [GTO] 13 [STO] [×] 1 [GTO] 08 [RCL] 1 [GTO] 00.

B.2.7.17 $A\uparrow + B \div \sqrt{}$I.

B.2.7.18 $A/B/+/$.

B.2.7.19 $D\uparrow C - 6 \div A + B + 2 \div$.

B.2.7.20 Aȝ.

B.2.7.21 $A\uparrow B$ ⌡ .

B.2.7.22 x [↑] [↑] [↑] 21 [+] [×] 157 [+] [×] 409 [+] [STO] [CLX] 2 [×] 45 [+] [×] 381 [+] [×] 1353 [+] [×] 1511 [+] [RCL] [÷] (42 keystrokes plus data).

The [CLX] after [STO] can be omitted on an HP-35 (why?).

x [↑] [↑] [↑] 9 [+] 8 [$x \gtrless y$] [÷] 7 [+] [+] 6 [$x \gtrless y$] [÷] 5 [+] [+] 4 [$x \gtrless y$] [÷] 3 [+] [+] [+] (24 keystrokes plus data).

For $x=2$, get 7.525153374. This example is from Fike (1968), p. 144.

B.2.7.23 (*a*) 3 [↑] 7 [×] 2 [+] 24 [×] 9 [+] 60 [×] 22 [+] 60 [×] 18 [+] (see 2,020,938 seconds).

(*b*) 18 [↑] 60 [÷] 22 [+] 60 [÷] 9 [+] 24 [÷] 2 [+] 7 [÷] 3 [+] (see 3.341498016 weeks).

These are called mixed-radix numbers; life contains all too many of them.

B.2.7.24 $1/(e-1) = /1, 1, 2, 1, 1, 4, 1, 1, 6, 1, 1, 8, \cdots/$. Whence this pattern?

See Knuth (1969), Vol. II, Section 4.5.3 and especially his exercise 16 on p. 336.

B.2.7.25 HP-21 SETUP: r [↑] 10 [÷] [B] [10^x] [↑] [B] [√x] 1 [+] [x≷y] 1 [−] [B] [√x] (see ε) [÷] n [1/x] [B] [y^x] [↑] [1/x] [STO]

LOOP: [−] 2 [÷] * k [↑] [↑] 1 [−] [+] n [÷] 90 [×] [↑] [R↓] [SIN] [CHS] [×] (see α_k) [R↓] [RCL] [+] [x≷y] [COS] [×] (see $\pm\omega_k$) [RCL] [1/x] [RCL] :|.

*If n is odd, see (minus) the real pole at this point, thus avoid going through the loop for $k=(n+1)/2$.

B.3.7.1 (*a*) From Figure 3.2.1, the slope of the straight line as drawn is

$$s=\frac{h(x_n)}{x_n-x_{n+1}}=\frac{x_n-f(x_n)}{x_n-F(x_n)}.$$

Then drop the subscript on x_n and get

$$s=\frac{x-f(x)}{x-[f(x)+xg(x)]/[g(x)+1]}=g(x)+1.$$

(*b*) The tangent to the curve has a slope of

$$h'(x)=\frac{d}{dx}[x-f(x)]=1-f'(x)=1+g(x),$$

provided $g(x)=-f'(x)$.

B.3.7.2 Third-order convergence means both $F'(a)=0$ and $F''(a)=0$. Substituting $x=a$, $f(a)=a$, and $g(a)=-f'(a)$ into the second derivative of equation 3.2.3 gives, after some manipulation,

$$F''(a)=\frac{f''(a)+2g'(a)}{1+g(a)},$$

which will be zero if

$$g'(a)=-\tfrac{1}{2}f''(a).$$

One possibility that satisfies this equation is

$$g(x)=-f'(x)+\frac{f''(x)[x-f(x)]}{2[1-f'(x)]},$$

which leads to Halley's method (see Grove, 1966, Section 1.13) as applied to $h(x)=0$.

B.3.7.3 (a) From Figure 3.3.1, the slope of the straight line as drawn is

$$s=\frac{h(x_n)}{\ln x_n-\ln x_{n+1}}=\frac{\ln x_n-\ln f(x_n)}{\ln x_n-\ln F(x_n)}.$$

Then drop the subscript on x_n and get

$$s=\frac{\ln x-\ln f(x)}{\ln x-\ln x-\alpha\left[\ln f(x)-\ln x\right]}=\frac{1}{\alpha}.$$

(b) The tangent to the curve has a slope of

$$\frac{dh(x)}{d\ln x}=\frac{d\left[\ln x-\ln f(x)\right]}{d\ln x}=1-\frac{xf'(x)}{f(x)}=\frac{1}{\alpha(x)},$$

provided α satisfies equation 3.3.5.

B.3.7.4 Equate $F(x)$ from equations 3.2.3 and 3.3.1 to give

$$x\left(\frac{f}{x}\right)^{\alpha}=\frac{f+xg}{1+g}.$$

Take ln of both sides,

$$\alpha\ln\left(\frac{f}{x}\right)=\ln[f+xg]-\ln[1+g]-\ln x,$$

and rearrange to get the expression for α in the problem.

B.3.7.5 (a) From Figure 3.4.1, the slope of the straight line as drawn is

$$s=\frac{h(x_n)}{1/x_n-1/x_{n+1}}=\frac{1/x_n-1/f(x_n)}{1/x_n-1/F(x_n)}.$$

Then drop the subscript on x_n and get

$$s=\frac{1/x-1/f(x)}{1/x-\beta/f(x)-(1-\beta)/x}=\frac{1}{\beta}.$$

(b) The tangent to the curve has a slope of

$$\frac{dh(x)}{d(1/x)}=\frac{d}{d(1/x)}\left[\frac{1}{x}-\frac{1}{f(x)}\right]=1-f'(x)\left[\frac{x}{f(x)}\right]^2=\frac{1}{\beta(x)},$$

provided β satisfies equation 3.4.5.

B.3.7.6 Equate $F(x)$ from equations 3.2.3 and 3.4.1 to give

$$\frac{xf}{f+\beta[x-f]}=\frac{f+xg}{g+1},$$

and rearrange to get equation 3.4.4.

B.3.7.7 Combine equation 3.6.5 (which is the derivative of equation 3.6.2) with equation 3.6.1 to give

$$f'(i)=\frac{in}{(1+i)\left[(1+i)^n-1\right]}\cong\frac{1}{1+i},$$

where the approximation comes from

$$(1+i)^n\cong 1+in, \qquad \text{for small } i.$$

For $i<0.1$, f' from this expression is $0.9\lesssim f'<1$.
Then from equation 3.6.3 derive

$$f'(i)=\frac{1-\left[1-i(n-1)\right](1+i)^{n-1}}{\left[1-(1+i)^n\right]^2}\cong\left[\frac{n-1}{n}\right]^2.$$

For $n<20$, f' from this expression is $f'\lesssim 0.9$.

B.3.7.8 (a) HP-45 SETUP: n [STO] 4 R [↑] S [÷] [↑] [↑] .1 [↑]
LOOP: 1 [+] [RCL] 4 [G] [y^x] 1 [−] [×] (see i_i) :|.

With the test case under algorithm 3.6.12, get i_i=0.1161478772, 0.1723915739, 0.6177934666, 1973.895851, ⋯ .

(b) From equation 3.6.14, get

$$f'(i)=\frac{R}{S}n(1+i)^{n-1}.$$

Then using equation 3.6.1,

$$f'(i)=\frac{in(1+i)^{n-1}}{(1+i)^n-1}\cong 1+i(n-1)>1.$$

(c) From equation 3.3.3, use

$$\alpha \cong \frac{1}{i(1-n)},$$

or

$$F(i) = i\left[\frac{(1+i)^n - 1}{iS/R}\right]^{1/[i(1-n)]}$$

HP-45 SETUP: R [↑] S [÷] [STO] 1 n [STO] 2 1 [$x \gtrless y$] [−] [STO] 3 .1
LOOP: [↑] [↑] [↑] 1 [+] [RCL] 2 [G] [y^x] 1 [−] [RCL] 1 [×] [$x \gtrless y$] [÷] [$x \gtrless y$] [RCL] 3 [×] [$1/x$] [G] [y^x] [×] (see i_i) :|.

With the same test case, get $i_i = 0.0939532846$, 0.0915217762, 0.0905719265, 0.0902058066, $\cdots$. So it converges slowly. Algorithm 3.6.16 is preferable for this problem.

B.3.7.9 Try $f(E) = M + e\sin E$, $f'(E) = e\cos E$. Now since $0 \leqslant e \leqslant 1$, and $|\cos E| \leqslant 1$, direct iteration on $f(E)$ will almost always converge. Convert to degrees by multiplying the equation by $180°/(\pi$ rads). Then for an HP-35:

SETUP: e [π] [÷] 180 [×] [STO] (M, °) [↑] [↑] [↑]; a good guess for E_0 is just M, so go on to
LOOP: [SIN] [RCL] [×] [+] (see E_i, °) :|.

A loop with only four keystrokes is hard to beat. For the test case, get $E_i = 69$, 79.69804365, 80.27442207, 80.29446667, 80.29514321, 80.29516602, 80.29516679, 80.29516681. Thus this algorithm takes seven cycles to converge to an answer with about nine significant figures, but with only 38 keystrokes total plus data. A g-method algorithm can be written for this problem and it converges in fewer iterations, but takes more than twice as many keystrokes for this test case. Only if e is very near unity would a more sophisticated algorithm be desirable.

B.3.7.10 Using trigonometric manipulation get

$$y = Al\sin y \equiv f(y).$$

Then

$$f'(y) = Al\cos y = -g(y),$$

and

$$F(y)=\frac{Al\sin y-yAl\cos y}{1-Al\cos y}.$$

Therefore a possible algorithm for an HP-21 would be
SETUP: (switch to RAD mode) A [CHS] [↑] l [×] [STO] y_0
LOOP: [↑] [↑] [TAN] [−] [$x \gtrless y$] [COS] [RCL] [×] [$1/x$] 1 [+] [÷] (see y_i) :|.

There are *approximately* Al/π solutions (bound energy levels) and a graphical approximation can be used to get first guesses for the algorithm. Plot both $\sin y$ and $y/(Al)$ for $y>0$; intersections are solutions. For $Al=10$, there are three solutions near $y=\pi$, 2π, and 3π, as given. This exercise is from Ball (1976a).

B.3.7.11 The problem is

$$x+y=10,$$

$$\frac{x}{y}+\frac{y}{x}=x \quad (\text{or } y).$$

After some manipulation, get

$$x^3-8x^2-20x+100=0.$$

Thus this is a cubic and can be solved by any one of several techniques (see Section A.5.3, *HP-45 Applications Book*, p. 74). The answer is $x=2.87836574$, $y=7.12163426$.

B.3.7.12 The problem is

$$x+y=10,$$

$$(x+\sqrt{x}\,)(y+\sqrt{y}\,)=A.$$

After some manipulation, get

$$x=\frac{A}{10-x+\sqrt{10-x}}-\sqrt{x}\equiv f(x),$$

and

$$f'(x)=\frac{A\left[1-1/(2\sqrt{10-x}\,)\right]}{(10-x+\sqrt{10-x}\,)^2}-\frac{1}{2\sqrt{x}}.$$

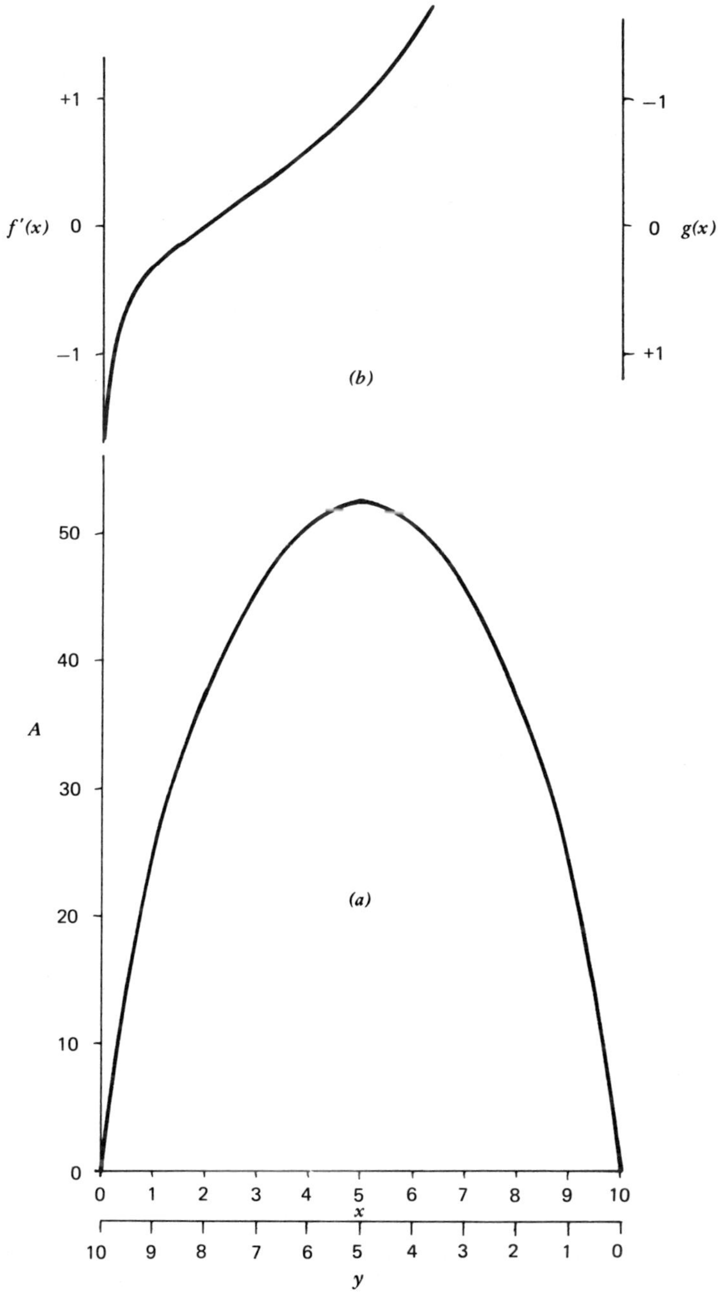

Figure B.3.1 (*a*) A as a function of x (or y). (*b*) $f'(x)$.

Figure B.3.1a is a plot of A as a function of x (or y) and Figure B.3.1b plots $f'(x)$, showing that direct iteration on $f(x)$ will work for some values of A, and $-f'(x)$ can be read off this plot to use for g=constant for other values. A possible algorithm for an HP-35 is then

SETUP: A [STO] x_0
LOOP: [↑] [↑] [↑] 10 [$x \gtrless y$] [−] [↑] [$\sqrt{x}$] [+] [RCL] [$x \gtrless y$] [÷] [$x \gtrless y$] [$\sqrt{x}$] [−] [$x \gtrless y$] g [×] [+] $(g+1)$ [÷] (see x_i) :|.

For $A=24$, use $g=0$, $x_0=2$; get $x_i=2$, 0.802174812, 1.066646477, 0.980261253, 1.006088122, 0.998145089, 1.000567276, 0.999826711, 1.000052954, etc.

Try again with $A=24$, $g=0.3$ (from Figure B.3.1b), $x_0=2$; get $x_i=2$, 1.078596009, 1.000330101, 0.9999986015, 1.000000006, 1.000000001, 1. For these and similar problems, see Kraitchik (1953).

B.3.7.13 $$g(x)=\frac{f(x)-f(f(x))}{f(x)-x}.$$

Note that this formula requires *two* evaluations of f but *no* evaluations of f'. This $g(x)$ is an approximation for $-f'(x)$ based on linearizing $f(x)$ between x and $x'=f(x)$.

REFERENCES: Isaacson and Keller (1966), Chapter 3; Section 2.4.

B.3.7.14 The torque due to the weight is $MgR\cos\theta$ and the two torques must be equal in equilibrium,

$$K\theta = MgR\cos\theta.$$

So set

$$f(\theta)=\frac{MgR}{K}\cos\theta \equiv C\cos\theta.$$

Use the g method with $g=-f'$ to get

$$F(\theta)=\frac{C\cos\theta+\theta C\sin\theta}{1+C\sin\theta}.$$

Thus an algorithm for the HP-21 is

SETUP: (set mode to RAD) C [STO] θ_0
LOOP: [↑] [↑] [TAN] [$1/x$] [+] [$x \gtrless y$] [SIN] [RCL] [×] [$1/x$] 1 [+] [÷] (see θ_i) :|.

For $C=3$ rad, starting from $\theta_0=1$ rad, get $\theta_i=1$, 1.176173146, 1.170126582, 1.170120950 rad.
This exercise is very similar to exercise 3.7.10—exchange cosine for sine. Spring scales are not usually made this way, but this version might be useful if one wanted a very nonlinear scale to weigh a wide range of masses.

B.4.4.1 Using the algorithm in Section A.6.1, get $a_1=2.365714286$, $a_0=-0.4571428583$, and $\sigma=0.3685234389$.

B.4.4.2 Using the algorithm in Section A.6.3, get $a=0.5785714286$, $b=0.9192857143$, $c=0.025$, $x_0=-0.7944444444$, $A=-0.3401607143$, $W=1.08437271$, and $\sigma=0.07503967215$.

B.4.4.3 Using the algorithm in Section A.6.4, get $A_x=0$, $A_y=2$, $C=2$, $A=2$, and $B=90°$.

B.4.4.4 Use the algorithm for straight-line fitting in Section A.6.1, but change x_i to θ_i [COS] [x^2] in the enter-data loop. Then $A=a_1$ and $B=a_0$.

B.4.4.5 Get $A=4$, $B=0$, and $\sigma=\sqrt{2}$.

B.4.4.6 Since

$$\cos(2\theta)=2\cos^2(\theta)-1,$$

the form to be fitted is equivalent to

$$f=2A\cos^2(\theta)+B\cos\theta+C-A.$$

Thus we can use the parabolic-fitting algorithm in Section A.6.3. Change x_i to θ_i [COS] in the enter-data loop. Then $A=a/2$, $B=b$, and $C=c+A$.

B.4.4.7 Get $A=2$, $B=2$, $C=2$.

B.4.4.8 Get 5.5 from equation 4.3.7, and 4.53 from equation 4.3.9.

B.4.4.9 Beginning with equation 4.3.10 and $F=\ln$, define

$$\varepsilon_i=\frac{Y_i-f_i}{Y_i}=1-\frac{f_i}{Y_i}=\frac{\Delta Y_i}{Y_i},$$

and note that ε_i is usually small, $|\varepsilon_i|\ll 1$.
So approximate

$$\ln(1-\varepsilon_i)\cong-\varepsilon_i,$$

and get

$$\Delta Y_i \cong w_i \ln\left(\frac{Y_i}{f_i}\right) = w_i \ln\left(\frac{1}{1-\varepsilon_i}\right) \cong w_i\varepsilon_i, \quad w_i \cong Y_i.$$

B.4.4.10 Using the algorithm in Section A.6.5, get $b=1.507986917$, $a=1.487057458$, and $\sigma=0.03321305233$ (excluding $x_i=0$).

B.4.4.11 Using the algorithm in Section A.6.7, get $W=3.146039603$, $x_0=3.139847725$, $A=3.137154187$, $\sigma=0.0000778271501$. This σ is rather smaller than one might expect for fitting to data given to only two decimal digits. The data were calculated from $W=x_0=A=\pi$ and were rounded to two decimal digits. However these π values give a σ more than 40 times larger!

B.4.4.12 Get $w_i \cong \exp(-Y_i)$, *but* this approximation is good only if $|\Delta Y_i| \ll 1$, a requirement very different from $|\varepsilon_i| \ll 1$ for exercise 4.4.9.

B.4.4.13 HP-45 INITIALIZE: [G] [CLEAR] [STO] 3 [STO] 4

ENTER DATA: x_i [↑] [↑] [↑] Y_i [CHS] [e^x] [STO] [+] 3 [×] [STO] [+] 4 [×] [$x \gtrless y$] [Σ+] (see i) :|

CODA: [RCL] 3 [RCL] 7 [×] [RCL] 4 [RCL] 5 [×] [−] [RCL] 3 [RCL] 8 [×] [RCL] 4 [x^2] [−] [÷] (see a_1) [RCL] 4 [×] [RCL] 5 [$x \gtrless y$] [−] [RCL] 3 [÷] (see a_0).

Then calculate the value of the curve at any x:

x [↑] a_1 [×] a_0 [+] [ln] (see Y).

B.4.4.14 Get $a_1=2.012671607$, $a_0=2.025952555$, and $\sigma=0.007954787674$. If the straight-line algorithm in Section A.6.1 is used, but with [e^x] after y_i, one gets a fit with a σ (in Y space) that is only slightly worse, namely, $\sigma=0.01$.

B.4.4.15

$$\ln Y_i - \ln f_i = \ln\left(\frac{Y_i}{f_i}\right) = \ln\left(\frac{1}{1-\Delta Y_i/Y_i}\right) = -\ln\left(1-\frac{\Delta Y_i}{Y_i}\right) \cong \frac{\Delta Y_i}{Y_i}.$$

B.4.4.16 The data could be fitted to either

$$I = I_0 \exp(UV),$$

with V as the independent variable, or, more reasonably, to

$$V = \frac{\ln I - \ln I_0}{U},$$

with I as the independent variable. The latter is preferable because I is given in presumably exact values, and because the fitting process to this second equation is easier. Use the straight-line fitting algorithm in Section A.6.1, but change x_i to I_i [ln], and y_i to V_i. Then interpret $a_1 = 1/U$ and $a_0 = -(\ln I_0)/U$. For the first half of the data get $a_1 = 0.02544965694$, $a_0 = 0.6805000053$, or $m = 1.001151044$, and $I_0 = 2.439900198 \times 10^{-12}$ A. For the second half of the data, get $a_1 = 0.02544965685$, $a_0 = 0.681100018$, or $m = 1.001151040$, and $I_0 = 2.383050086 \times 10^{-12}$ A. Thus the m factor does not seem to vary very much.

B.4.4.17 Write

$$S_\nu = S_1 \nu^{-\alpha},$$

where S_1 is the flux at $\nu = 1$ MHz (the units are arbitrary). This is a power curve as in equation 4.3.12 and Section A.6.5. The fit using this algorithm gives $\alpha = 0.6659978463$, $S_1 = 34352.30504$, and $\sigma = 2.702340997$. Using the algorithm in the *HP-45 Applications Book*, p. 86 (no w_i) gives $\alpha = 0.6645659613$, $S_1 = 34018.88430$, and $\sigma = 3.130083748$. Either fit is acceptable.

B.4.4.18 Choose $w_i = P_i^2$ (why?), then

HP-45 INITIALIZE: turn calculator off, then on

ENTER DATA: P_i [↑] [↑] [STO] [+] 2 [×] [STO] [+] 1 t_i [↑] [R↓] [×] [STO] [+] 4 [R↓] [×] [STO] [+] 3 [x^2] [STO] [+] 5 :|

CODA: [RCL] 1 [RCL] 5 [×] [RCL] 4 [x^2] [−] [RCL] 4 [RCL] 2 [×] [RCL] 1 [RCL] 3 [×] [−] [÷] (see K) [RCL] 2 [×] [RCL] 4 [+] [RCL] 1 [÷] (see t_d).

From the given data get $K = 185476.7563$ million years, and $t_d = 2021.553359$ A.D. ≅ July 21, 2021. From somewhat different population data, von Foerster et al. (1960) calculated "Doomsday: Friday, 13 November, A.D. 2026." See also von Hoerner (1972).

B.4.4.19 The weighting factor is $w_i \cong \cos\alpha_i$ (why?) where $\alpha_i \equiv \sin^{-1} Y_i$. Then, for $0 \leqslant \alpha_i \leqslant 90°$:

INITIALIZE: Turn calculator off, then on.

ENTER DATA: Y_i [G] [SIN^{-1}] [↑] [↑] [COS] [STO] [+] 1 [×] [STO] [+] 2 t_i [↑] [R↓] [×] [STO] [+] 3 [R↓] [COS] [×] [STO] [+] 4 [×] [STO] [+] 5 :|

CODA: [RCL] 1 [RCL] 3 [×] [RCL] 4 [RCL] 2 [×] [−] [RCL] 1 [RCL] 5 [×] [RCL] 4 [x^2] [−] [÷] (see ω) [RCL] 4 [×] [RCL] 2 [$x \gtrless y$] [−] [RCL] 1 [÷] (see θ).

The "not" answers given in the exercise result from omitting the weighting factors. Writing a sine curve fitting algorithm that works over all quadrants is much more difficult.

B.5.7.1 $\pi/2 = 1.570796327$. Simpleton's formula (Section A.3.1) with $n \geqslant 2$ is exact for this integral. (Why?)

B.5.7.2 This is the gamma function $\Gamma(x)$, and $\Gamma(4) = 3! = 6$. The two-point Gauss-Laguerre formula (Section A.3.2) is exact for this integral. (Why?) This is another way to calculate a gamma function, but it is not very precise except for x a small integer, and then the answer is known anyway.

REFERENCES: HMF 6.1; and Section A.5.8.

B.5.7.3

$$\int_1^{1.7} y(x)\,dx \cong 0.0551359259,$$

$$y(1.32) \cong 0.02812 \qquad \text{or} \qquad 0.0279056,$$

$$y'(1.32) \cong 0.283.$$

These data were calculated from $y = (x-1)^{\pi}$, and the answers can be compared with the exact answers if there were no round-off.

B.5.7.4 $\pi^2/4 = 2.467401100$. The Gauss-Chebyshev formula (Section A.3.2) is exact for this integral. (Why?)

B.5.7.5 Because it is a periodic function integrated over a full period (why?), Simpleton's formula (Section A.3.1) works as well as any. For $x = 2$ and $n = 6$, get 0.2238907837. The precise answer is Bessel function $J_0(2) = 0.2238907791$. This is another way to calculate a Bessel function.

REFERENCES: HMF 9.1.21; Section 21.5 in Hamming (1973); and Section A.5.9.

B.5.7.6 For $x = 1$, get 0.8426900188 from the three-point Gauss-Legendre algorithm (Section A.3.2), or 0.8427003240 from the five-point Lobatto-Radau algorithm (Section A.3.2). The precise answer is error function $\text{erf}(1) = 0.8427007929$. This is another way to calculate an error function.

REFERENCES: HMF 7.1; HMF Table 7.1; Section A.5.10.

B.5.7.7 For $x = 1$, get 0.5379904675 from the three-point Gauss-Legendre algorithm (Section A.3.2), or 0.5380817906 from the five-point Lobatto-Radau algorithm (Section A.3.2). The precise answer is 0.5380795069. This is Dawson's integral.

REFERENCE: HMF Table 7.5.

B.5.7.8 For $x=1$, get 0.4380405866 from the three-point Gauss-Legendre algorithm (Section A.3.2), 0.4382750839 from the five-point Lobatto-Radau algorithm (Section A.3.2), 0.4385009355 from the five-point Newton-Cotes algorithm (Section A.3.1), or 0.4384277565 (!) from the five-point Simpleton algorithm (Section A.3.1). The precise answer is Fresnel's integral $S(1)=$ 0.4382591474.

REFERENCES: HMF 7.3.2; HMF Table 7.7.

B.5.7.9 This is Airy's differential equation and the particular solution from these initial values is Airy function $\mathrm{Ai}(-x)$. Representative values are as follows:

x	$\mathrm{Ai}(-x)$	Approximate y
0.5	0.47572809	0.4757279328
1.0	0.53556088	0.5355595556
1.5	0.46425658	0.4642527094
2.0	0.22740743	0.2274026793
2.3	0.02670633	0.02670483264
2.4	−0.04333414	−0.04333355958

The third column in this tabulation is from the modified third-order Runge-Kutta algorithm in Section A.4.2 with $h=0.1$. The error plot for this case has holes at $x \cong 1.85$ for y' and at $x \cong 2.35$ for y.

Integrating the differential equation is the preferred method for evaluating the Airy integral,

$$\mathrm{Ai}(x)=\frac{1}{\pi}\int_0^\infty \cos\left(\frac{t^3}{3}+xt\right)dt.$$

"This quadrature is not feasible," says John Todd (1962), p. 71.

REFERENCES: HMF 10.4; HMF Table 10.11.

B.5.7.10 Using second-order Runge-Kutta from Section A.4.1 (HMF 25.5.6), with $h=0.1$, get $y(2)=0.3548042050$. Or using third-order Runge-Kutta from Section A.4.1 (HMF 25.5.8), with $h=$

0.1, get $y(2) = 0.3542997875$. The precise answer is the Goodwin-Staton function, $y(2) = 0.3543359288$.

REFERENCES: Hartree (1958), p. 119; HMF 27.6

B.5.7.11 This is a tricky integral. The straightforward approach with [...f...] = [×] [−] [e^x] [$x \gtrless y$] x [+] [÷] in the two-point Gauss-Laguerre algorithm gives 0.420748815, which is pretty far from the truth. The substitution $s = t^2$ leads to [...f...] = [$\sqrt{x}$] x [×] [+] [$1/x$] 2 [÷], which gives 0.2119397662—even farther from the precise value given in the preceding answer. Higher order Gauss-Laguerre formulas give better values of course, but the improvement is slow. Goodwin and Staton tried several methods and decided that integrating the differential equation, as in the preceding problem, may be the easiest way to obtain values for this integral. It can also be expressed in terms of Dawson's integral and the exponential integral. See the references in B.5.7.10.

B.5.7.12 With third-order Runge-Kutta from Section A.4.1 with $h = 0.1$, some representative values are as follows:

t	θ (rad)
0.5	0.259422134
1.0	0.487260681
1.5	0.657309365
2.2	0.766128187
2.3	0.767986555 (sic)

Note that the last value is larger than $\theta_0 = 0.767944871$ rad. Integrating the first-order differential equation near θ_0 is very difficult. (Why?) The period is approximately $4 \times 2.3 = 9.2$. The period from the approximate formula given in the problem is 8.885765875. The precise period is

$$P = 4\sqrt{l/g}\, K\left(\sin^2\left(\frac{\theta_0}{2}\right)\right),$$

where $K(m)$ is the complete elliptic integral of the first kind (see Section A.5.7). This precise formula gives $P = 9.224796844$.

REFERENCES: MacMillan (1958), p. 311; Kopal (1955), p. 236.

B.5.7.13 For the sum, HP-45:

SETUP: [G] [CLEAR] [Σ+] θ [↑] [↑] [↑] 1
LOOP: [×] [SIN] [RCL] 5 [x^2] [÷] [Σ+] (check: is display 7? If so, go to coda.) :|
CODA: [RCL] [Σ+] (see Σ· · ·).

With $\theta = 50°$ get 0.9847817128. For the integral use [···f···] = 2 [÷] [SIN] 2 [×] [ln] in the three-point Gauss-Legendre formula (Section A.3.2) and get $54.°72469811 \times \pi/180° = 0.9551261644$. Neither of these answers is very precise. The sum converges slowly and the singularity at the end of the range of integration makes its numerical evaluation difficult. One way out of this difficulty uses the definite integral (CRC SMT p. 467)

$$\int_0^{\pi/2} \ln(\sin x)\,dx = -\frac{\pi}{2}\ln 2,$$

to write

$$-\int_0^{\theta} \ln\left(2\sin\frac{t}{2}\right)dt = (\pi - \theta)\ln 2 + \int_{\theta}^{\pi} \ln\left(\sin\frac{t}{2}\right)dt.$$

Then use the three-point Gauss-Legendre formula (Section A.3.2) to get 1.001241512, or the five-point Lobatto-Radau formula to get 1.000748143. This form of the log-sin integral is called Clausen's integral, and the precise value for $\theta = 50°$ is 1.000790624.

REFERENCE: HMF 27.8.1.

B.5.7.14 Using the three-point Gauss-Legendre formula (Section A.3.2) with [···f···] = [ln] [1/x], get 170.6687071, and with the five-point Lobatto-Radau formula (Section A.3.2) get 229.3720681. This integral can be written in terms of exponential integrals, namely, $P(n) = \text{Ei}(\ln n) - \text{Ei}(\ln 2)$. The precise value of $P(1000)$ is 176.5644944, and there are 168 primes less than 1000.

REFERENCES: Hunter and Madachy (1963), p. 7; HMF 5.1.3; HMF 5.1.50 (for a plot of the error in $P(n)$); Section A.5.11.

B.6.6.1 x [↑] a [×] b [+] [×] c [+] [×] d [+] [×] e [+]. This is the same as the algorithm in Section 2.4, except that with the ȼ system one need not fill up the stack ([↑] [↑] [↑] after x), provided the stack is initially clear.

B.6.6.2 1 [↑] [↑] [↑] R_1 [÷] [↔] R_2 [÷] [+] [↔] R_3 [÷] [+] [÷].

B.6.6.3 In the Lagrange three-point interpolation algorithm in Section A.2, change [$x \gtrless y$] to [↔], and change [x^2] to [↑] [×].

B.6.6.4 The tabular quadrature algorithms in Section A.3.1 will all work on the model A with no changes.

Each of the following answers works if the model B initially has the r stack active and auto-enter enabled, or else $\not c$ in X_r and X_i.

B.6.6.5 a [S] b [S] c [S] d [÷] [$\sqrt{x}$] 1 [−] (see u) [S] (see v).

B.6.6.6 1 [S] 2 [↑] [CHS] [×] 1 [S] 3 [÷] 2 [S] 4 [÷] 5 [↑] 14 [S] 2 [÷] [−] (see $u=0$) [S] (see $v=0$).

B.6.6.7 1 [CHS] [↑] 3 [$1/x$] [y^x] (see -1) [S] (see 0), thus a cube root of -1 is -1. Then [RCL] 0 [×] (see $\frac{1}{2}$) [S] (see $-\frac{1}{2}\sqrt{3}$), therefore another cube root of -1 is $\frac{1}{2}-i\frac{1}{2}\sqrt{3}$. Then [RCL] 0 [×] (see $\frac{1}{2}$) [S] (see $\frac{1}{2}\sqrt{3}$), thus the third cube root of -1 is $\frac{1}{2}+i\frac{1}{2}\sqrt{3}$.

B.6.6.8 a [S] b [S] c [S] d [y^x] [RCL] 0 [$x \gtrless y$] [÷] (see u) [S] (see v).

B.6.6.9 Probably. See Demidovich and Maron (1973), Section 4.11.

B.6.6.10 HP-45 or HP-35:

[EEX] 9 [↑] .5 [+] (see 1000000001.) [EEX] 9 [↑] .4999999999 [+] (see 1000000000.).

So far, so good. But,

[EEX] 9 [↑] .5 [+] .5 [−] (see 1000000001.) .5009999999 [−] (see 1000000001.) .501 [−] (see 1000000000.)!

APPENDIX C

References and Bibliography

Woe be to him who reads but one book.

—*George Herbert (1593–1633)*

Abramowitz, M., and Stegun, I. A. See HMF.

AENA. *The American Ephemeris and Nautical Almanac*. Government Printing Office, Washington, D.C. (yearly). "The Blue Book."

Allen, C. W. *Astrophysical Quantities*. 2nd ed. Athlone Press, London, 1973.

Ambartsumyan, V. A. (Ed.). *Theoretical Astrophysics*. Pergamon Press, New York, 1958.

Applications Programs (for HP-25), quarterly. The Viola Grange, Route 2, Box 965, Estacada, Ore. 97023.

Ball, J. A. *Some FORTRAN Subprograms Used in Astronomy*. MIT Lincoln Laboratory Technical Note TN1969-42, Lexington, Mass., July 16, 1969. Parts of TN1969-42 were reproduced in Appendix A of *Methods of Experimental Physics*, Vol. 12C, M. L. Meeks, Ed. Academic Press, New York, 1976.

Ball, J. A. "Reverse-Polish or Algebraic Entry." *Electron. Des.* **23-2**, 50, 1975, **23-15**, 7, 1975a.

Ball, J. A. *Algorithms for the HP-45 and HP-35*. Center for Astrophysics Technical Report, Cambridge, Mass., rev. ed. March 9, 1975b.

Ball, J. A. "Computations in Radio-Frequency Spectroscopy." In *Methods in Computational Physics*, Vol. 14, B. Alder, S. Fernbach, and M. Rotenberg, Eds. Academic Press, New York, 1975c.

Ball, J. A. "Comment on 'Simple Iterative Procedures for Solving Transcendental Equations with the Electronic Slide Rule.'" *Am. J. Phys.* **44/5**, 488, 1976a.

Ball, J. A. "Measurements with Radio-Frequency Spectrometers." In *Methods of Experimental Physics*, Vol. 12C, M. L. Meeks, Ed. Academic Press, New York, 1976b.

Ball, J. A. "Pointing and Range on a Pocket Calculator." Submitted to *Ham Radio Mag.*, 1977.

Born, M. *Einstein's Theory of Relativity*. Dover, New York, 1962.

Bracewell, R. N. *The Fourier Transform and Its Applications*. McGraw-Hill, New York, 1965.

Collins, P., and Cherniak, J. *SAO FORTH—A Primer*. Technical Report, Smithsonian Astrophysical Observatory, Cambridge, Mass., 1974.

Conrad, V., and Pollak, L. W. *Methods in Climatology*. Harvard University Press, Cambridge, Mass., 1950.

CRC HCP. *Handbook of Chemistry and Physics*. R. C. Weast, Ed. Chemical Rubber Company, Cleveland (yearly). The page numbers in the text refer to the 56th edition (1975–1976), but almost all the references are also in other editions, perhaps with different page numbers.

CRC HTM. *Handbook of Tables for Mathematics*. S. M. Selby, Ed. Chemical Rubber Company, Cleveland, 1967, and other editions.

CRC SMT. *Standard Mathematical Tables*. S. M. Selby, Ed. Chemical Rubber Company, Cleveland, 1973, and other editions.

Davidon, W. C. "Fast Least-Squares Algorithms." *Am. J. Phys.* **45/3**, 260, 1977.

Demidovich, B. P., and Maron, I. A. *Computational Mathematics*. Mir Publishers, Moscow, 1973.

Dicke, R. H., and Wittke, J. P. *Introduction to Quantum Mechanics*. Addison-Wesley, Reading, Mass., 1960.

Eisberg, R. M. *Applied Mathematical Physics with Programmable Pocket Calculators*. McGraw-Hill, New York, 1976.

Elmore, W. C. "Straight-Line Fitting with an HP-45 Calculator." *Am. J. Phys.* **42**, 253, 1974a.

Elmore, W. C. "Complex Number Arithmetic with a Pocket Calculator." *Am. J. Phys.* **42**, 340, 1974b.

Elmore, W. C. "Perpetual Calendar." *Am. J. Phys.* **44/5**, 482, 1976.

ESE. *Explanatory Supplement to the Ephemeris*. Her Majesty's Stationery Office, London, 1961. "The Green Book."

Euclid. *The Thirteen Books of Euclid's Elements*. c. 300 B.C.

Everything You've Always Wanted to Know About RPN but Were Afraid to Pursue—Comprehensive Manual for Scientific Calculators. T. K. Enterprises, 16611 Hawthorne Blvd., Lawndale, Calif. 90260, 1976.

Feynman, R. P., Leighton, R. B., and Sands, M. *The Feynman Lectures on Physics*. 3 vols. Addison-Wesley, Reading, Mass., 1963.

FFP. *Fundamental Formulas of Physics*. 2 vols. D. H. Menzel, Ed. Dover, New York, 1960.

Fike, C. T. *Computer Evaluation of Mathematical Functions*. Prentice-Hall, Englewood Cliffs, N.J., 1968.

FORTH. See Moore, 1974; Collins and Cherniak, 1974; Rather, Moore, and Hollis, 1974; and FORTH, Inc., 815 Manhattan Ave., Manhattan Beach, Calif. 90266.

Fowler, R. W. *A Collection of HP-25 Routines*, Published by the author at 12104 Arbie Road, Silver Spring, Md. 20904, 1976.

Grove, W. E. *Brief Numerical Methods*. Prentice-Hall, Englewood Cliffs, N.J., 1966.

Hamming, R. W. *Numerical Methods for Scientists and Engineers*. McGraw-Hill, New York, 1962 and 1973.

Hartree, D. R. *Numerical Analysis*. 2nd ed. Oxford University Press, London, 1958.

Haurwitz, B. *Dynamic Meteorology*. McGraw-Hill, New York, 1941.

Hildebrand, F. B. *Introduction to Numerical Analysis*. McGraw-Hill, New York, 1956.

HMF. *Handbook of Mathematical Functions*. M. Abramowitz and I. A. Stegun, Eds. Government Printing Office, Washington, D.C., 1964 and 1972; also Dover, New York, 1965.

HP-21 Applications Book. Hewlett-Packard Company, 1000 N.E. Circle Blvd., Corvallis, Ore. 97330, 1975.

HP-25 Applications Programs. Hewlett-Packard Company, Corvallis, Ore., 1975.

HP-35 Math Pac. Hewlett-Packard Company, Corvallis Ore., 1974.

HP-45 Applications Book. Hewlett-Packard Company, Corvallis, Ore., 1974.

HP-55 Mathematics Programs. Hewlett-Packard Company, Corvallis, Ore., 1974.

HP-55 Statistics Programs. Hewlett-Packard Company, Corvallis, Ore., 1974.

HP-65 and *HP-67/97 Pacs*. Hewlett Packard Company, Corvallis, Ore., various dates. These are 15 or more sets of programs and writeups for the HP-65 or HP-67 and HP-97, grouped by subject matter.

HP-65 Users' Library; Catalog of Contributed Programs. Hewlett-Packard Company, Corvallis, Ore.

HP-67/97 Users' Library; Catalog of Contributed Programs. Hewlett-Packard Company, Corvallis, Ore.

HP Key Notes. A quarterly newsletter for HP-65 and HP-67/97 users. Hewlett-Packard Company, Corvallis, Ore.

Humphreys, W. J. *Physics of the Air*. McGraw-Hill, New York, 1940.

Hunter, J.A.H., and Madachy, J.S. *Mathematical Diversions*. Dover, New York, 1975.

IBM GC20-8011. *Random Number Generation and Testing*. IBM, White Plains, N.Y., 1959 and 1969.

IBM SSP. *Scientific Subroutine Package* (Programmer's Manual). H20-0205, IBM, White Plains, N.Y., 1968.

Isaacson, E., and Keller, H. B. *Analysis of Numerical Methods*. Wiley, New York, 1966.

Knuth, D. E. *The Art of Computer Programming*. Vol. I, *Fundamental Algorithms*. 2nd ed., 1973. Vol. II, *Seminumerical Algorithms*, 1969. Addison-Wesley, Reading, Mass.

Kopal, Z. *Numerical Analysis*. Chapman & Hall, London, 1955.

Kopchenova, N. V., and Maron, I. A. *Computational Mathematics*. Mir Publishers, Moscow, 1975.

Kraitchik, M. *Mathematical Recreations*. 2nd ed., Dover, New York, 1953.

Lanczos, C. *Applied Analysis*. Prentice-Hall, Englewood Cliffs, N.J., 1956.

Lang, K. R. *Astrophysical Formulae*. Springer-Verlag, Berlin, 1974.

Lewart, C. R. "Short HP-25 Program Does Combinations Faster." *Electron. Des.* **24-15**, 11, 1976; **24-22**, 12, 1976. See also *Applications Programs* 1-21 to 1-24.

Lilley, A. E., Palmer, P. "Tables of Radio-Frequency Recombination Lines." *Astrophys. J. Suppl.* **16-144**, 1968.

Lubkin, Y. J. *Filter Systems and Design: Electrical, Microwave, and Digital*. Addison-Wesley, Reading, Mass., 1970.

Lukasiewicz, J. *Aristotle's Syllogistic from the Standpoint of Modern Formal Logic*. Oxford University Press, London, 1951.

McKelvey, J. P. "Simple Iterative Procedures for Solving Transcendental Equations on the 'Electronic Slide Rule.'" *Am. J. Phys.* **43**, 331, 1975.

MacMillan, W. D. *Statics and the Dynamics of a Particle*. Dover, New York, 1958.

Menzel, D. H. See FFP.

Milne, W. E. *Numerical Calculus*. Princeton University Press, Princeton, N.J., 1949.

Milne, W. E. *Numerical Solution of Differential Equations*. Wiley, New York, 1953.

Moore, C. H. "FORTH: A New Way to Program a Mini-Computer." *Astron. Astrophys. Suppl.* **15-3**, 497, 1974. See also Rather, Moore, and Hollis, 1974; and FORTH.

Nelson, R. J., see *65 Notes*.

Péter, R. *Playing with Infinity*. Atheneum, New York, 1964; Dover, New York, 1976.

Pólya, G. *How to Solve It*. Princeton University Press, Princeton, N.J., 1973.

Rather, E. D., Moore, C. H., and Hollis, J. M. *Basic Principles of FORTH Language as Applied to a PDP-11 Computer*. Computer Division Technical Report Number 17, National Radio Astronomy Observatory, 2010 North Forbes Boulevard, Tucson, Ariz. 85705, March 1974.

Rice, S. O. "Mathematical Analysis of Random Noise." In *Selected Papers on Noise and Stochastic Processes*, N. Wax, Ed. Dover, New York, 1954.

Ruze, J., *The Effect of Aperture Errors on Antennas*. MIT RLE Technical Report 248, Cambridge, Mass., October 1952.

Scarborough, J. B. *Numerical Mathematical Analysis*. 6th ed. Johns Hopkins Press, Baltimore, 1966.

Schaffer, H. E. "Beating the Standard Set for the HP-25." *Electron. Des.* **24-7**, 7, 1976.

Selby, S. M. See CRC SMT and CRC HTM.

65 Notes. The Newsletter of the HP-65 Users' Club; c/o Richard J. Nelson, 2541 West Camden Place, Santa Ana, Calif. 92704.

Slater, J. C. *Introduction to Chemical Physics*. McGraw-Hill, New York, 1939; Dover, New York, 1970.

Smart, W. M. *Text-Book on Spherical Astronomy*. 5th ed. Cambridge University Press, Cambridge, 1962.

Smith, J. M. *Scientific Analysis on the Pocket Calculator*. Wiley, New York, 1975.

Strunk, W. Jr., and White, E. B. *The Elements of Style*. Macmillan, New York, 1972.

Thuesen, H. G. *Engineering Economy*. Prentice-Hall, Englewood Cliffs, N.J., 1950.

TN1969-42. See Ball, 1969.

Todd, J. (Ed.). *Survey of Numerical Analysis*. McGraw-Hill, New York, 1962.

Traub, J. F. *Iterative Methods for the Solution of Equations*. Prentice-Hall, Englewood Cliffs, N.J., 1964.

von Foerster, H., Mora, P., and Amiot, L. "Doomsday: Friday, 13 November, A.D. 2026." *Science* **132**, 1291, 1960.

von Hoerner, S. "Population Explosion and Interstellar Expansion." In *Einheit und Vielheit*, E. Scheibe and G. Süssman, Eds. Vandenhoeck & Ruprecht, Göttingen, 1972.

Wall, H. S. *Analytic Theory of Continued Fractions*. Van Nostrand, New York, 1948; Chelsea, Bronx, New York, 1967.

Weast, R. C., see CRC HCP.

Whittaker, Sir E. T., and Robinson, G. *The Calculus of Observations*. 4th ed. Blackie, London, 1944.

INDEX

This index contains references to both section numbers (identifiable by the imbedded decimal points) and to page numbers. Thus "4." means (all of) Chapter 4, and "4" means page 4. References beginning with A or D are section numbers in the corresponding appendix.

Whenever the reference is to an exercise, see also the corresponding answer in Appendix B. But using Appendix B when doing an exercise is no fair.